AF609221

Chemistry and Technology of Lubricants

Chemistry and Technology of Lubricants

Edited by

R.M. MORTIER
Technology Development Manager

and

S.T. ORSZULIK
Senior Technologist

Castrol Ltd
Reading

Springer Science+Business Media, LLC

10 9 8 7 6 5 4 3 2 1

Originally published by Blackie and Son Ltd in 1992

British Library Cataloguing in Publication Data

Chemistry and technology of lubricants.
I. Mortier, R.M. II. Orszulik, S.T.
665

ISBN 978-0-216-92921-0

Library of Congress Cataloging-in-Publication Data

Chemistry and technology of lubricants / edited by R.M. Mortier and S.T. Orszulik.
p. cm.
Includes bibliographical references and index.
ISBN 978-0-216-92921-0 ISBN 978-1-4615-3272-9 (eBook)
DOI 10.1007/978-1-4615-3272-9
1. Lubrication and lubricants. I. Mortier, R.M. II. Orszulik, S.T.
TJ1077.C418 1992
665.5'385—dc20 91-44485
CIP

Photosetting by Advanced Filmsetters (Glasgow) Ltd

Preface

The use of lubricants began in ancient times and has developed into a major international business through the need to lubricate machines of increasing complexity. The impetus for lubricant development has arisen from need, so lubricating practice has preceded an understanding of the scientific principles. This is not surprising as the scientific basis of the technology is, by nature, highly complex and interdisciplinary. However, we believe that the understanding of lubricant phenomena will continue to be developed at a molecular level to meet future challenges. These challenges will include the control of emissions from internal combustion engines, the reduction of friction and wear in machinery, and continuing improvements to lubricant performance and life-time.

More recently, there has been an increased understanding of the chemical aspects of lubrication, which has complemented the knowledge and understanding gained through studies dealing with physics and engineering. This book aims to bring together this chemical information and present it in a practical way. It is written by chemists who are authorities in the various specialisations within the lubricating industry, and is intended to be of interest to chemists who may already be working in the lubricating industry or in academia, and who are seeking a chemist's view of lubrication. It will also be of benefit to engineers and technologists familiar with the industry who require a more fundamental understanding of lubricants.

Throughout the book the range of uses of liquid lubricants, the base fluid types, and the various classes of additives available are covered. In the chapters on lubricant technology, the authors have been given the opportunity to draw on their extensive industrial experience. Although it has not been possible to cover all aspects of such a broad subject, the aim is to provide an insight into the more important aspects of the chemistry of lubricants, together with an indication of how lubricants are formulated to meet the needs of lubrication technology.

We would like to express our thanks to the authors for their contributions and for their patience during the editing process. The additional contribution by Tony Lansdown after the very late withdrawal of one of our authors is particularly appreciated. Thanks are also due to the publishers for the layout of the book and for guiding us through the editing. Finally, we would welcome comments, criticisms and suggestions.

R.M.M.
S.T.O.

Contributors

Mr C.I. Betton	Castrol Research Laboratories, Whitchurch Hill, Pangbourne, Reading RG8 7QR, UK
Mr M. Brown	ICI Chemicals and Polymers Ltd, Wilton, Cleveland TS6 8JE, UK
Mr B.H. Carter	Castrol Research Laboratories, Whitchurch Hill, Pangbourne, Reading RG8 7QR, UK
Mr C.C. Colyer	The Lubrizol Corporation, 29400 Lakeland Blvd, Wickliffe, Ohio, 44092, USA
Dr J. Crawford	Adibis, 36/44 High Street, Redhill, Surrey RH1 1RW, UK
Professor D. Dowson	Department of Mechanical Engineering, University of Leeds, Leeds LS2 9JT, UK
Mr W.C. Gergel	The Lubrizol Corporation, 29400 Lakeland Blvd, Wickliffe, Ohio, 44092, USA
Mr G. Gow	AB Axel Christiernsson, PO Box 19, S-440 41 Nol, Sweden
Mr T.J. Hoyes	Castrol Research Laboratories, Whitchurch Hill, Pangbourne, Reading RG8 7QR, UK
Professor C. Kajdas	Technical University of Radom, ul. Malczewskiego 29, 26-600 Radom, Poland
Dr A.R. Lansdown	10 Havergal Close, Caswell, Swansea SA3 4RL, UK
Dr C.M. Lindsay	Castrol Research Laboratories, Whitchurch Hill, Pangbourne, Reading RG8 7QR, UK

Mr A.J. Mills	Castrol Research Laboratories, Whitchurch Hill, Pangbourne, Reading RG8 7QR, UK
Dr R.M. Mortier	Castrol Research Laboratories, Whitchurch Hill, Pangbourne, Reading RG8 7QR, UK
Dr S.T. Orszulik	Castrol Research Laboratories, Whitchurch Hill, Pangbourne, Reading RG8 7QR, UK
Mr R.J. Prince	Castrol Ltd, Swindon SN3 1RE, UK
Dr A. Psaila	OIS Research Laboratory, Adibis, Saltend, Hedon, Hull HU12 8DS, UK
Dr S.J. Randles	ICI Chemicals and Polymers Ltd, Wilton, Cleveland TS6 8JE, UK
Dr M. Rasberger	Ciba-Geigy Ltd, Additives Division AD7.1, Building R-1032.4.58, CH-4002, Basle, Switzerland
Mr R.L. Stambaugh	Rohm and Haas, Research Laboratories, 727 Norristown Road, Springhouse, Philadelphia, 19477, USA
Dr P.M. Stroud	ICI Chemicals and Polymers Ltd, Wilton, Cleveland TS6 8JE, UK

Contents

Introduction

D. DOWSON

The recorded use of lubricants dates back almost to the birth of civilisation (Dowson, 1979), with early historical developments being concerned with the use of lubricants of animal or vegetable origin in transportation or machinery. During the Middle Ages (AD 450–1450) there was a steady development in the use of lubricants, but it was not until AD 1600–1850 (particularly the industrial revolution in AD 1750–1850) that the value of lubricants in decreasing friction and wear was recognised.

The classical studies of friction (Amontons, 1699; Coulomb, 1785) suggested that surface roughness played a major role in determining the resistance to sliding between two surfaces. It was therefore thought that lubricants were effective because they filled up the hollows in surfaces and reduced the roughness. This view was widespread and long sustained, but the classical experimental investigations by Tower (1883) and Petrov (1883) clearly indicated that in successful bearings the rotating journal was completely separated from the bearing by a coherent film of lubricant. The principle of fluid-film lubrication was fully recognised when Reynolds (1886) analysed the slow, viscous flow of lubricants in plain bearings and derived the differential equation for pressure that underpins bearing design procedures to the present day.

When the principle of fluid-film lubrication had been established for almost 40 years, Hardy (1922) drew attention to another form of protection now known as boundary lubrication. In this condition—normally associated with high loads, low speeds and low viscosities—the surfaces cannot be separated by coherent fluid films that exceed in considerable measure the composite surface roughness. The friction and wear characteristics in this regime are thus determined by the properties of surface films, often of molecular proportions formed on the solids, or generated by adsorption or by chemical reaction between constituents of the lubricant and the solids. The recognition of this major mode of lubrication provided a sound basis for the spectacular development of certain forms of additives later in the 20th century.

Mineral oil was first produced commercially in the 18th and early 19th centuries, but this ubiquitous lubricant really established itself towards the end of the 19th century, following the drilling of Drake's Well at Titusville, Pennsylvania, USA in 1859. Production started at a similar time in Russia and Roumania, but it was the involvement of the Nobel brothers after 1873 that established the Baku field as a major producer. An account of lubricants and lubrication in the 19th century has been presented by Dowson (1974).

Mineral oil rapidly overtook oils of animal and vegetable origin as the essential lubricant of the 20th century. The disparate nature of the mineral oils derived from different sources of crude oil is outlined in chapter 1. Mineral oils readily oxidise at temperatures above about 100°C and are reluctant to flow at temperatures lower than about −20°C. As the aerospace industry developed in the 20th century it became necessary to introduce alternative, synthetic base oils. Synthetics still represent a small volume proportion of the base oils used today, but their range of applications is extending steadily and they have enabled machinery to operate under conditions that could not have been accommodated by mineral oils. An account of synthetic base oils is given in chapter 2.

Fluid-film lubrication is associated with the physical rather than the chemical nature of lubricants, and modern bearings rely substantially on this exceptional mechanism for separating sliding solids. However, all bearings start and stop, and the surfaces of the sliding solids come into contact with each interruption of movement. Furthermore, the major economic pressure for improved efficiency in most forms of machinery in the latter years of the 20th century has caused designers to adopt higher mean bearing pressures and lower viscosity fluids, thus reducing the effective minimum film thickness in bearings. In many dynamically loaded bearings and most severely-taxed, lubricated machine elements such as gears, cams and piston rings, the friction and life of the machine is greatly influenced by the protection afforded by films of molecular proportions formed on the solids by additives. Such substances, added to mineral oil, have contributed in a major way to the spectacular development of lubricants in the second half of the 20th century. Additives are also added to oils to restrict oxidation of the lubricant, to act as rust inhibitors, and to perform a role as detergents. Furthermore, they can be used to modify the viscosity–temperature characteristics of lubricants and to depress the pour point. Detailed accounts of the roles of additives are given in chapters 3, 4, 5, 6 and 12.

The greatly improved ability to analyse and design lubricated machine elements in the latter stages of the 20th century has progressed alongside the formulation of additive-containing mineral oils and the introduction of synthetic lubricants. It is therefore particularly helpful to have chapters of the present book devoted to the nature of lubricants developed specifically for the internal combustion engine (chapter 7), general industrial machinery (chapter 8) and aviation and marine applications (chapters 9 and 10). Finally, chapters 11 and 13 deal with greases and environmental aspects of lubricants, both topics of major significance. The response to ever more demanding pressures for improved efficiency and reliability of equipment, for machinery to operate in severe environments has been quite remarkable. If the engineer or tribologist can draw some satisfaction from the fact that improved analysis, design and manufacture have permitted most lubricated components to function well with effective fluid films of thickness 0.1–1 μm, the chemist or lubricant technologist can certainly claim that none of this would have been possible if additive packages had not produced thin protective layers of molecular proportions on the surfaces of the bearing solids.

References

Amontons, G. (1699) De la resistance caus'ee dans les machines. *Mémoires de l'Académie Royale* **A** 251–282. (Chez Gerard Kuyper, Amsterdam, 1706).

Coulomb, C.A. (1785) Théorie des machines simples, en ayant égard au frottement de leurs parties, et a la roideur des cordages. *Mém. Math. Phys.* (Paris) **X** 161–342.

Dowson, D. (1974) Lubricants and lubrication in the nineteenth century. *Joint Institution of Mechanical Engineers—Newcomen Society Lecture*, pp. 1–8.

Dowson, D. (1979) *History of Tribology*. Longman Group Limited, London, pp. 1–677.

Hardy, W.B. (1922) *Collected Scientific Papers of Sir William Bate Hardy (1936)*. Cambridge University Press, Cambridge, pp. 639–644.

Petrov, N.P. (1883) Friction in machines and the effect of the lubricant. *Inzh. Zh. St. Peterb* **1** 71–140; **2** 277–279; **3** 377–436; **4** 535–564.

Reynolds, O. (1886) On the theory of lubrication and its application to Mr Beauchamp Tower's experiments, including an experimental determination of the viscosity of olive oil. *Phil. Trans. R. Soc.* **177** 157–234.

Tower, B. (1883) First report in friction experiments (friction of lubricated bearings). *Proc. Instn. Mech. Engrs.* November 1883, 632–659; January 1984, 29–35.

1 Base oils from petroleum

R.J. PRINCE

1.1 Introduction

Modern lubricants are formulated from a range of base fluids and chemical additives. The base fluid has several functions but primarily it is the lubricant, providing a fluid layer separating moving surfaces or removing heat and wear particles while keeping friction at a minimum. Many of the properties of the lubricant are enhanced or created by the addition of special chemical additives to the base fluid. For example, stability to oxidation and degradation in an engine oil can be improved by the addition of antioxidants while extreme pressure (EP) anti-wear properties needed in gear lubrication are created by the addition of special EP additives. The base fluid also functions as the carrier for these additives and must therefore be able to keep the additives in solution under all normal working conditions.

The majority of lubricant base fluids are produced by the refining of crude oil. Bromilow (1990) has estimated that 24.1 million tonnes of petroleum base oils were used in the world in 1989 (excluding the USSR, China and Eastern Europe which are thought to use some 15 million tonnes of lubricant). The reasons for the predominance of refined petroleum base oils are simple and obvious—performance, availability and price. Large scale oil refining operations can produce base oils which have excellent performance in modern lubricant formulations at economic prices. Non-petroleum base fluids find application where special properties are necessary, where petroleum base oils are in short supply or where substitution by natural products is practicable or desirable.

This chapter concerns base oils made from crude oil petroleum. Crude oil is an extremely complex mixture of organic chemicals ranging in size from simple gaseous molecules, such as methane, to very high molecular weight asphaltic components. Obviously only some of these crude oil constituents are desirable in a lubricant base fluid and so a series of physical refining steps are needed to separate the good from the bad. Other process steps involving chemical reactions may also be used to enhance properties of the oil. Different types of base oils are produced at refineries; oils which have different viscosities or chemical properties are needed for different applications.

1.2 Base oil composition

Crude oil is the end result of physical and chemical processes acting over many millions of years on the buried remains of plants and animals. Although crude oil is usually formed in fine grained source rocks, it can migrate into more permeable reservoir rocks and large accumulations of petroleum can be accessed by drilling.

Each accumulation or oilfield contains a different type of crude oil, varying in chemical composition and physical properties. Some crudes have a low sulphur content and flow easily. Some may contain wax and only flow when heated while others contain large amounts of very high molecular weight asphalt. Table 1.1 shows some basic properties of a number of different crude oils.

Despite the wide range of hydrocarbons and other organic molecules that are found in crude oils, the main differences between crudes are not in the types of molecules but rather in the relative amounts of each type that occur in the oil.

Table 1.1 Variations in crude oil properties.

Source	North Sea	Indonesia	Venezuela	Middle East
Sulphur content (% wt)	0.3	0.2	5.5	2.5
Pour point (°C)	−3	39	9	−15
Viscosity at 40 °C (cSt)	4	12	19 000	8

1.2.1 *Components of crude oil*

The components of crude oil can be classified into a few broad categories. Some of these components have properties which are desirable in a lubricant while others have properties which are detrimental.

1.2.1.1 *Hydrocarbons* Hydrocarbons (organic compounds composed exclusively of carbon and hydrogen) predominate in all crude oils and can be further subdivided:

Alkanes (also known as paraffins). These have saturated linear or branched chain structures.

Alkenes (also known as olefins). These are unsaturated molecules, but they are comparatively rare in crude oils. Certain refining processes, however, do produce large amounts of alkenes by cracking or dehydrogenation.

Alicyclics (also known as naphthenes). These are saturated cyclic structures based on five- and six-membered rings.

Aromatics These are cyclic structures with conjugated double bonds, based on the six-membered benzene ring.

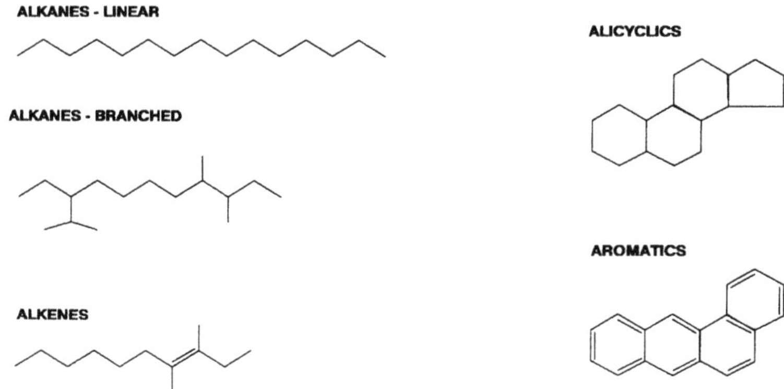

Figure 1.1 Examples of hydrocarbons.

Examples of these types are shown in Figure 1.1.

This classification is a simplification because there are many hydrocarbons which are combinations of these classes, for example alkyl-substituted cyclics or mixed polycyclics containing both aromatic and fully saturated rings.

1.2.1.2 *Non-hydrocarbons* Many organic compounds in crude oil incorporate other elements sometimes within ring structures or as functional groups attached to a hydrocarbon structure. Organosulphur compounds are generally much more prevalent than nitrogen- or oxygen-containing molecules, while organometallics are usually present as traces.

Within the boiling range appropriate to lubricant base oils, almost all of the organosulphur and organonitrogen compounds are heterocyclic molecules (see Figure 1.2 for examples). In contrast, the principal oxygen-containing molecules are carboxylic acids; either saturated aliphatic acids or cycloalkanoic acids (naphthenic acids). Traces of phenols and furans may also occur.

Finally, there are very high molecular weight resins and asphaltenes which contain a variety of aromatic and heterocyclic structures. Resins are the lower

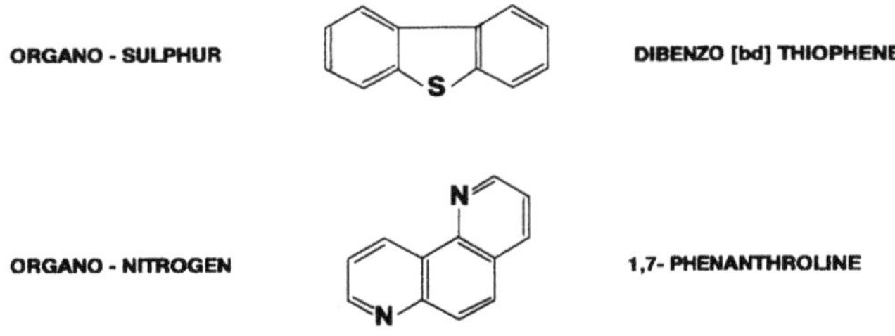

Figure 1.2 Non-hydrocarbon examples.

molecular weight (< 1000) species while asphaltenes are the result of the linking together of many other structures and have exceptionally high molecular weights.

1.2.2 *Characteristics of the hydrocarbons for lubricant performance*

In this section, only hydrocarbon properties will be discussed because most of the non-hydrocarbons are prone to oxidation or degradation and are deleterious to lubricant performance. However, organosulphur molecules are known to act as naturally occurring antioxidants and it is frequently desirable to retain some of these in the refined base oil.

Alkanes, alicyclics and aromatics of the same molecular weight have markedly different physical and chemical characteristics. Physical characteristics will affect the viscometrics of the lubricant, and the chemical stability of each class to oxidation and degradation while in use is also very important.

Alkanes have relatively low density and viscosity for their molecular weight and boiling point. They have good viscosity/temperature characteristics (i.e. they show relatively little change in viscosity with change in temperature—see viscosity index in section 1.3.1) compared to cyclic hydrocarbons. However, there are significant differences between isomers as the degree of branching of the alkane chain increases (see Figure 1.3).

The linear alkanes (normal paraffins) in the lubricant boiling range have good viscosity/temperature characteristics but their high melting points cause them to crystallise out of solution as wax. In contrast, highly-branched alkanes are not waxy but have less good viscosity/temperature characteristics. There is a compromise region in which acceptable viscosity index (VI) and acceptable low-temperature properties are achieved simultaneously.

In general, alkanes also have good viscosity/pressure characteristics, are reasonably resistant to oxidation and have particularly good response to oxidation inhibitors.

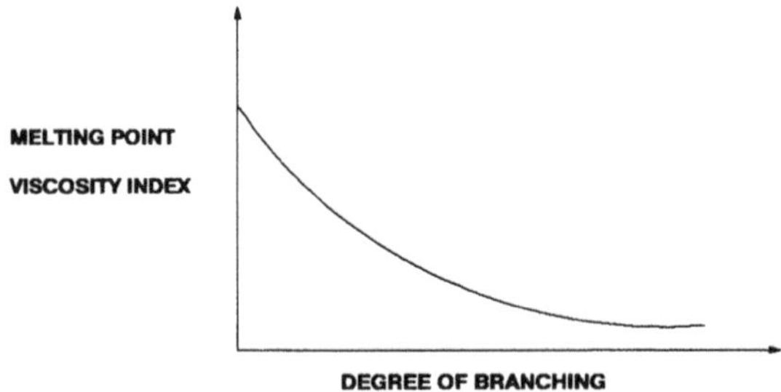

Figure 1.3 Variation in properties of alkane isomers.

Alicyclics have rather higher density and viscosity for their molecular weight compared to the alkanes. An advantage which alicyclics have over the alkanes is that they tend to have low melting points and so do not contribute to wax. However, one disadvantage is that alicyclics have inferior viscosity/temperature characteristics. Single ring alicyclics with long alkyl side chains, however, share many properties with branched alkanes and can in fact be highly-desirable components for lubricant base oils. Alicyclics tend to have better solvency power for additives than pure alkanes but their stability to oxidative processes is inferior.

Aromatics have densities and viscosities which are still higher. Viscosity/temperature characteristics are in general rather poor but melting points are low. Although they have the best solvency power for additives, their stability to oxidation is poor. As for alicyclics, single ring aromatics with long side chains (alkylbenzenes) may be very desirable base oil components.

1.2.3 *Crude oil selection for base oil manufacture*

Different crude oils contain different proportions of these classes of organic components and they also vary in the boiling range distribution of their components. The main factors affecting crude selection for manufacture of base oils are:

- Content of material of suitable boiling range for lubricants
- Yield of base oil after manufacturing processes
- Base oil product properties, physical and chemical

Crude oil assay data will reveal the potential content of lubricant boiling range material. The manufacturing process at a base oil refinery consists of a series of steps to separate the desirable lube components from the bulk of the crude oil. These steps are described in detail in section 1.4, but briefly, their aims are as follows:

Distillation This step removes both the components of too low boiling point and too high boiling point, leaving the lubricant boiling range distillates.
Aromatics removal This step leaves an oil that is high in saturated hydrocarbons and improves VI and stability.
De-waxing This step removes wax and controls the low-temperature properties of the base oil.
Finishing This step removes traces of polar components, and improves the colour and stability of the oil.

The yield of base oil after the application of these processes depends on the amount of desirable components in the lube boiling range. Lube distillates made from different crudes can have radically different properties and some examples are shown in Table 1.2.

Table 1.2 Comparison of distillates from a range of crude oils.

Crude Source	North Sea (Forties)	Middle East (Arabian)	Nigeria (Forcados)	Venezuela (Tia Juana)
Viscosity at 40 °C (cSt)	16	14	18	23
Pour point (°C)	25	19	18	−48
Viscosity index	92	70	42	10
Sulphur (% wt)	0.3	2.6	0.3	1.6
Aromatics (% wt)	20	18.5	28	21

Both the Forties and Arabian distillates have relatively high VI and high pour point, because they are rich in alkanes, and are examples of paraffinic crude oils. Paraffinic crudes are preferred for the manufacture of base oils where viscosity/temperature characteristics are important (e.g. automotive lubricants which have to operate over a wide temperature range). However, there is a big difference in sulphur content between these two crudes and this has an effect on the base oil composition and its chemical properties, especially natural oxidation stability. Careful control of the manufacturing processes can minimise some of these differences.

The Nigerian and Venezuelan distillates are examples of naphthenic products because they are relatively low in alkane content. In particular, the Venezuelan distillate is wax-free and so no de-waxing step is required. Although naphthenic products have inferior viscosity/temperature characteristics, they have other beneficial properties which are particularly useful in industrial applications.

These examples are all crudes that are regularly used to make base oils, but many other crudes do not contain sufficient useful lubricant components and cannot economically be used for conventional base oil production. However, in section 1.5, a modern catalytic process is described which can upgrade distillates of less suitable origin and so create desirable lubricant components.

1.3 Products and specifications

Lubricants are formulated by blending base oils and additives to meet a series of performance specifications. These specifications relate to the physical properties of the oil when it is new and also ensure that the oil continues to function and protect the engine or machinery in service. Self-evidently, lubricant performance is determined by the base oils and the additives that are used in the formulation.

When selecting the appropriate base oil to use in a formulation, there is a range of properties that can be measured and used to predict performance. Many of these properties are also used as quality control checks in the manufacturing process to ensure uniformity of product quality. Although many of these properties are modified or enhanced by the use of additives, knowledge of the base oil characteristics, especially any limitations, is vital to effective formulation of any lubricant.

The complexity of chemical composition of the base oils requires that most measurements are of bulk physical or chemical properties which indicate the average performance of all the molecular types in the base oil. Many tests are empirically based and are used to predict or correlate with the real field performance of the lubricant. Although perhaps not rigorously scientific, the importance of such tests should not be underestimated.

In the early days of the oil industry, a wide range of tests was developed by different companies and different countries. Nowadays, many tests are standardised and controlled on an international basis by organisations such as:

USA	American Society for Testing and Materials (ASTM)
UK	Institute of Petroleum (IP)
Germany	Deutsches Institut für Normung (DIN)
International	International Organisation for Standardisation (ISO)

1.3.1 *Physical properties*

1.3.1.1 *Viscosity* Viscosity is a measurement of the internal friction within a liquid; the way the molecules interact to resist motion. It is a vital property of a lubricant because it influences the ability of the oil to form a lubricating film or to minimise friction and reduce wear.

Newton defined the absolute viscosity of a liquid as the ratio between the shear stress applied and the rate of shear which results. Imagine two plates of equal area A which are separated by a liquid film of thickness D, as shown in Figure 1.4. The shear stress is the force (F) applied to the top plate (causing it to move relative to the bottom plate) divided by the area (A) of the plate. Shear rate is the velocity (V) of the top plate divided by the separation distance (D).

The unit of absolute viscosity is the pascal second (Pa.s), although the

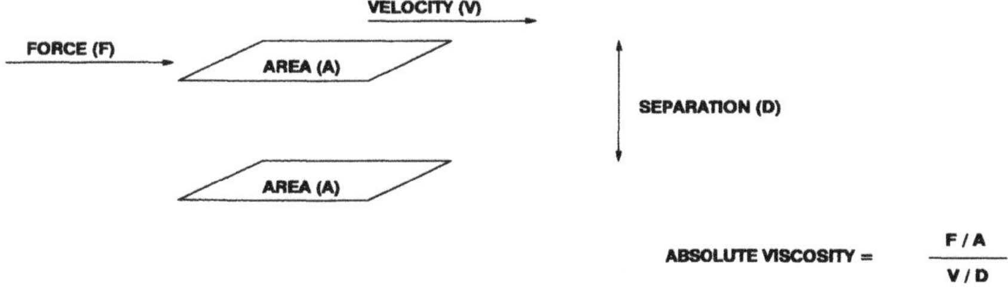

Figure 1.4 Definition of absolute viscosity.

centipoise (cP) is often used as an alternative unit (1 Pa.s = 10^3 cP). Absolute viscosity is usually measured with rotary viscometers in which a rotor spins in a container of the fluid and the resistance to rotation (torque) is measured.

Another method of defining viscosity is to measure the rate of flow of the liquid through a capillary under the influence of the constant force of gravity. This is the kinematic viscosity and is defined as follows:

$$\text{Kinematic viscosity} = \frac{\text{Absolute viscosity}}{\text{Liquid density}}$$

The unit of kinematic viscosity is m^2/s, but for practical reasons it is more common to use the centistoke (cSt) (1 cSt = $10^{-6}\,m^2/s$).

There are other, empirical, scales in use, such as SUS (Saybolt Universal Seconds) or the Redwood scales, and conversion tables are available. Base oil grades are sometimes referred to by their SUS viscosities.

Kinematic viscosity is routinely measured with ease and great precision in capillary viscometers suspended in constant temperature baths. Standard methods are ASTM D445, IP 71 and a number of standard temperatures are used. Measurement of the kinematic viscosity at more than one temperature allows the viscosity/temperature relationship to be determined.

Absolute viscosity is an important measurement for the lubricating properties of oils used in gears and bearings. However, it cannot be measured with the same degree of simplicity and precision as kinematic viscosity.

1.3.1.2 *Viscosity/temperature relationship—viscosity index* The most frequently used method for comparing the variation of viscosity with temperature between different oils is by calculation of a dimensionless number, known as the viscosity index (VI). The kinematic viscosity of the sample is measured at two different temperatures (40 °C, 100 °C) and the viscosity change is

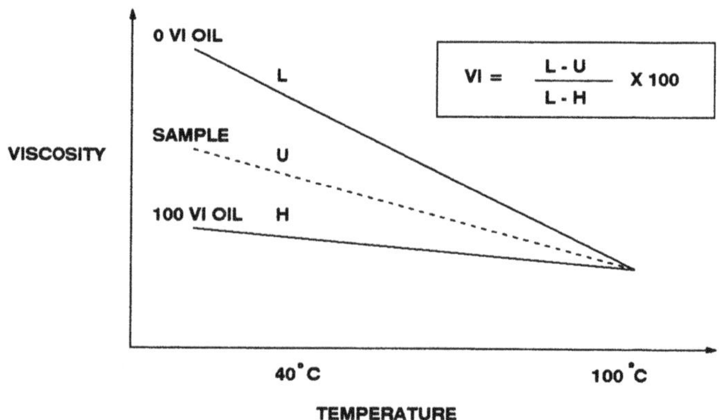

Figure 1.5 Definition of viscosity index.

compared with an empirical reference scale. The original reference scale devised by Dean and Davis (1929) was based on two sets of lubricant oils derived from separate crude oils—Pennsylvania crude, arbitrarily assigned a VI of 100 and a Texas Gulf crude assigned a VI of 0. The higher the VI number, the less the effect of temperature on the viscosity of the sample.

Full definitions of the methods of calculation are given in ASTM or IP manuals (ASTM D2270, IP 226) and a summary is shown in Figure 1.5. In this figure L is the viscosity at 40 °C of an oil of 0 VI which has the same viscosity at 100 °C as the sample under test; H is the viscosity at 40 °C of an oil of 100 VI which has the same viscosity at 100 °C as the sample under test; and U is the viscosity at 40 °C of the oil sample. L and H are obtained from standard tables. A modified procedure applies to oils of VI above 100 or to oils of high viscosity.

The VI scale is a useful tool in comparing base oils, but it is vital to recognise that it is arbitrarily based and has limitations. Extrapolation outside the measured temperature range of 40–100 °C may result in false conclusions, especially as wax crystals form at low temperatures.

VI is also used as a convenient measure of the degree of aromatics removal during the base oil manufacturing process, but comparison of VIs of different oil samples is only realistic if they are derived from the same distillate feedstock. Great care should be exercised in applying a measurement of VI as an indication of base oil quality.

1.3.1.3 *Low-temperature properties* When a sample of oil is cooled, its viscosity increases in a predictable manner until wax crystals start to form. The matrix of wax crystals becomes sufficiently dense with further cooling to cause an apparent solidification of the oil, but it has not undergone a true phase change in the way a pure compound, such as water, freezes. Although the 'solidified' oil does not pour under the influence of gravity, it can be moved if sufficient force is applied, e.g. by applying torque to a rotor suspended in the oil. Further decrease in temperature causes more wax to form, increasing the complexity of the wax/oil matrix and requiring still more torque to turn the rotor.

Many lubricating oils have to be capable of flow at low temperatures and a number of properties should be measured.

Cloud point is the temperature at which the first signs of wax formation can be detected. A sample of oil is warmed sufficiently to be fluid and clear and then cooled at a specified rate. The temperature at which a haziness is first observed is recorded as the cloud point. The oil sample must be free of water because this can interfere with the test (ASTM D2500, IP 219).

Pour point is the lowest temperature at which the sample of oil can be made to flow by gravity alone. The oil is warmed and then cooled at a specified rate and the test jar is removed from the cooling bath at intervals to see if the sample is still mobile. This procedure is continued until movement does not occur. The pour point is the last temperature before movement ceases, not the temperature at which solidification occurs. Oils of high viscosity may cease to flow because their viscosity becomes too high at low temperatures rather than because of wax formation. In these cases, the pour point will be higher than the cloud point (ASTM D97, IP 15).

The *cold crank simulator* test measures the apparent viscosity of the oil sample at low temperatures and high shear rate. It relates to the cold starting characteristics of engine oils and should be as low as possible. The oil sample fills a chamber between the rotor and stator of an electric motor. When the equipment has been cooled to the test temperature, the motor is started and the speed of rotation gives an indication of the apparent viscosity of the oil. The test is used for comparing oil samples rather than accurate prediction of the absolute performance in a specific engine (ASTM 2602, IP 383).

The *Brookfield viscosity* test measures the low-temperature viscosity of gear oils and hydraulic fluids under low-shear conditions. Brookfield viscosities are measured in centipoise units using a motor driven spindle immersed in the cooled oil sample (ASTM D2983).

1.3.1.4 *High-temperature properties* The high-temperature properties of an oil are governed by the distillation or boiling range characteristics of the oil.

Volatility is important because it is an indication of the tendency of an oil to be lost in service by vaporisation (e.g. in a hot engine). Several methods are used to characterise volatility and these include:

- Distillation curve, measured by vacuum distillation (ASTM D1160) or simulated by gas chromatography (ASTM D2887)
- Thermogravimetric analysis
- Noack volatility, where the sample is heated for an hour at 250 °C and the weight loss is measured (DIN 51581).

The *flash point* of an oil is important from a safety point of view because it is the lowest temperature at which auto-ignition of the vapour occurs above the heated oil sample. Different methods are in use and it is essential to know which type of equipment has been used when comparing results (ASTM D92, D93).

1.3.1.5 *Other physical properties* Various other physical properties may be measured, most of them relating to specialised lubricant applications. Since there is insufficient space to describe them in detail, a list of some of the more important measurements is given:

Density important, because oils may be formulated by weight, but measured by volume.

Demulsification ability of oil and water to separate.

Foam characteristics tendency to foam formation and stability of the foam that results.
Pressure/viscosity characteristics.
Thermal conductivity important for heat transfer fluids.
Electrical properties resistivity, dielectric constant.
Surface properties surface tension, air separation.

1.3.2 *Chemical properties*

1.3.2.1 *Oxidation* Degradation of lubricants by oxidative mechanisms is potentially a very serious problem. Although the formulated lubricant may have many desirable properties when new, oxidation can lead to a dramatic loss of performance in service by reactions such as:

- corrosion due to formation of organic acids
- formation of polymers leading to sludge and resins
- viscosity changes
- loss of electrical resistivity

A variety of different stability tests has been devised to measure resistance to oxidation under different conditions which correlate with different service uses of lubricants.

Since oxidation inhibitors are frequently added to base oils, the response of the base oil to standard inhibitors is an important property to measure. Therefore some tests are carried out in the presence of standard doses of antioxidants. Other tests include catalysts to cause accelerated ageing of the oil and reduce the duration of testing to manageable periods.

The sulphur content of base oils is often regarded as a useful indicator of natural oxidation resistance. This is because many naturally occurring organosulphur compounds in crude oil are moderately effective in destroying organic peroxide intermediates and breaking the oxidation chain mechanism. However, the effectiveness of these natural inhibitors is usually rather inferior to synthesised additives which can be much more specific in their action.

1.3.2.2 *Corrosion* The lubricant base oil should not contain components which promote corrosion of metal parts in an engine or machine. The problems of oxidation products leading to corrosion have been mentioned above.

Corrosion tests usually involve bringing the oil sample into contact with a metal surface (copper and silver are often used) under controlled conditions. Discoloration of the metal, changes in surface condition or weight loss may be used to measure the corrosion tendency of the oil. Other tests have been devised to measure corrosion protection properties of the oil under adverse conditions (e.g. in the presence of water, brine or acids formed as combustion products), however these tests are more applicable to formulated lubricants rather than base oils.

1.3.2.3 *Carbon residue* This test is used to measure the tendency of a base oil to form carbonaceous deposits at elevated temperatures. Tests such as the Conradson carbon residue test (ASTM D189) determine the residue which remains after pyrolytic removal of volatile compounds in the absence of air.

1.3.2.4 *Seal compatibility* Lubricants are often used in machines where they come into contact with rubber or plastic seals. The strength and degree of swell of these seals may be affected by interaction with the oil. Various tests have been devised to measure the effects of base oils on different seals and under different test conditions.

1.3.3 *Base oil categories*

1.3.3.1 *Paraffinics* Paraffinic base oils are made from crude oils that have relatively high alkane contents. Typical crudes are from the Middle East, North Sea, US mid-continent. This is not an exclusive list, nor does it follow that all North Sea crudes, for example, are suitable for production of paraffinic base oils. The manufacturing process requires aromatics removal (usually by solvent extraction) and de-waxing.

Paraffinic base oils are characterised by their good viscosity/temperature characteristics, i.e. high viscosity index, adequate low-temperature properties and good stability. In oil industry terminology they are frequently referred to as solvent neutrals (SN), where solvent means that the base oil has been solvent-refined and neutral means that the oil is of neutral pH. An alternative designation is high viscosity index (HVI) base oil.

Most of the base oils produced in the world are paraffinics and they are available in the full range of viscosities, from light spindle oils to viscous brightstock. Some examples of a range of paraffinic base oils from typical refinery production are given in Table 1.3.

Paraffinic base oils of very high viscosity index can also be manufactured by severe hydrotreatment or hydrocracking processes (see section 1.5) in

Table 1.3 Paraffinic base oils—typical properties (Arabian crude).

Grade	Spindle	150 SN	500 SN	Brightstock
Density at 20 °C (g ml^{-1})	0.85	0.87	0.89	0.91
Viscosity at 40 °C (cSt)	12.7	27.3	95.5	550
Viscosity at 100 °C (cSt)	3.1	5.0	10.8	33
Viscosity index	100	103	97	92
Pour point (°C)	−15	−12	−9	−9
Sulphur content (% wt)	0.4	0.9	1.1	1.5

which iso-alkanes are created by chemical reaction and crude oil origin is of reduced importance.

1.3.3.2 *Naphthenics* Naphthenics are made from a more limited range of crude oils than paraffinics, and in smaller quantities, at a restricted number of refineries. Important characteristics of naphthenic base oils are their naturally low pour points, because they are wax-free, and excellent solvency powers. Their viscosity/temperature characteristics are inferior to paraffinics (i.e. they have low to medium VI), but they are used in a wide range of applications where this is not a problem.

Since naphthenic crudes are free of wax, no de-waxing step is needed, but solvent extraction or hydrotreatment is nowadays often used to reduce aromatic content and especially to remove polycyclic aromatics which may present a health hazard in untreated oils.

The main producers of naphthenics are in North and South America, because most of the world's supply of naphthenic lube crudes are to be found there.

1.3.3.3 *Other base oil categories* Base oil refineries may produce a range of other products besides their main output of paraffinic or naphthenic base oils. These products may be either by-products of the main base oil refinery processes or speciality products made by additional process steps or by more severe processing. The main types are:

White oils These are very highly refined oils which consist entirely of saturated components, all aromatics having been removed by treatment with fuming sulphuric acid or by selective hydrogenation. Their name reflects the fact that they are virtually colourless and the most highly refined medicinal white oils are used in medical products and in the food industry.

Electrical oils Oils are used in industrial transformers for electrical insulation and heat transfer. They must have low viscosity and very good low-temperature properties. They are made, therefore, either from naphthenic crudes or by urea or catalytic de-waxing from paraffinic crudes.

Process oils Lightly refined base oils or the highly aromatic extracts which are a by-product of base oil manufacture are often used in various industrial products, for example as plasticisers in automotive tyres, in printing inks and in mould release oils.

1.4 Conventional base oil manufacturing methods

1.4.1 *Historic methods*

Very early lubricants were made by simple distillation of petroleum to recover the lower boiling gasoline and kerosine fractions and leave a residue which was useable as a lubricant. It was found that lubricant quality could be improved by additional very simple processing to remove some of the less desirable components, such as asphalt, wax and aromatics. In this era, lubricants relied on the inherent properties of the base oil because virtually no additives were used.

Distillation under vacuum allowed the separation of lube distillates from the crude oil, leaving the asphalt behind in the distillation residue. Wax was removed by chilling the lube distillate and filtering in plate and frame presses. Aromatics were reduced by treating the oil with sulphuric acid and separating the acid tar phase. Finally, finishing treatments such as adsorption of acid residues and impurities by activated clays gave further improvement in product quality.

These processes were mainly batch operations, labour intensive and characterised by their hazardous nature. They were not suitable for the great expansion in production capacity which the industry was being called upon to supply. New technology was developed which allowed continuous operation so that plants became much larger and could make more consistent quality products at lower cost. These new process methods were based on the use of solvents: continuous selective solvent extraction for aromatics removal was the process which replaced acid treatment and continuous solvent dewaxing replaced the very labour-intensive cold-pressing technique.

Technology has developed further in the last 30 years. Catalytic hydrogenation processes have become the normal method for finishing of base oils and in a more severe form can be used as an alternative to solvent extraction for the control of aromatics content.

With the exception of these newer hydrotreatment processes, all other processes used in modern base oil plants are physical separation techniques, i.e. all the eventual constituents of the finished base oil were present in the original crude oil and processing methods are used to concentrate the desirable components by removing the less desirable components as by-products.

1.4.2 *Base oil manufacture in a modern refinery*

Most base oil plants are integrated with mainstream oil refineries which produce a range of transportation and heating fuel products. Overall production capacity for lubricant base oils is a very small part of total refinery throughputs, in America amounting to less than 1.5% according to Thrash (1991).

Figure 1.6 indicates where a lubricant base oil plant fits into the process flow-scheme of a typical refinery—if there is such a thing. Although the scheme is simplified, the inter-relationship between the base oil plant and other process units and product streams is evident. In a sense, the base oil plant and fuels upgrading plant, such as the catalytic cracker, compete for

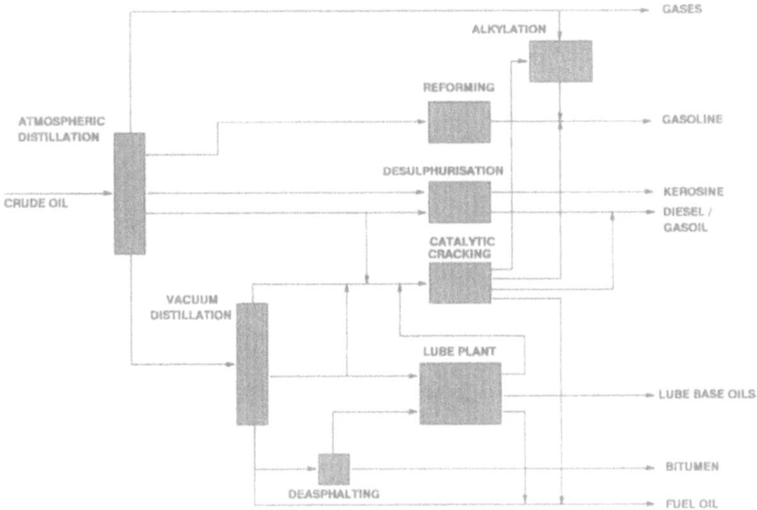

Figure 1.6 Simplified refinery flow-scheme.

feedstock from vacuum distillation. By-products from base oil manufacture are largely incorporated into fuels production streams. These interactions are very important to the logistics and production economics of making base oils.

Base oil manufacture produces large quantities of by-products, the unwanted components of the crude oil. Figure 1.7 is a typical base oil production flow-scheme in which the numbers indicate the relative amounts of

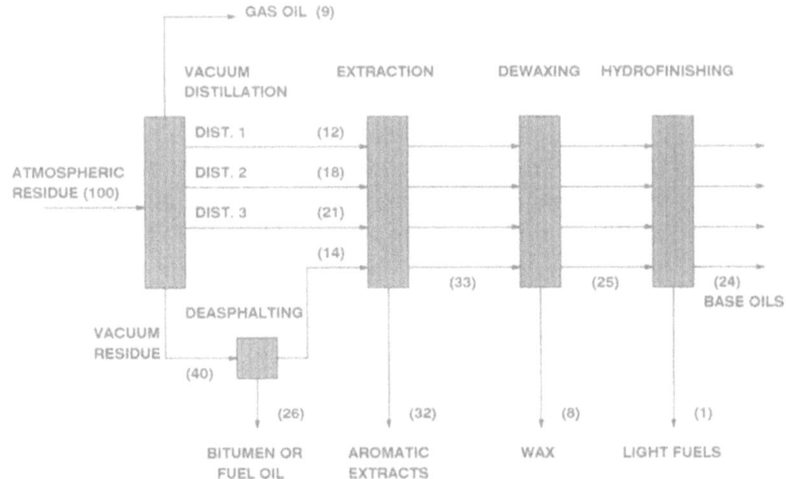

Figure 1.7 Base oil production flow-scheme.

intermediate and final products throughout the manufacturing process. The basis for the scheme is the processing of the residue from atmospheric distillation of a good quality Middle East crude. Starting with 100 parts of residue (which itself represents only about 50% of the original crude oil), even when the maximum possible amount of each base oil grade is produced, only 24 parts of base oil result. In practice, demand for different base oil grades is unlikely to match the potential output of each grade and surplus distillates or residue will be returned to the main fuels part of the refinery. It is quite normal for the actual output of base oil to be less than 10% of the crude oil purchased for making base oil.

Since the choice of crude oil is restricted when making base oils, the production of relatively small volumes of base oil actually makes a large imposition on the crude purchasing requirements for a refinery. If suitable crudes are only available at a premium price, there is an economic penalty for the refinery. Consequently, refining companies have gone to a lot of trouble in recent years to expand the portfolio of crude oils which they can use to make satisfactory base oils so that they have more flexibility in crude oil purchasing.

1.4.3 *Base oil production economics*

Each oil refinery is different. They have different process units with different relative production capacities arranged in different schemes to make different product ranges. Thus a view of production economics must of necessity be rather generalised. Production costs can be divided into several categories:

- Net feedstock or hydrocarbon cost of making base oil
- Variable operating costs (e.g. energy, chemicals)
- Fixed operating costs (e.g. wages, maintenance, overheads)
- Costs of capital (e.g. depreciation, interest)

Production cost per tonne of base oil is calculated by dividing the total annual costs by the total annual production of base oils.

Net feedstock cost can be calculated in several ways, but it will not necessarily be identical to the cost of crude oil. Since the base oil plant in a sense competes with fuels production units for feedstock, the basic feedstock cost to the lube complex should be determined by the alternative value of that feedstock if it were used to make mainstream fuels products.

The by-products of base oil manufacture also have values for blending into fuels streams or in some cases for direct sale as speciality products (e.g. waxes, bitumen). Credit must be given for these products so that the net value of the hydrocarbon content of the base oil can be calculated. Refineries use sophisticated linear programming computer models to optimise refinery operations based on different crude oil input, process yields, market prices, production targets, etc.

Variable and fixed operating costs are usually well defined, but when these costs are divided by the relatively small output of base oil, they are seen to be significant. If the base oil plant operates below maximum capacity then the fixed costs have to be shared over an even smaller volume and overall production costs rise in proportion. Energy costs are high because of the number of process steps needed and the energy intensive nature of equipment such as refrigeration plant and solvent recovery systems. Obviously energy use will vary between refineries, but consumptions as high as 0.4 tonnes fuel oil equivalent per tonne of base oil product are not uncommon.

The costs of capital tend to relate to the age of the base oil plant. A brand new plant has to be financed and since base oil production plant is very expensive to build, depreciation and interest charges will be considerable. Most of the present day base oil plant is at least 15 years old and so by now is almost fully depreciated. Therefore, for many base oil refineries, the cost of making base oil is limited to the hydrocarbon value and operating costs which makes it generally a profitable activity.

1.4.4 *Distillation*

Distillation is the primary process for separating the useful fractions for making lubricant base oils from crude oil. Crude oil is distilled at atmospheric pressure into components essentially boiling below 350 °C (gases, naphtha, kerosine and gas oil) and a residue containing lube boiling range components. Thermal decomposition is increasingly likely to occur at higher temperatures and so further separation of the atmospheric residue into lube distillates is carried out at reduced pressure in a vacuum distillation unit of the type shown in Figure 1.8.

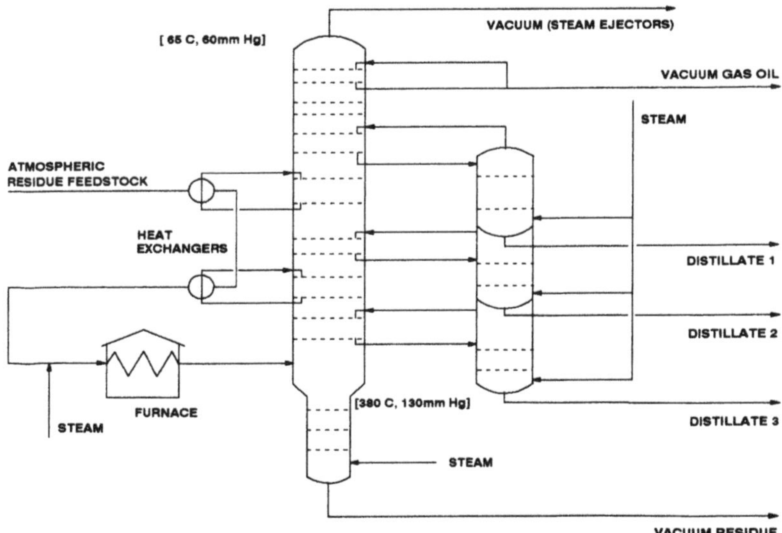

Figure 1.8 Lube vacuum distillation unit.

Atmospheric residue feedstock is injected with steam and pre-heated in a furnace before entering the lower part of the vacuum column. Inside the column, a variety of different internal mechanical arrangements are used to assist separation of different boiling range fractions:

- Trays, placed at intervals, with holes, bubble caps or valves to allow rising vapour and falling liquid to contact and come to equilibrium.
- Packing with random oriented rings or other particles, giving a high surface area for liquid/vapour contact.
- Packing with geometrically structured mesh, giving excellent contact and distribution of liquid and vapour.

Vacuum is provided at the top of the column, normally by steam ejectors in which condensing steam creates a vacuum, or sometimes by mechanical pumps. Typical top pressures are 60–80 mm Hg (absolute pressure) while, after allowing for pressure drops down the column, base pressure in the flash zone is likely to be in range 100–140 mm Hg. Injection of superheated steam helps to reduce the partial pressure of hydrocarbons in the flash zone. This reduction aids separation of the heavy distillate from the residue and helps to restrict overheating.

From the flash zone, the mixture of vaporised hydrocarbons and steam passes upwards while condensed liquid descends. A temperature gradient through the column from about 140 °C near the top to about 360 °C at the base is created by taking several sidestreams from selected trays at different levels in the column and cooling the streams before re-injection at a higher level.

The required lube distillates are also withdrawn as sidestreams and are steam stripped to provide the best possible separation between each fraction. A residue, typically boiling above 550 °C, is drawn from the column base.

Distillation provides a limited number of fractions (usually 3 lube distillates and a residue), each of which has viscosity and boiling range defined within quite a narrow range. The quality and consistency of fractionation has considerable impact on all the subsequent process steps. Careful design and operation of the vacuum column should achieve the following desirable results:

- Minimum overlap in boiling range between fractions (note, however, that some overlap is inevitable)
- Avoidance of entraining high molecular weight asphaltic components in the heaviest distillate fraction
- Ability to take a very heavy distillate fraction, rather than losing this material in the distillation residue
- Flexibility to run different crudes and still achieve design specifications for the properties of each lube fraction
- Minimum energy usage

Use of structured packings in recent years together with good design of the flash zone region of the column has helped to achieve these aims on modern base oil plants. Re-vamping and modernisation of older columns has also given substantial benefits.

The lube distillates and residue streams are run to heated intermediate tankage from where they are drawn to feed downstream process units.

1.4.5 *De-asphalting*

The residue from vacuum distillation is a black and very viscous material because it contains large amounts of asphaltic and resinous components. When these are removed, a useful high viscosity base oil fraction, known as brightstock, is available. Low molecular weight hydrocarbons are effective for dissolving the more desirable components while leaving the asphaltic material as a separate phase. Liquefied propane is by far the most frequently used solvent for de-asphalting of residues to make lubricant bright stock, whereas butane or pentane produce lower grade de-asphalted oils which are more suitable for feeding to fuels upgrading units.

The propane is kept close to its critical point and, under these conditions, raising the temperature increases selectivity. A temperature gradient is set up in the extraction tower to facilitate separation. Solvent to oil ratios are kept high because this enhances rejection of asphalt from the propane/oil phase. Counter-current extraction takes place in a tall extraction tower of the type shown in Figure 1.9. Vacuum residue enters near the top of the tower while

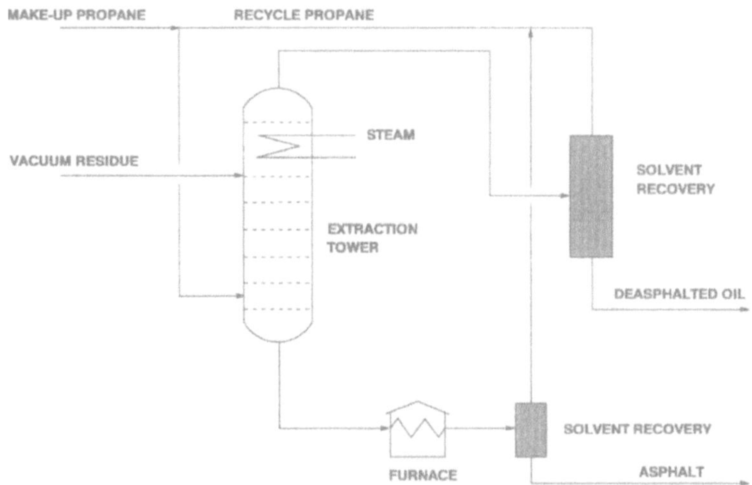

Figure 1.9 Propane de-asphalting unit.

propane enters near the base. The de-asphalted oil/propane phase, being lower in density, is taken from the top of the tower and the heavy asphalt phase leaves at the bottom. Steam heating coils provide the temperature gradient within the tower. Typical operating conditions are as follows:

• Propane/vacuum residue volume ratio	5–10:1
• Pressure	25–35 bar
• Top temperature	60–80 °C
• Base temperature	30–40 °C

Good contacting of feedstock and propane is essential and a variety of tower internals can be used to achieve this target:

- Tower packed with random particles e.g. ceramic rings
- Baffles or trays
- Mechanical rotating disc contactors

The bulk of the de-asphalting plant is actually taken up with equipment for solvent recovery for the de-asphalted oil and asphalt phases, and cooling and compression for recycling of the solvent. The de-asphalted oil product is a viscous, waxy material and requires solvent extraction and de-waxing before it can be used as a base oil. The asphalt can be a valuable feedstock for making bitumen grades or it may be blended into fuel oil.

1.4.6 *Solvent extraction*

Solvent extraction replaced acid treatment as the method for improving oxidative stability and viscosity/temperature characteristics of base oils. The solvent selectively dissolves the undesired aromatic components (the extract), leaving the desirable saturated components, especially alkanes, as a separate phase (the raffinate). Choice of solvent is determined by a number of factors:

- selectivity (i.e. to give good yields of high quality raffinate)
- solvent absorption power (to minimise the solvent/oil ratio)
- ease of separation of extract and raffinate phases
- ease of solvent recovery (boiling point must be below that of the raffinate or extract)
- desirable solvent properties (such as stability, safety, low toxicity, ease of handling, cost).

Solvents in commercial use include sulphur dioxide (historically important, but rare nowadays), phenol (use is in decline), furfural (the most widely used) and *N*-methylpyrrolidone (increasing in importance). *N*-methylpyrrolidone is gaining in popularity for new units and conversions because it has the lowest toxicity and can be used at lower solvent/oil ratios which saves energy.

Each distillate or bright stock stream is processed separately, because different process conditions are needed to obtain optimum results for each base oil grade. The main factors in operation of such a plant are:

Solvent/oil ratio. Increasing the solvent/oil ratio will allow deeper extraction of the oil, removing more aromatics and, of course, decreasing the raffinate yield. Over-extraction should be avoided because good lubricant components may be lost.

Extraction temperature. Solvent power increases with temperature, but selectivity decreases until feed and solvent become miscible. Clearly this extreme of complete miscibility must not be allowed to happen. The use of temperature gradients in extraction towers aids selectivity.

Solvent/oil contacting. The solvent and oil streams must be brought into contact, mixed efficiently and then separated into the extract and raffinate phases. The principal methods used are:

- Multi-stage mixing vessels, arranged in series so that flows of solvent and oil run in counter-current fashion.
- Extraction towers packed with ceramic rings or with sieve trays. Flows of solvent and oil are counter-current as described in section 1.4.5 on the de-asphalting plant. A temperature gradient is maintained within the tower.
- Extraction towers using a vertically-mounted rotating disc contactor. The spinning discs alternate with wall-mounted baffles and create a high shear mixing regime around the discs and allow excellent mixing. Rotor speed can be used as a control mechanism.
- Multi-stage centrifugal extractors both mix the incoming solvent and oil streams and separate the raffinate and extract products. They have advantages of small size and small hold-up volume.

Whichever contacting method is used, the end result is two product streams. The raffinate stream is mainly extracted oil containing a limited amount of solvent while the extract stream is a mixture of solvent and aromatic components. The streams are handled separately during solvent recovery and the recombined recovered solvent stream is recycled within the plant. A large proportion of an extraction plant is allocated to solvent recovery and it is an energy intensive part of the process.

As mentioned above, feedstocks are run individually (so-called blocked operation) so that the required properties for each base oil grade can be met economically. If a very wide boiling range feedstock were to be solvent extracted, then at one extreme of the boiling range over-extraction occurs, while at the other end under-extraction results. The result is a poor yield of indifferent quality product. Normal refinery operating procedure is to process each base oil grade in turn, drawn from intermediate product tanks, avoiding changes between dissimilar grades i.e. processing up and down the viscosity range. This procedure avoids major changes in operating conditions and minimises wastage of mixed fractions during change-overs.

1.4.7 *Solvent de-waxing*

The material which crystallises out of solution from lube distillates or raffinates is known as wax. Wax content is a function of temperature, because as the temperature is reduced, more wax appears. Sufficient wax must be removed from each base oil fraction to give the required low-temperature properties for each base oil grade. Naphthenic feedstocks, of course, are relatively free of wax and do not normally require de-waxing.

The molecular types within the wax fraction change as the boiling range of the feedstock increases. Linear alkanes crystallise easily in the form of large crystals and these are the predominant constituent of wax in the lighter distillates. Iso-alkane waxes form smaller crystals and these predominate in the heavier fractions. In addition, as the temperature of de-waxing decreases, the molecular composition of the wax which crystallises out of solution also changes; the highest melting point components crystallise first. Different grades of wax can be separated from different viscosity feedstocks at different temperatures.

The original de-waxing method involved cooling the waxy oil and filtering in large plate and frame presses. Pressures of up to 20 bar could be applied to the wax cake to force out the oil. However, the process had severe drawbacks; it was very labour intensive and oils of high viscosity could not be filtered at low temperatures. Filtration efficiency could be greatly improved by diluting the oil with solvents such as naphtha, but the selectivity for wax removal was reduced.

Improved solvent systems have been developed to give better de-waxing performance and important factors in the choice of solvent are:

- Good solubility of oil in the solvent and low solubility of wax in the solvent
- Small temperature difference between de-waxing temperature and the pour point of the de-waxed oil.
- Minimum solvent/oil ratios.
- Formation of large wax crystals which are easily filtered.
- Ease of solvent recovery (i.e. low boiling point).
- Desirable properties (such as stability, safety, low toxicity, ease of handling and cost).

Solvents in commercial use include propane, methyl isobutylketone, and mixed solvents such as methylethylketone/toluene or methylene chloride/dichloromethane. The use of paired solvents helps to control the oil solubility and wax crystallisation properties better than use of a single solvent.

A simplified flow scheme for a modern solvent de-waxing plant is shown in Figure 1.10. Solvent and oil are mixed together, then progressively chilled to the required temperature for filtration (this will be several degrees lower than the desired pour point). The rate of chilling influences the size and form of the wax crystals and the subsequent ease of filtration. Chilling takes place in

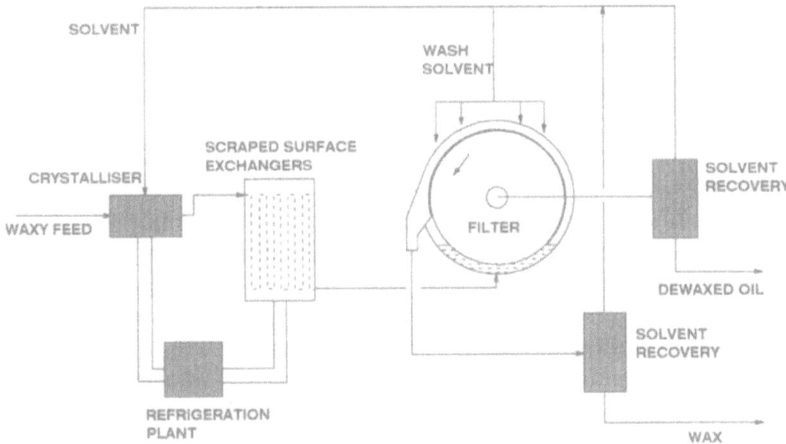

Figure 1.10 Solvent de-waxing plant.

special heat exchangers which have mechanically driven scrapers to keep the pipe walls clear of solidifying wax, aid heat transfer and ensure that the oil/wax/solvent slurry remains mobile.

Filtration is carried out in large rotary drum filters. Suction is applied to the inside of the horizontally mounted drum which slowly rotates with the lower part of the drum immersed in the chilled slurry. As oil passes through the filter cloth, a layer of wax (about 0.5 cm thick) builds up on the cloth and is removed by a scraper blade or blown off by inert gas. The de-waxed oil/solvent stream and the crude wax are handled as separate streams for solvent recovery. The wax contains an appreciable amount of oil because filtration pressures are very low and this oil can be recovered in a second filtration stage which also yields high quality wax products.

If very low pour points are necessary, the costs of de-waxing become very high and the yield correspondingly low. A variant of solvent de-waxing, urea de-waxing, is effective for making low pour point base oils from lower boiling lube feedstocks. Urea de-waxing relies upon the fact that urea, as it crystallises, forms crystals containing linear channels which can trap linear alkanes (i.e. wax). The urea–wax adduct is removed by filtration to leave a very low pour point oil. Urea and solvent are recovered and recycled.

1.4.8 *Finishing*

Despite the intensive series of process steps carried out so far, trace impurities may still be present in the base oil and a finishing step is needed to correct problems such as:

- poor colour
- poor oxidation or thermal stability

- poor demulsification properties
- poor electrical insulating properties

These undesirable components tend to be nitrogen-, oxygen- or, to a lesser extent, sulphur-containing molecules.

In the past, selective adsorbents such as clay or bauxite were used to remove impurities, but these processes were messy and gave waste disposal problems. Today, hydrofinishing has almost completely taken over.

Hydrofinishing differs from all the process steps used so far because it is not a physical separation procedure. It depends on the selective, catalysed hydrogenation of the impurities to form harmless products and is carried out under relatively mild conditions. Yields of finished base oil are high (at least 95%) and costs are quite low. Hydrofinishing should be effective for removing organonitrogen molecules because they are largely responsible for poor colour and stability of base oils, while organosulphur molecules should be retained because they tend to impart natural oxidation stability to the base oil.

A simplified flow diagram of a hydrofinishing plant is shown in Figure 1.11. Oil and hydrogen are pre-heated and then allowed to trickle downwards through a reactor filled with catalyst particles where the hydrogenation reactions take place. The oil product is separated from the gaseous phase and then stripped to remove traces of dissolved gases or water. Typical reactor operating conditions for hydrofinishing are as follows:

Catalyst temperature	250–350 °C
Operating pressure	20–60 bar
Catalyst type	Ni, Mo supported on high surface area alumina particles

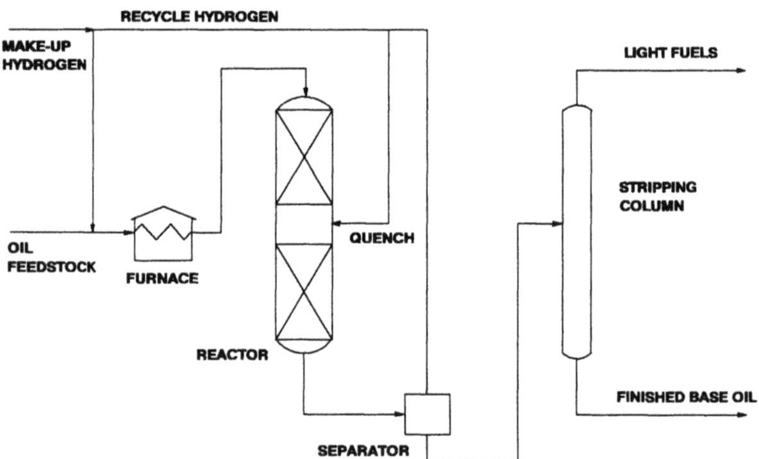

Figure 1.11 Hydrofinishing unit.

1.5 Modern catalytic processes

In recent years, some of the solvent-based separation processes have acquired competition from new processes based on catalytic hydrogenation as an alternative means of removing unwanted components from the base oil. Hydrogenation may offer economic advantages over solvent processes and gives products that are clearly differentiated from conventional solvent-refined base oils. In fact, some of the processes can go a stage further and actually create new and highly desirable components, so that the resulting base oils have characteristics which are superior to anything that could be made by conventional solvent-refining technology relying on physical separation processes.

The types of reaction that occur in catalytic hydrogenation processes are:

- Hydrogenation of aromatics and other unsaturated molecules
- Ring-opening, especially of multi-ring molecules
- Cracking to lower molecular weight products
- Isomerisation of alkanes and alkyl side-chains
- Desulphurisation
- Denitrogenation
- Reorganisation of reactive intermediates, e.g. to form traces of stable polycyclic aromatics.

The extent to which each of these reaction types occurs is determined by the type of catalyst used, the process conditions and the feedstock composition.

1.5.1 *Severe hydrotreatment*

A mild version of this process has already been described in section 1.4.8 as the hydrofinishing step at the end of a conventional base oil production scheme. Under much more severe operating conditions, hydrogenation of aromatics and ring opening reactions become important and the result is to substantially reduce the aromatic content of the lube distillate. Reactions, however, are not limited to hydrogenation and ring opening. Chain-breaking or hydrocracking reactions which lead to molecular weight reduction are also very important. The distillate feedstock is converted to a range of lower boiling point products (naphtha, kerosine, gas oil) in addition to material which remains within the lube boiling range.

The lube products have high VI and are analogous to the products made by solvent extraction of distillates, but there are also important differences. Denitrogenation and desulphurisation reactions lead to products of extremely low sulphur and nitrogen content. Severe hydrotreatment chemically changes the molecular composition, destroying some kinds of molecules and creating other kinds which have good VI properties. Thus the chemical properties and some physical properties of severely hydrotreated base oils are not quite the same as solvent refined base oils.

Severe hydrotreatment actually decreases the range of molecular types within the base oil compared to solvent extracted base oils and so hydrotreated base oils produced from different crude oils are more consistent in their properties than extracted oils made from different crudes. Since the hydrotreating reactions create high VI molecules, it becomes possible to produce base oils from crude oils that have an inherently low content of higher VI components and so are normally unsuitable for conventional solvent refining. This benefit is discussed by Farrell and Zakarian (1986).

Production of base oils by this route is sometimes described as lube oil hydrocracking because it is really a variant of the common refinery process of hydrocracking to make light fuels products from vacuum distillate feedstocks. It is not a complete process for making base oils. Distillation, de-waxing and usually also hydrofinishing steps are needed, just as in a conventional lube plant.

The catalysts used for severe hydrotreatment are specialised types of hydrocracking catalyst. Normally they comprise sulphides of metals from Groups VI and VIII of the Periodic Table (Mo, W, Ni, Co), supported on a high surface area, high acidity base (alumina or silica–alumina). Although aluminosilicate zeolites are often used as supports for hydrocracking catalysts, they are preferred for processes which make light fuels products rather than lube products.

The catalysts are manufactured as mechanically strong particles by extrusion, tableting, or spheridisation so that they can be packed by the tonne to make a porous catalyst bed inside the reactor vessel.

A severe hydrotreating plant will have a similar flow-scheme to the hydrofinishing unit shown in Figure 1.11. Hydrocracking is a highly exothermic reaction so cold hydrogen has to be injected at several points in the catalyst bed to moderate the temperature rise. Operating conditions are severe:

Pressure	100–180 bar
Reactor temperature	350–420 °C

Because the hydrocracking process causes a change in the boiling range of the oil, the feedstock does not need to be a narrow range distillate as is desirable for conventional solvent processing. It must, however, contain a substantial proportion of high boiling components or there will be a very poor yield of lube boiling range product.

1.5.2 *Special base oils from hydrocracking*

The severe hydrotreatment process described above is an alternative route to the manufacture of high VI base oils from conventional solvent extraction

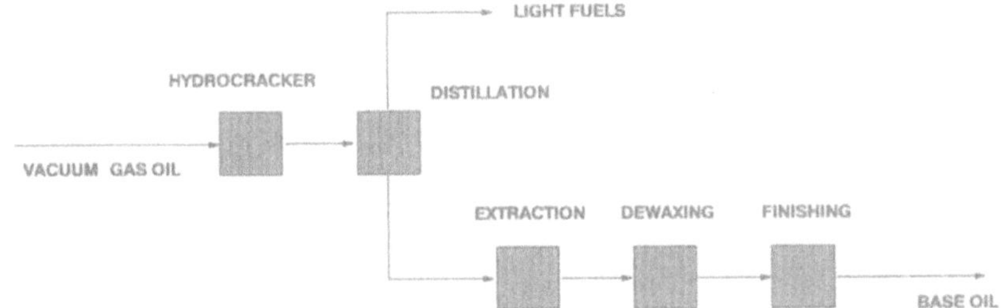

Figure 1.12 Scheme for hydrocracked base oils.

technology. If the hydrotreatment is still more severe, then hydrocracking becomes the dominant mechanism and the yield of material still within the useful lube boiling range falls from 40–70% to 5–15%. Under these conditions, aromatics destruction is largely complete and the potential base oil product is highly paraffinic. Such a base oil has advantages over ordinary hydrotreated or extracted base oils: the viscosity index is higher and volatility is lower than for a conventionally produced oil of the same viscosity. These are important benefits for formulating automotive engine lubricants. Although the natural oxidation stability of these special very high VI base oils is unremarkable, the response to additives such as inhibitors and viscosity index improvers is particularly good. In many respects, these base oils approach the desirable characteristics of synthetic polyalphaolefin base oils (see chapter 2), but are considerably less expensive to manufacture.

An existing lube hydrocracker can be operated at higher severity to make this special product, but the sharp reduction in yield may not be attractive for the base oil plant economics. However, an alternative source of hydrocracked base oil is available from a few of the many fuels hydrocrackers in existence. These hydrocrackers are important refinery conversion units and are used to make a range of fuels products from vacuum distillate feedstocks. Some plants do not fully convert the feed in one pass to low boiling products and the limited amount of residue which remains (5–10%) can be recycled within the plant, used as a fuel oil blending component or be upgraded to make these special base oils.

Further processing of the hydrocracker residue is needed and Figure 1.12 indicates the extra steps. Extraction and hydrotreatment are desirable to remove traces of polycyclic aromatics and improve product quality. Dewaxing is essential because the hydrocracker residue is invariably waxy and distillation is needed to adjust boiling range and viscosity of the base oil.

The economics of making special base oils from fuels hydrocracker residue are determined both by the hydrocracker operation and additional processing at a conventional base oil plant, which is often at a separate site.

Table 1.4 Base fluid comparisons.

	Solvent refined	Hydrocracked	Wax isomerised	Polyalphaolefin
Viscosity at 100 °C (cSt)	5.2	5.6	5.0	5.8
Viscosity index	98	125	146	137
Pour point (°C)	−15	−15	−18	−60

1.5.3 *Special base oils by wax isomerisation*

Another variant of the severe hydrotreatment process is the substitution of wax for lube distillate as feedstock. The wax recovered from conventional solvent de-waxing units is essentially a pure alkane feedstock containing a high proportion of linear alkanes. With this type of feedstock and under appropriate operating conditions, the isomerisation reaction can be made to predominate over cracking reactions. Unconverted wax can be removed by conventional methods to yield a base oil that is exclusively composed of iso-alkanes and which resembles synthetic polyalphaolefin base fluids more closely than the hydrocracked base oils described in section 1.5.2. A comparison of some of these base fluid properties is shown in Table 1.4.

Synthetic polyalphaolefins are composed of a very limited number of branched alkane isomers, all having the same molecular weight and they are also completely wax-free. The wax isomerisation product has a wider spread of isomers and covers a broad band of different molecular weights. It also contains some wax and so these products cannot match the low temperature properties of the polyalphaolefins.

A process flow scheme for making these base oils is shown in Figure 1.13. The wax feedstock reacts over a catalyst in a hydrogen atmosphere but, despite control of conditions to favour isomerisation, a significant amount of cracking to lighter products is inevitable. Products are separated by distil-

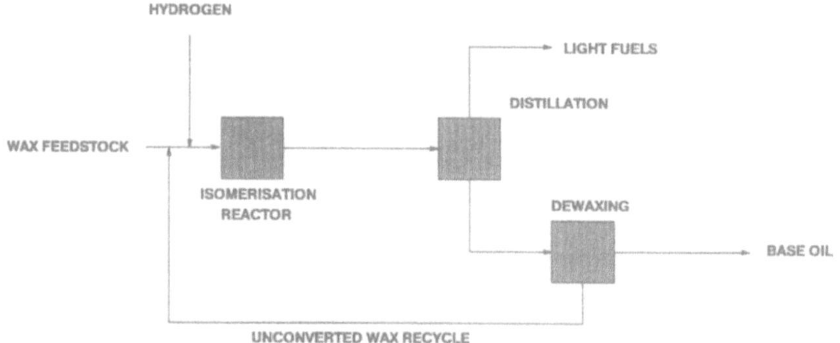

Figure 1.13 Scheme for wax isomerised base oils.

lation and the lube boiling range material is conventionally de-waxed. It is possible to recycle the unconverted wax to increase the overall base oil yield.

The isomerisation reactor may be part of a purpose-built plant or it may be possible to use an existing lube hydrocracker, as described by Bull and Marmin (1979). Such a dual purpose plant will be operated for some periods as a severe hydrotreater to make normal VI base oils, and for other periods as an isomerisation unit to make these special base oils.

1.5.4 *Catalytic de-waxing*

In catalytic de-waxing, special shape selective catalysts are used to selectively hydrocrack only the straight chain alkanes to low boiling point by-products. Since it is the linear alkanes that comprise the bulk of the waxy components in the lube boiling range, the oil is effectively de-waxed. There are differences in the composition and properties of base oil de-waxed by catalytic or solvent processes because of different selectivities.

De-waxing catalysts are based on molecular sieves (zeolites, silicalite, etc.) which have a highly porous structure based on a regular arrangement of channels. The channel openings have diameters of 5–7 Å which will admit only linear or very lightly branched alkanes. Non-waxy branched alkanes and cyclic structures cannot enter the pore structure and so are not converted, although long alkyl side chains on aromatic or alicyclic nuclei can be stripped off. Side chain removal may lead to the loss of some beneficial molecules which have good lubricant performance and are still non-waxy. In contrast, solvent de-waxing separates only those molecules which have crystallised under the conditions of de-waxing. Since there is always some good oil trapped in the filter cake of wax, there is also a loss of useful molecules in solvent de-waxing, but these molecules are not the same as those lost by the catalytic route.

The cracking of alkanes initially produces unsaturated, low molecular weight by-products which can polymerise and, through coke formation, cause a rapid loss of catalyst activity. This problem can be minimised by use of a zeolite with low coke forming tendencies (e.g. ZSM-5) or by incorporation of a hydrogenation function (e.g. Pt, Pd) within the de-waxing catalyst. In the latter case, the oil is simultaneously de-waxed and hydrofinished whereas in the former case a separate hydrotreatment step is necessary to give a stable product, therefore a separate reactor and catalyst system must be provided. Components such as platinum require greater protection from catalyst poisons, but the molecular sieve action of the catalyst automatically tends to protect the active catalytic sites from many potential poisons which are too large to enter the pores.

Catalytic de-waxing is highly selective for linear alkanes and, since these molecules have the highest VI characteristics, it is usual for catalytically de-waxed base oils to have lower VIs than solvent de-waxed oils of the same pour point. However, it is interesting to observe that the directly measured low-temperature properties such as cold crank simulator and Brookfield viscosities are actually superior for the catalytic products. This is explained by the fact that the molecules of highest melting point and tendency to form wax gels, are removed more selectively by the catalytic process. The catalytic process is especially effective for the production of very low pour point oils from paraffinic crudes that can be replacements for scarce naphthenic oils or for special winter grade lubricants.

A catalytic de-waxing plant resembles other hydrotreating plants, and operating conditions do not need to be particularly severe. Operating costs can be significantly below solvent de-waxing costs, especially for low pour point oils where refrigeration costs become prohibitive. The commercialisation of this technology has been described by Hargrove *et al.* (1979) and Smith *et al.* (1980).

1.6 Future developments

According to Vlemmings (1988), world lubricant consumption is approximately 40 million tonnes annually. Once the additive content and use of re-refined and synthetic oils is subtracted, the demand for petroleum derived base oils is likely to be about 36 million tonnes annually. The majority of base oil plants are of the conventional solvent type and were built at least 15 years ago. Since 1975 there has been significant rationalisation within the oil refining industry and many of the smaller, older plants have closed. However, a few new plants have been built and many others have undergone modernisation programmes and capacity increase through de-bottlenecking. As a result, there has been a modest increase in production capacity over this period.

Future developments in petroleum-derived base oils will be driven by marketing demands for improved base oils and developments in manufacturing technology. Health and safety constraints have reduced demand for naphthenic base oils, but the application of hydroprocessing technology has largely solved this problem.

Automotive lubricant performance trends are towards lower viscosity and lower volatility base oils with improved stability and better low temperature properties. These technical demands can be achieved by the use of relatively expensive synthetic fluids or, increasingly, by use of the special hydrocracked and wax isomerisation types of base oils. There are also concerns about interchangeability of base oils made by different refineries.

Base oil refineries are under pressure to make the base oils that the lubricant marketers require in addition to meeting volume demands and economic pressures to reduce costs. These pressures will be met by modernisation programmes and, to a lesser extent, by construction of new plant. New base oil capacity is very expensive to build and can only be justified at present where there are special economic reasons.

Examples of modernisation developments that are already happening include:

- Revamping of old vacuum distillation units by replacing trays with structured packing. This allows more efficient operation which reduces costs and, very importantly, improves separation of the lube distillates. This in turn assists operation of downstream processing units and can reduce costs further or increase capacity while also improving product performance.
- De-bottlenecking an existing solvent de-waxing plant by adding a catalytic de-waxing unit to process the most difficult base oil grades. It is possible to convert redundant hydrotreating units which may already exist at a refinery at much lower cost than building a new plant.

New base oil capacity will be needed, and most of the projects that are being evaluated or planned are for the regions with highest growth in demand, such as Asia. Any new plant proposals should consider taking advantage of the most modern technologies with the aims of:

- making high quality products
- minimising plant operating costs
- achieving maximum flexibility to run different crude oils

Although hydroprocessing plant tends to be more expensive to build compared to conventional solvent process plant, the advantages of feed flexibility and operating costs make such a route increasingly attractive.

Finally, there is potential to tap into the significant volumes of fuels hydrocracker residues which are available at refineries around the world and upgrade these at existing base oil plants to make more of the special higher VI base oils that are in increasing demand.

References

Bromilow, I.G. (1990) *Supply and demand of lube oil: an update of the global perspective.* AM-90-27 presented at the 1990 NPRA Annual Meeting.

Bull, S. and Marmin, A. (1979) Lube oil manufacture by severe hydrotreatment PD 19. *10th World Petroleum Congress*, Bucharest.

Dean, E.W. and Davis, G.H.B. (1929) *Chem. Metall. Eng.* **36** 618–619.

Farrell, T.R. and Zakarian, J.A. (1986) Lube facility makes high quality lube oil from low quality feed. *Oil and Gas Journal*, May 19, 47–51.

Hargrove, J.D., Elkes, G.J. and Richardson, A.M. (1979) New dewaxing process proven in operations. *Oil and Gas Journal*, Jan 15, 103–105.

Smith, K.W., Starr, W.C. and Chen, N.Y. (1980) *A new process for dewaxing lube base stocks: Mobil lube dewaxing.* API 45 Midyear refining meeting May 1980.

Thrash, L.A. (1991) Annual refining survey. *Oil and Gas Journal*, March 18, 84–105.

Vlemmings, J.M.L.M. (1988) *Supply and demand of lube oils. A global perspective.* AM-88-19 presented at the 1988 NPRA Annual Meeting.

2 Synthetic base fluids

S.J. RANDLES, P.M. STROUD, R.M. MORTIER,
S.T. ORSZULIK, T.J. HOYES and M. BROWN

2.1 Introduction

Synthetic lubricants have been used for many years. In the early 1930s, synthetic hydrocarbon and ester technologies were being simultaneously developed in Germany and the US. In the US, the development of a process for the catalytic polymerisation of olefins led to the formulation of automotive crankcase lubricants with improved low temperature performance (Sullivan *et al.*, 1931; Sullivan and Vorhees, 1934). These products were not commercialised due to the inherent cost of the new synthetic base fluids and to improvements in the performance of lubricants based on mineral oils. In Germany, low temperature performance was one of the driving forces behind the development of similar products by Zorn (Gunderson and Hart, 1962) although the main objective was to overcome the general shortage of petroleum base stocks.

With the exception of the special circumstances of World War II, synthetic lubricants did not become commercially significant until after the war. In general, the improved properties of lubricants achieved with the early synthetic base stocks could be obtained more cost effectively by improved formulations based on mineral oils. However, the requirements for lubricants, particularly military and aero-engine lubricants, to perform over increasing temperature ranges, has stimulated the continuing development of synthetic lubricant technology. Synthetic lubricants are now found in all areas of lubrication such as automobiles, trucks, marine diesels, transmissions and industrial lubricants, as well as aviation and aerospace lubricants.

Many compounds have been investigated as possible base stocks for synthetic lubricants. Gunderson and Hart (1962) identified over 25, of which seven types are of major importance:

- polyalphaolefins
- alkylated aromatics
- polybutenes
- aliphatic diesters

- polyolesters
- polyalkyleneglycols
- phosphate esters

Other materials such as silicones, borate esters, perfluoroethers and polyphenylene ethers are also of importance, but their applications are restricted due either to high cost or to performance limitations.

2.2 Polyalphaolefins

The term polyalphaolefin, or PAO, when used for lubricant base stocks, refers to hydrogenated oligomers of an α-olefin, usually α-decene. Several methods have been investigated for the oligomerisation of α-olefins for use as lubricant base stocks, the most important being free radical processes, Ziegler catalysis, and cationic, Friedel–Crafts catalysis.

2.2.1 *Free radical oligomerisation*

Compared with the other methods, very little work has been done on the free radical initiated oligomerisation of α-olefins. This term includes the thermal oligomerisation of α-olefins which is presumed to occur by a free radical process initiated by adventitious peroxide. The process has a high activation energy (Seger *et al.*, 1950) and gives low yields of poor quality products. The use of benzoyl peroxide or a di-tertiary-alkylperoxide allows oligomerisation to take place at lower temperatures than the thermal reaction (Garwood, 1960). Garwood reported that oligomerisation of α-decene using di-tertiary-butylperoxide gave a product with similar rheological properties to the product of a BF_3 catalysed oligomerisation. However, in general, the free radical process is not amenable to control of the degree of polymerisation and gives low product yields. Furthermore, the products tend to have poor viscosity/temperature characteristics due to skeletal isomerisation during oligomerisation.

2.2.2 *Ziegler catalysed oligomerisation*

Ziegler catalysts studied for the oligomerisation of α-olefins have tended to be based on modified first generation catalysts, i.e. triethylaluminium/titanium tetrachloride. The ratio of aluminium to titanium could have a marked effect on the properties of the product (Beynon *et al.*, 1962, 1967). At a ratio of less than 0.8:1 liquids were produced whereas at ratios above 1:1 waxy products were obtained. This was believed to relate to a change in the mechanism of catalysis from cationic to anionic with higher proportions of aluminium.

A disadvantage of standard Ziegler catalysts is the tendency to give products with a broad distribution of oligomers. Antonsen *et al.* (1963) used a propylene oxide-modified alkylaluminium/titanium tetrachloride catalyst to produce a series of oligomers based on C_2 to C_{12} α-olefins. The C_8-based trimer was claimed to be an outstanding hydrocarbon fluid with a wide range of operating temperatures. Other modified Ziegler systems are also capable of giving high yields of high quality base stocks (Isa, 1986). A more recent example is described in a US patent (White, 1985) which covers the use of an alkylaluminium halide/alkoxide zirconium halide catalyst. It is reported that by changing the temperature of reaction of this system, the degree of oligomerisation of α-decene can be controlled.

Ziegler catalysts have also been used to produce base stocks by co-oligomerisation of ethene with other α-olefins (Gates *et al.*, 1969; Kashiwa and Toyota, 1986). Useful products were obtained using a catalytic complex of di-isobutylaluminium chloride and di-ethoxychlorovanadate in the presence of hydrogen to control the molecular weight of the products.

Although improvements to the Ziegler catalyst have been made, the requirement for a solvent, and difficulties with handling, separation and recovery of the catalyst have led to the use of cationic, Friedel–Crafts type catalysts. These systems also tend to give higher monomer conversions, faster cycle times and greater control over oligomer distribution.

2.2.3 *Friedel–Crafts catalysed oligomerisation*

The use of aluminium trichloride for the polymerisation of α-olefins is described by Sullivan *et al.* (1931). The reaction mechanism requires addition of a proton to the double bond (equation 2.1) therefore a co-catalyst such as water is used (Evans and Polanyi, 1947; Evans *et al.*, 1946a, b).

$$AlCl_3 + H_2O \longrightarrow AlCl_3OH_2 \longrightarrow (AlCl_3OH)^- + H^+ \qquad (2.1)$$

Whilst such catalysts are very reactive, they can lead to cracking and isomerisation which can give products with complex structures and wide molecular weight distributions (Fontana, 1963). The oligomerisation can be controlled to some extent by the use of Lewis bases such as ethyl acetate or polyhydric alcohols in place of water as the co-catalyst (Higashimura *et al.*, 1982). The difficulty in controlling the oligomerisation and the need for low viscosity base fluids for automotive lubricants and aircraft hydraulic oils has led to the predominance of BF_3 based catalyst systems for α-olefin oligomerisation.

As with $AlCl_3$, BF_3 requires a co-catalyst for effective oligomerisation. The type of co-catalyst, which can be water, or an alcohol, acid, ether or ketone, has a significant effect on the degree of oligomerisation. The most common co-catalysts are alcohols, especially *n*-propanol and *n*-butanol.

It has been suggested that the co-catalyst is needed in a less than molar equivalence to the Lewis acid (Brennan, 1980). The following initiation step was proposed to account for this finding:

$$CH_2{=}CHR + R'OH.BF_3 + BF_3 \longrightarrow \left[\begin{matrix} \overset{H}{HC} \cdots H \cdots \overset{R'}{O}BF_3 \\ \| \qquad\qquad \vdots \\ \underset{H}{RC} \qquad\qquad BF_3 \end{matrix}\right] \longrightarrow \begin{matrix} CH_3 \\ | \\ \underset{H}{RC^{\oplus}} \end{matrix} \quad \begin{matrix} \ominus\overset{R'}{O}BF_3 \\ BF_3 \end{matrix} \qquad (2.2)$$

The propagation step proposed by Whitmore (1934) involves the insertion of monomer between the olefinic cation and its associated gegenion. Termination occurs by proton transfer from the oligomeric cation:

$$R{-}CH_2{-}CH{=}CH_2 \underset{\text{Termination}}{\overset{\text{Initiation}}{\rightleftharpoons}} R{-}CH_2{-}\overset{\oplus}{C}H{-}CH_3 \xrightarrow{RCH_2CH=CH_2} R{-}CH_2{-}\overset{CH_3}{\overset{|}{C}H}{-}CH_2{-}\overset{\oplus}{C}H{-}CH_2R \rightarrow \text{etc.}$$

$$R{-}CH_2{-}\overset{CH_3}{\overset{|}{C}H}{-}CH_2{-}\overset{\oplus}{C}H{-}CH_2R \xrightarrow{\text{Termination}} RCH_2{-}\overset{CH_3}{\overset{|}{C}H}{-}CH{=}CH{-}CH_2R \qquad (2.3)$$

Whilst the BF_3 catalysed oligomerisation of α-olefins has many advantages over other methods, such as allowing control of the degree of oligomerisation, high conversions, lack of solvent and a short time-cycle for the reaction, there is a problem in the inability to recycle the catalyst. The catalyst is normally removed by washing the reaction mixture with a solution of ammonia. A method of overcoming this problem with the use of a heterogeneous silica-BF_3-water catalyst, which allows the BF_3 to be easily recovered by distillation, has been proposed (Madgavkar, 1983; Madgavkar and Barlek, 1981; Madgavkar and Swift, 1981, 1983).

The mechanism of the BF_3 catalysed oligomerisation of α-olefins is not fully understood. In particular, the mechanism proposed by Whitmore (1934) does not account for the number of methyl groups in the products. Shubkin *et al.* (1979, 1980) considered this problem and postulated that methyl group migration in the dimer might explain the difference. However, the proposed mechanism requires the formation of an intermediate protonated cyclopropyl compound which is energetically unfavourable.

Analysis of the monomer fraction product of an oligomerisation reaction has been considered as a route to understanding the reaction mechanism (Brennan, 1980; Onopchenko *et al.*, 1982, 1983). The polymerisation of α-decene to high conversion gave a monomer fraction which contained 60% α-decene, 35% internal alkenes and 5% methylnonenes. It was concluded that, in addition to methyl group migration, isomerisation of the double bond could also take place, even with the monomer. Since polyalphaolefins derived from internal olefins usually give products with inferior temperature/viscosity properties, double bond isomerisation can be a problem.

The analysis of products from the oligomerisation of long chain alkenes can be difficult, therefore the oligomerisation of low molecular weight alkenes has been used as a model. Infra-red analysis of the products from the oligomerisation of propene, but-1-ene and but-2-ene has been used to identify the structural units (Puskas *et al.*, 1979). Only internal alkenes were found and the structures obtained were believed to derive from intramolecular 1,2-hydride shifts and some tail-to-tail coupling of monomer units.

Developments in ^{13}C nmr spectroscopy have allowed unambiguous assignments of chemical shifts in a number of hydrocarbon structures. Such information has allowed specific structures within the hydrogenated dimer and trimer fractions of propene and butene polymerisations to be identified. This method has led to the proposal of 1,2-hydride and methanide shifts, hydride transfer and proton elimination to explain the structure of the products (Corno *et al.*, 1979; Ferraris *et al.*, 1980).

Analysis of the dimer and trimer fractions of the polymerisation products of propene and but-1-ene by gas chromatography–mass spectrometry also suggested intramolecular hydride and methanide shifts as the source of the isomers formed. A possible mechanistic scheme for the oligomerisation of propene has been proposed (Audisio *et al.*, 1988):

$$H_2C{=}CH(CH_3) \xrightarrow{H^+} H_3C{-}\overset{+}{C}H(CH_3) \xrightarrow{H_2C{=}CH(CH_3)} H_3C{-}CH(CH_3){-}CH_2{-}\overset{+}{C}H(CH_3)$$

$$\xrightarrow{H^- \text{ shift}} \underset{(A)}{H_3C{-}CH(CH_3){-}\overset{+}{C}H{-}CH_2(CH_3)} \xrightarrow{CH_3^- \text{ shift}} H_3C{-}\overset{+}{C}H{-}CH(CH_3){-}CH_2(CH_3)$$

$$\xrightarrow{-H^+ (+H_2)} \text{3-methylpentane}$$

$$(A) \xrightarrow{CH_3^- \text{ shift}} H_3{-}CH(CH_3){-}CH(CH_3){-}\overset{+}{C}H_2 \xrightarrow{H^- \text{ shift}}$$

$$H_3C{-}CH(CH_3){-}\overset{+}{C}(CH_3){-}CH_3 \xrightarrow{-H^+ (+H_2)} \text{2,3-dimethylbutane} \qquad (2.4)$$

In most of the above studies, the products have been analysed after hydrogenation. However, a study by Chaffee *et al.* (1987), which described the analysis of the C_7 product of a propene/but-1-ene copolymerisation before hydrogenation, again confirmed the role of hydride and methanide shifts in determining the structures of the products.

Analysis of the 'simple' structures obtained from the dimer and trimer fractions of short chain alkenes has shown the complexity of the oligomerisation reaction. It is not surprising, therefore, that oligomerisation of the longer chain alkenes such as dec-1-ene should give complex mixtures of isomers. However, it should be remembered that the rearrangements which give rise to the mixture of isomers are all intramolecular thus giving products with predictable molecular weights.

Developments in PAO technology have included the use of acidic, aluminosilicate shape-selective catalysts. The Mobil Oil Corporation (Chen and Shuihua, 1989) reported the use of a chromium on silica gel catalyst, and a shape-selective metallosilicate to produce olefin oligomers. These were suitable intermediates for reaction with an enophile to produce polar synthetic base stocks.

2.3 Alkylated aromatics

2.3.1 *Introduction*

Three types of alkyl benzenes are available for use as or in lubricants: (i) dialkyl benzenes produced as a by-product in the manufacture of linear dodecyl or tridecyl benzene, a sulphonation feedstock for anionic detergent manufacture; (ii) dialkyl benzenes produced from these detergent alkylates by reacting further with α-olefins; and (iii) synthesised alkyl aromatics purpose-built from benzene and short chain olefins such as propylene. Characteristics of the materials which either individually or in combination make them suitable for particular applications include sulphur free chemistry, low pour point, good thermal stability (when compared to mineral oils), and good solvency.

2.3.2 *Production and general properties*

The lowest cost (di)alkyl benzenes are produced as a by-product in the manufacture of linear monoalkylates used extensively in detergent manufacture. Synthesis is usually by Friedel–Crafts type reactions, either by the reaction of the alkyl chloride and benzene with anhydrous aluminium chloride as catalyst, or by the catalysed reaction of the appropriate length α-olefin and benzene. The more specialised alkyl benzenes are synthesised from benzene and propylene, with catalysts and conditions selected to give some control over the molecular weight and structure, and hence the physical and chemical properties, of the alkyl component.

Properties of synthetic alkyl benzenes vary widely depending on chemical type, and are therefore difficult to generalise. However, a typical structure is shown in Figure 2.1. A typical value of x would be 3, giving a viscosity at

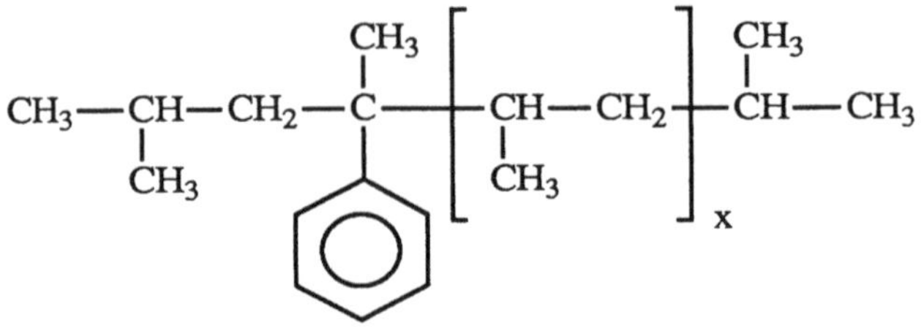

Figure 2.1 Synthetic alkyl benzene.

100 °C of 4 cSt and a viscosity index of 100. With this structure, pour points of −40 °C are attainable as the material is essentially wax free. The presence of the aromatic ring donates a fairly high degree of polarity to the molecule resulting in good solvating powers for most lubricant additives.

2.3.3 *Applications*

The lower cost dialkyl benzenes are used in a wide variety of industrial and metalworking products. In particular, their sulphur-free chemistry has led to their extensive use in rolling and drawing oils for copper. The synthesised alkyl benzenes, even though optimised for their chemistry, generally exhibit poorer properties than PAOs. However, their excellent solvency and low pour point make them suitable for lubricants designed to operate in extremely low temperature conditions, such as arctic greases, gear oils, hydraulic and power transmission fluids. The most widespread use of the tailored alkylates is, however, as refrigerator oils. When carefully designed, the oils are fully compatible with fluorinated refrigerants (such as R22 and R502) whereas mineral oils suffer from wax precipitation and PAOs give phasing problems. The oil may be carried through the entire refrigeration circuit and can experience extremes of temperature. Good thermal stability is therefore essential. The natural solvency of the alkylates is also of benefit in minimising the formation of sludges, varnishes and other deposits. Normally no supplementary additives are required, but in applications where operating conditions are particularly severe, anti-wear compounds such as triaryl phosphates have been successfully used.

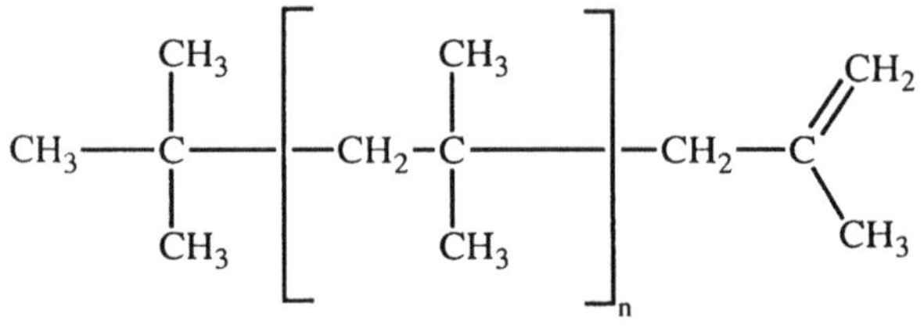

Figure 2.2 Unhydrogenated polyisobutylene.

A final application of the lower molecular weight alkyl benzenes (alkyl chain length C_9 to C_{16}), either neat or blended with mineral oils, is in transformer oils where exceptionally high resistance to gas evolution is required.

2.4 Polybutenes

2.4.1 *Introduction*

Polybutenes used in lubricants are usually composed mainly of isobutylene, and are therefore also referred to as polyisobutylenes or PIBs. Polybutenes are available in a wide variety of molecular weights and physical properties, ranging from low viscosity fluids to rubbers.

Amongst the polyolefins, PIB shows substantially different properties to the PAO-type lubricants, which are generally made from higher molecular weight straight chain α-olefins. In lubricant terms, therefore, it is treated separately. PIB has commonly been used in the adhesives and sealants industries. In the lubricant industry it is used in the manufacture of dispersant additives and viscosity index (VI) improvers, where it has been particularly exploited for its clean burning characteristics. Lower molecular weight fluids, though rarely used as true base fluids, have found application where their ability to depolymerise and burn completely without leaving any deposits is advantageous. Being hydrocarbons, polybutenes are also compatible with mineral oils, and confer good metal-wetting properties and improved film strength.

2.4.2 *Production and general chemical properties*

Polybutenes are generally produced by polymerising a mixed C_4 stream from a catalytic cracking process. A typical feedstock would be a butane–butene mixture containing both *n*-butene and isobutylene. As with PAO manufacture, a cationic polymerisation process (e.g. $AlCl_3/HCl$ or BF_3/methanol), in which isobutylene is selectively polymerised in the presence of the other C_4 compounds, is usually employed. The resulting PIB, containing some polymerised 1- and 2-butene, is used in lubrication in preference to polymers of pure isobutylene, which are also available commercially. Unlike PAOs, PIB is normally sold in the olefin form, i.e. the residual double bond is not hydrogenated, but there are some examples of hydrogenated PIB. Figure 2.2 shows PIB with a double bond in the terminal position, though other structures are also possible. The ability of PIBs to depolymerise and oxidise on heating limits their use as base fluids, but is an important property in applications where low deposits are required, particularly in metalworking and two-stroke lubricant formulations, and as VI improvers.

2.4.3 *Application of PIB fluids*

Lower viscosity PIB fluids in the molecular weight range 400–1300 are used as synthetic lubricants, whilst very viscous PIBs, of molecular weights generally over 10 000, find use as shear-stable VI improvers. Lubricants blended with PIB, which is more tacky than mineral oil, form a tough lubricating film even under severe conditions. In two-strokes the use of PIBs alleviates problems associated with deposits, such as port blocking and smoke formation, whilst providing excellent protection against wear and scuffing. Furthermore, the superior lubricating properties of PIBs leads to lower oil:fuel ratio requirements (Souillard *et al.*, 1971; Suigara and Kagaya, 1977).

PIBs are used to lubricate high-pressure ethylene compressors in polyethylene manufacture, and as lubricants in high temperature applications, such as ovens, dryers, furnaces and metalworking, where their ability to burn off cleanly is essential. Polyisobutylene is non-biodegradable and this may limit its use in some applications where environmental issues are significant.

2.5 Synthetic esters

2.5.1 *Introduction*

Prior to the early 19th century, the main lubricants were natural esters contained in animal fats such as sperm oil and lard oil, or in vegetable oils such as rapeseed and castor oil. During World War II a range of synthetic oils was developed. Amongst these, esters of long chain alcohols and acids proved to be excellent for low temperature lubricants. Following World War II, the further development of esters was closely linked to that of the aviation gas turbine. In the early 1960s, neopolyol esters were used in this application because of their low volatility, high flash points and good thermal stabilities. Esters are now used in many applications including automotive and marine engine oils, compressor oils, hydraulic fluids, gear oils and grease formulations. The inherent biodegradability of ester molecules offers added benefits to those of performance.

2.5.2 *Ester types*

The direct effect of the ester group on the physical properties of a lubricant is to lower the volatility and raise the flash point. This is due to strong dipole moments, called the London forces, binding the lubricant together. The presence of the ester group also affects other properties such as:

- thermal stability
- hydrolytic stability

- solvency
- lubricity
- biodegradability

These properties will be discussed more fully later in the chapter. The major types of esters and their feedstocks are reviewed in Table 2.1. Table 2.2 summarises the physical properties of these esters.

2.5.3 *Manufacture of esters*

The manufacturing process of esters consists of three distinct processes: esterification, filtration and distillation. The fundamental reaction process is that of acid + alcohol → ester + water. This reaction is reversible, but is driven to completion by the use of excess alcohol and removal of water as it forms. The use of an azeotroping agent, e.g. toluene, to aid water removal is optional.

The acid and alcohol can be reacted thermally, usually in the presence of a catalyst in an esterification reactor. Possible catalysts include sulphuric acid, *p*-toluene sulphonic acid, tetra alkyl titanate, anhydrous sodium hydrogen sulphate, phosphorus oxides and stannous octanoate. After the ester has been formed, unreacted acid is neutralised using sodium carbonate or calcium hydroxide and removed by filtration.

Typical reaction conditions are 230 °C and 50–760 mmHg pressure. A significant amount of alcohol vaporises along with the water and must be recovered. This is accomplished by condensing the reactor vapours and separating the resulting two-phase liquid mixture. The alcohol is then returned to the reactor.

Polyol esters are made by reacting a polyhydric alcohol, such as neopentyl glycol (NPG), trimethylol propane (TMP) or pentaerythritol (PE), with a monobasic acid to give the desired ester. When making neopolyol esters, excess acid is used because the acid is more volatile than the neopolyol and is therefore easy to recover from the ester product.

2.5.4 *Physicochemical properties of ester lubricants*

Mineral oil base stocks are derived from crude oil and consist of complex mixtures of naturally occurring hydrocarbons. Synthetic ester lubricants, on the other hand, are prepared from man-made base stocks having uniform molecular structures, and therefore well-defined properties that can be tailored to specific applications.

Many lubricant requirements are translated into specific properties of an oil measurable by conventional laboratory tests, e.g. viscosity, evaporation, flash point, etc. Other, more critical requirements are related to the chemical properties of the lubricant, and many of these can only be measured satisfac-

Table 2.1 Ester types

Diesters (dioates)

$R'OOC(CH_2)_nCOOR''$

R',R'' = linear, branched or mixed alkyl chain

n = 4 = adipates
n = 7 = azelates
n = 8 = sebacates
n = 10 = dodecanedioates

C_{36} dimer acid esters

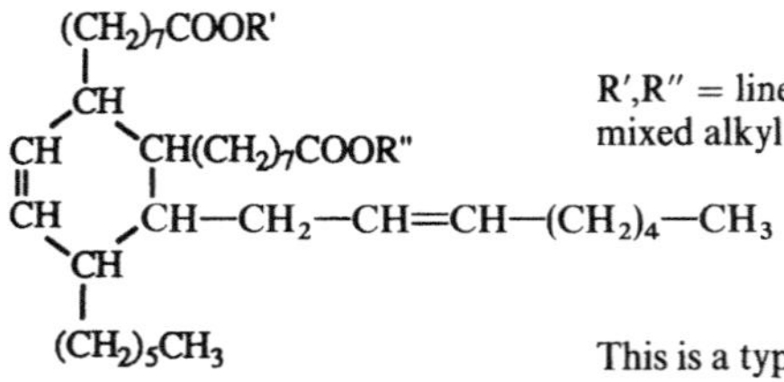

$(CH_2)_7COOR'$
CH
CH $CH(CH_2)_7COOR''$
CH $CH—CH_2—CH{=}CH—(CH_2)_4—CH_3$
CH
$(CH_2)_5CH_3$

R',R'' = linear branched or mixed alkyl chain

This is a typical structure encountered in dimer acids, the ester can also be fully hydrogenated

Trimellitate esters (1,2,4-benzene tricarboxylate)

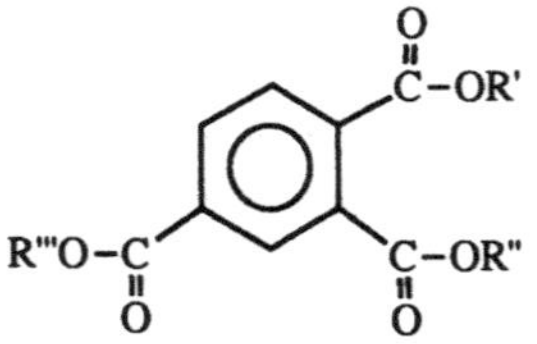

$R'''O-C(=O)-C_6H_3(-C(=O)-OR')(-C(=O)-OR'')$

R',R'',R''' = linear, branched or mixed alkyl chain

Phthalate esters (1,2-benzene dicarboxylate)

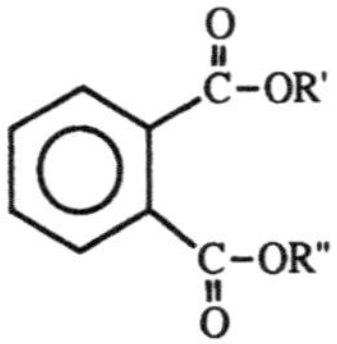

$C_6H_4(-C(=O)-OR')(-C(=O)-OR'')$

R',R'' = linear, branched or mixed alkyl chain

Polyols (hindered esters)

$C(CH_2OCOR)_4$	Pentaerythritol esters
$CH_3CH_2C(CH_2OCOR)_3$	Trimethylolpropane esters
$(CH_3)_2C(CH_2OCOR)_2$	Neopentylglycol esters

R = Branched, linear or mixed alkyl chain

Table 2.2 Summary of ester properties

	Diesters	Phthalates	Trimellitates	C_{36} dimer esters	Polyols	Polyoleates
Viscosity at 40 °C	6 to 46	29 to 94	47 to 366	13 to 20	14 to 35	8 to 95
Viscosity at 100 °C	2 to 8	4 to 9	7 to 22	90 to 185	3 to 6	10 to 15
Viscosity index	90 to 170	40 to 90	60 to 120	120 to 150	120 to 130	130 to 180
Pour point (°C)	−70 to −40	−50 to −30	−55 to −25	−50 to −15	−60 to −9	−40 to −5
Flash points	200 to 260	200 to 270	270 to 300	240 to 310	250 to 310	220 to 280
Thermal stability	Good	Very good	Very good	Very good	Excellent	Fair
Conradson carbon	0.01 to 0.06	0.01 to 0.03	0.01 to 0.40	0.20 to 0.70	0.01 to 0.10	?
% Biodegradability	75 to 100	46 to 88	0 to 69	18 to 78	90 to 100	80 to 100
Costs (PAO = 1)	0.9 to 2.5	0.5 to 1.0	1.5 to 2.0	1.2 to 2.8	2.0 to 2.5	0.6 to 1.5

torily by elaborate and expensive rigs specially developed to simulate performance.

A wide variety of raw materials can be used for the preparation of ester type base fluids and this can affect a number of lubricant properties including:

- viscosity
- flow properties
- lubricity
- thermal stability
- hydrolytic stability
- solvency
- biodegradability

2.5.4.1 *Viscosity* The viscosity of an ester lubricant can be altered by:

- increasing the molecular weight of the molecule by
  —increasing the carbon chain length of the acid
  —increasing the carbon chain length of the alcohol
  —increasing the number of ester groups
- increasing the size or degree of branching
- including cyclic groups in the molecular backbone
- maximising dipolar interactions

One disadvantage of very long chain molecules is their tendency to shear into smaller fragments under stress.

2.5.4.2 *Flow properties* The viscosity index (VI) of an ester lubricant can be increased by:

- increasing the acid chain length
- increasing the alcohol chain length
- increasing the linearity of the molecule
- not using cyclic groups in the backbone, which lowers the VI even more than aliphatic branches
- molecular configuration—viscosity indices of polyol esters tend to be somewhat lower than their diester analogues due to the more compact configuration of the polyol molecule

The pour point of the lubricant can be decreased by:

- increasing the amount of branching
- the positioning of the branch—branching in the centre of the molecule gives better pour points than branches near the end
- decreasing the acid chain length
- decreasing the internal symmetry of the molecule

As can be seen from the above lists, there is a natural trade-off between viscosity index and pour point. For instance by increasing the linearity of the ester the viscosity index improves but the pour point increases. Esters made from mixtures of normal and branched acid (having the same carbon number) have viscosity indices between those of the normal and branched acid esters, but have lower pour points than esters used from either branched or normal acids.

2.5.4.3 *Lubricity* Ester groups are polar and will therefore affect the efficiency of anti-wear additives. When a too polar base fluid is used, it, and not the anti-wear additives, will cover the metal surfaces. This can result in higher wear characteristics. Consequently, although esters have superior lubricity properties compared to mineral oil, they are less efficient than anti-wear additives.

Esters can be classified in terms of their polarity, or non-polarity by using the following formula (Van der Waal, 1985):

$$\text{Non-polarity index} = \frac{\text{total number of C atoms} \times \text{molecular weight}}{\text{number of carboxylic groups} \times 100}$$

Generally, the higher the non-polarity index, the lower the affinity for the metal surface. Using the above formula it can be seen that as a general rule, increasing molecular weight improves overall lubricity. Esters terminated by normal acids or alcohols have better lubricities than those made from branched acids/alcohols, while esters made from mixed acids/alcohols have lubricities intermediate between esters of normal acids/alcohols and esters of branched acids/alcohols.

2.5.4.4 *Thermal stability* The ester linkage is an exceptionally stable one; bond energy determinations predict that it is more thermally stable than the C–C bond.

The advantage in thermal stability of polyol esters compared to diesters is well documented and has been investigated on a number of occasions. It has been found that the absence of hydrogen atoms on the beta-carbon atom of the alcohol portion of an ester leads to superior thermal stability. The presence of such hydrogen atoms enables a low energy decomposition mechanism to operate via a six-membered cyclic intermediate producing acids and 1-alkenes (see Figure 2.3a). When beta-hydrogen atoms are replaced by alkyl groups this mechanism cannot operate and decomposition occurs by free radical mechanism. This type of decomposition requires more energy and can only occur at higher temperature (see Figure 2.3b).

Short linear chains generally give better thermal stability than long branched chains, whilst esters made from normal acids generally have higher flash points than those made from branched acids. Increasing molecular weight also increases flash points.

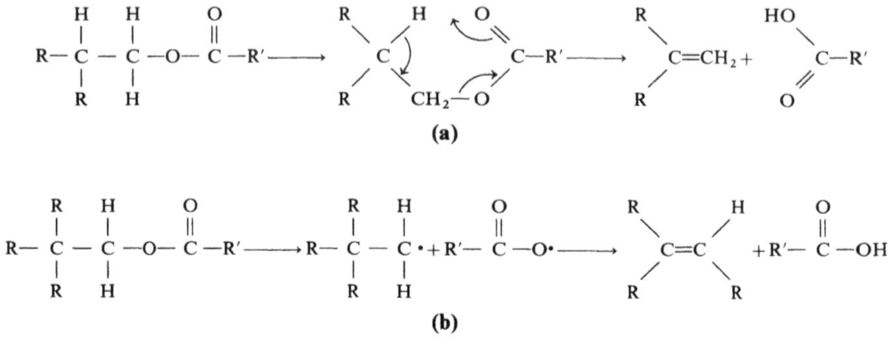

$$R{-}C(H)(R){-}C(H)(H){-}O{-}C({=}O){-}R' \longrightarrow \text{[cyclic transition state]} \longrightarrow R_2C{=}CH_2 + HO{-}C({=}O){-}R'$$

(a)

$$R{-}C(R)(R){-}C(H)(H){-}O{-}C({=}O){-}R' \longrightarrow R{-}C(R)(R){-}C(H)(H)\cdot + R'{-}C({=}O){-}O\cdot \longrightarrow R_2C{=}C(H)R + R'{-}C({=}O){-}OH$$

(b)

Figure 2.3 Thermal decomposition of (a) esters with beta hydrogens (e.g. dibasic acid esters) and (b) esters without beta hydrogens (e.g. neopolyol esters).

2.5.4.5 *Hydrolytic stability* The hydrolytic stability of esters depends on two main features:

- processing parameters
- molecular geometry

If the final processing parameters of esters are not tightly controlled they can have a major effect on the hydrolytic stability of the esters. Such processing parameters include:

- acid value
- degree of esterification
- catalyst used during esterification and the level remaining in the ester after processing

Esters must have a low acid value, a very high degree of esterification and a low ash level before the effects of molecular geometry will begin to assert themselves.

Molecular geometry can affect hydrolytic stability in several ways. By sterically hindering the acid portion of the molecule (hindrance on the alcohol portion having relatively little effect) hydrolysis can be slowed down. To this purpose, geminal di-branched acids (e.g. neoheptanoic acids) have been used. However, when using these feedstocks, there are penalties to be paid, namely very long reaction times to achieve complete esterification, and poor pour points. The hydrolytic stability of neopolyol esters can generally be regarded as superior to that of dibasic esters.

2.5.4.6 *Solvency* This can be divided into compatibility with additives and other lubricants, and elastomer compatibility.

(i) *Compatibility with additives and other lubricants* Esters are generally fully compatible with mineral oils. This gives them three major advantages. First, there are no contamination problems therefore esters can be used in machinery that previously used mineral oil. In addition, they can be blended with mineral oil (semi-synthetics) to boost their performance. Second, most additive technology is based on mineral oil and this technology is usually directly applicable to esters. Third, esters can be blended with other synthetics such as polyalphaolefins (PAOs). This gives esters great flexibility, whilst blending with other oils gives unrivalled opportunities to balance the cost of a lubricant blend against its performance.

(ii) *Elastomer compatibility* Elastomers brought into contact with liquid lubricants will undergo an interaction with the liquid diffusing through the polymer network. There are two possible kinds of interaction, chemical (which is rare) and physical. During physical interactions two different processes occur:

- absorption of the lubricant by the elastomer, causing swelling
- extraction of soluble components out of the elastomer, causing shrinkage

The degree of swelling of elastomeric material can depend on:

- size of the lubricant—the larger the lubricant the smaller the degree of swelling
- molecular dynamics of the lubricant—linear lubricants diffuse into elastomers quicker than branched or cyclic lubricants
- closeness of the solubility parameters of the lubricant and the elastomer. The 'like-dissolves-like' rule is obeyed
- polarity of the lubricant. It is known that some elastomers are sensitive to polar ester lubricants. The non-polarity index can be used to model elastomeric seal-swelling trends for specific ester types.

Several polar esters are well known industrial plasticisers. Non-polar basestocks, such as PAOs, have a tendency to shrink and harden elastomers. By carefully balancing these compounds with esters, lubricants which are neutral to elastomeric materials can be formulated.

2.5.4.7 *Environmental aspects* Growing environmental awareness has turned the threat to our waters into a major issue. The environment can become polluted in many ways, for example oils and oil-containing effluent can have devastating consequences for fish stocks and other water fauna.

(i) *Ecotoxicity* In Germany materials are classified according to their water endangering potential or Wassergahrdungklasse (WGK). Substances are given a ranking of between 0 and 3.

WGK 0	Not water endangering
WGK 1	Slightly water endangering

WGK 2	Water endangering
WGK 3	Highly water endangering

Esters generally have the the following rankings:

Polyols, polyoleates, C_{36} dimer esters, diesters	0
Phthalates and trimellitates	0 to 2

This shows esters to have a low impact on the environment.

(ii) *Biodegradability* The general biochemistry of microbial attack on esters is well known and has been well reviewed. The main steps of ester hydrolysis (Macrae and Hammond, 1982), beta-oxidation of long chain hydrocarbons (Wyatt, 1982) and oxgenase attack on aromatic nuclei (Cerniglia, 1984) have been extensively investigated. The main features which slow or reduce microbial breakdown are:

- position and degree of branching (which reduces β-oxidation)
- degree to which ester hydrolysis is inhibited
- degree of saturation in the molecule
- increase in molecular weight of the ester

Figure 2.4 shows the biodegradabilities of a wide range of lubricants as measured using the CEC-L-33-T-82 test (Randles *et al.*, 1989).

2.5.5 *Application areas*

2.5.5.1 *Engine oils* It is now widely accepted that synthesised fluids, such as polyalphaolefin/ester blends, offer a number of inherent performance advantages over conventional petroleum based oils for the formulation of modern automotive engine oils. Practical benefits which may derive from their use include improved cold starting, better fuel and oil economy, together with improved engine cleanliness, wear protection and viscosity retention during service. Fluid types used in the development of automotive crankcase oils, either commercialised or considered for commercialisation, include polyalphaolefins (PAOs)—more correctly hydrogenated olefin oligomers, organic dibasic esters, polyolesters, alkylated aromatic hydrocarbons, and polyglycols. Experience from numerous laboratories of engine bench and vehicle test programmes conducted over the last ten years has shown that a blend of PAO and an organic ester provides an excellent base fluid for the formulation of synthesised crankcase oils (O'Connor and Ross, 1989; Krulish *et al.*, 1977).

Low temperature viscosity is perhaps the single most important technical feature of a modern crankcase lubricant. Cold starts are a prime cause of engine wear which can be mitigated only by immediately effective lubricant circulation. Furthermore, motor vehicles are increasingly required to operate reliably in arctic conditions. Esters provide this essential low temperature

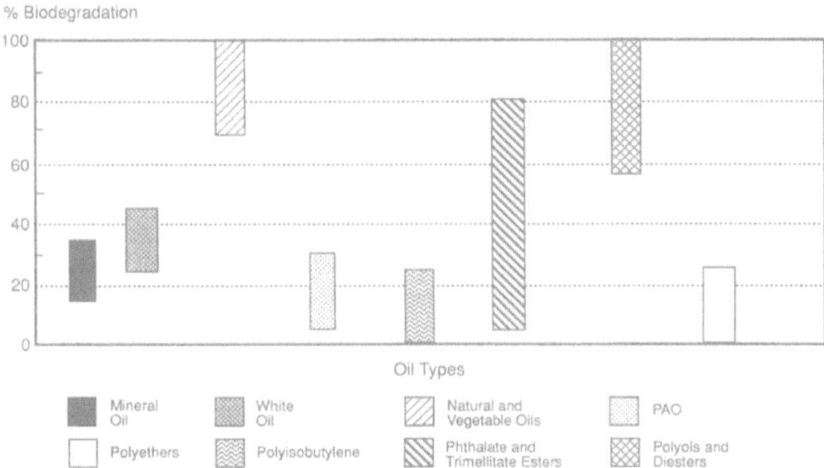

Figure 2.4 Biodegradability of lubricants as measured by the CEC-L-33-T-82 test.

fluidity and, because of their low volatility, do so without any sacrifice of lubricant efficiency at high operating or ambient temperatures. Low volatility is especially important in the context of the modern trend towards smaller sump capacities and longer oil change intervals.

2.5.5.2 *Two-stroke oils* Ester lubricants (such as C_{36} dimer esters and polyoleates) offer a number of advantages over mineral oils as the lubricant component of two-stroke engine mixtures. First, the clean-burn characteristics result in less engine fouling with much reduced ring stick and lower levels of dirt build-up on ring grooves, skirts and undercrowns. Ignition performance and plug life are also enhanced. Second, due to their polar nature, esters are more efficient lubricants than mineral oils. Mineral oil has oil:fuel dilution ratios of 50:1 whereas esters can be used at 100:1 and even 150:1. This higher dilution factor results in reduced oil emissions which is a benefit in environmentally-sensitive applications such as marine outboard engines and chainsaw motors. Third, in some applications, such as engines used to power snowmobile-type vehicles, low temperature performance is important. In these applications, esters with low pour point (down to $-56\,°C$) are very suitable.

Finally a 25% decrease in the amount of PAH (polyaromatic hydrocarbons) in the exhaust emissions of a two-stroke engine has been found when a carboxylic ester has been used in place of a mineral oil (Cosmacki *et al.*, 1988). PAHs have been found to be one of the major contributors to the carcinogenic nature of exhaust emissions. Esters can also be used to reduce the level of smoke emitted by the engine.

2.5.5.3 *Compressor oils* This sector of the market covers a wide range of compressor types, used for a number of different gases. Diesters and phthalates have found their major application in air compressor lubricants, but are also used in compressors handling natural gas. In reciprocating compressors, where oils of rather higher viscosity are preferred, trimelliate esters can be used. Diesters and polyol esters may also be blended with PAOs for use in the various compressor types.

Diesters have inherently good oxidation resistance and low volatilities (3–10%, according to viscosity) when compared to mineral oil. Coupled with their higher flash and auto-ignition temperatures, and low order of toxicity for vapour inhalation, ingestion and skin irritation, these properties make them considerably safer lubricants to use than mineral oil. Their low ecotoxicity and high biodegradabilities can also lessen their environmental impact. Diesters generally have high viscosity indices, giving them a wide temperature range without the use of viscosity improvers. (The latter can shear in this application.) A further advantage of esters is their good thermal conductivity which allows them to conduct heat away from heat sources more effectively than mineral oils. Specific heat values of 5–10% higher than mineral oils enable esters to 'soak' up heat and allow the compressor to operate at cooler temperatures (Wits, 1989).

2.5.5.4 *Aviation oils* The bulk of aviation lubricant demand is for gas turbine lubricants for both military and civilian use. The requirements placed on jet engine oils, namely lubricating, oxidation and ageing stability, cannot be met by hydrocarbon oils. The first generation of oils (Type 1) were diesters but, over the last 25 years, these have slowly lost ground to the more expensive (Type 2) polyol esters. Some diesters are still used in less-demanding applications, e.g. for small private aircraft, turbo-prop engines, etc. Type 2 aviation gas turbine lubricants are produced to a viscosity of 5 cSt (at 100 °C). For some military applications, where operability at low temperatures is vital, the corresponding viscosity is reduced to 3 cSt.

2.5.6 *Future trends*

The requirement for lubricants to operate at high temperature is causing a move away from mineral oil to esters. In particular, due to the better temperature stability of polyols, there is a growing tendency to use these in preference to diesters. In response to increased environmental pressure, the chemistry of esters is being modified so as to produce compounds which have high biodegradabilities, low toxicity, and clean engine emissions.

2.6 Polyalkylene glycols

2.6.1 *Introduction*

The term 'polyalkylene glycol' covers an extensive range of polymeric molecules which, depending on chemical structure, can exhibit quite different physical properties. For example, polyalkylene glycols can be solid or liquid, water soluble or water insoluble, and can be produced to give almost any viscosity required. Polyalkylene glycols have the following inherent chemical and physical properties which make them suitable for a large number of engineering and lubrication applications:

- wide viscosity range
- excellent viscosity/temperature characteristics
- low pour points
- good thermal stability
- high flash points
- good lubricity
- low toxicity
- good shear stability
- water solubility or insolubility
- non-corrosive to common metals
- volatile or soluble breakdown products
- little effect on rubbers
- practically non-flammable in aqueous solution

Typical application areas are industrial gear oils and greases, compressor lubricants, metal working fluids, aqueous quenching fluids, fire-resistant hydraulic fluids, textile lubricants, heat transfer fluids, etc. Section 2.6.5 offers a detailed account of these applications.

2.6.2 *Production*

Polyalkylene glycols are produced by reacting compounds containing active hydrogen atoms (e.g. alcohols, water) with alkylene oxides (also referred to as oxiranes or epoxides), usually in the presence of a basic catalyst (e.g. sodium or potassium hydroxide or tertiary amines. The commonly used alkylene

(a)  $H_2C \overset{O}{\frown} CH_2$ (b) 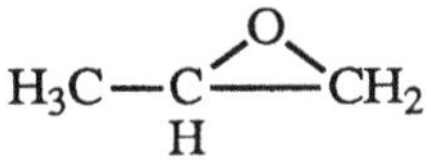 $H_3C{-}\underset{H}{C} \overset{O}{\frown} CH_2$

Figure 2.5 Structures of (a) ethylene oxide and (b) propylene oxide.

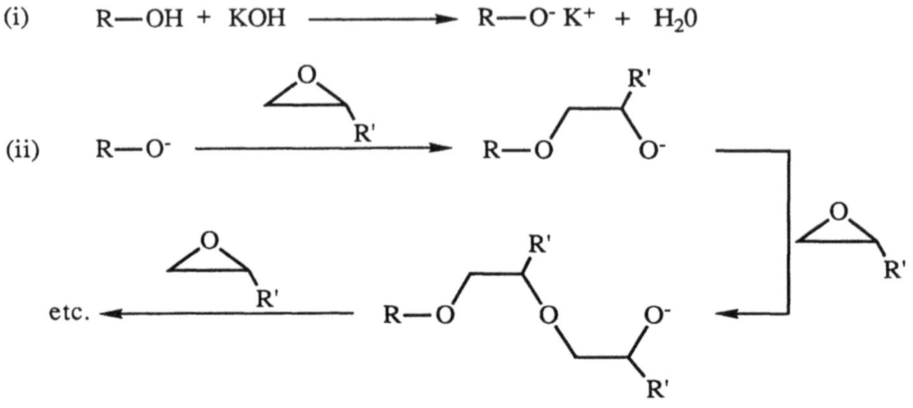

Figure 2.6 Typical polymerisation reaction for an alcohol with KOH as a catalyst.

oxides are ethylene oxide and propylene oxide (see Figure 2.5). Alkylene oxides are very reactive and show a propensity for polymerisation. Reactions are highly exothermic.

Under basic polymerisation conditions, ethylene oxide yields products with exclusively primary terminal hydroxyl groups whereas propylene oxide preferentially results in secondary terminal hydroxyl groups. Figure 2.6 depicts a typical polymerisation reaction for an alcohol (ROH) with KOH as catalyst. Although only a small number of alkoxide ions (i.e. RO^-) will be present in the reaction mixture at any time (only a catalytic quantity of KOH is added), continuous fast exchange of protons results in products of relatively narrow molecular weight distributions. Products are normally characterised by an average molecular weight or by their viscosity. When polymerisation is complete the residual catalyst can be (i) neutralised, e.g. by the addition of phosphoric acid; (ii) demineralised, e.g. amphoteric aluminosilicates; or (iii) left in the product.

2.6.3 *Chemistry*

The ether linkages in polyalkylene glycols are responsible for the unique properties possessed by these polymers. The carbon–oxygen bond of an ether is stronger than the carbon–carbon bond of a hydrocarbon (351 kJ mol^{-1}cf. 348 kJ mol^{-1} respectively) and the electron-rich oxygen atoms offer many sites for co-ordination. In fact it is hydrogen bonding to these sites that confers water solubility properties to many of these molecules. The polarity resulting from the oxygen atoms also confers quite different gas solubilities. Obvious routes to controlling product properties are:

- variation of the starting molecule (initiator)

- variation of the alkylene oxide(s) used
- the molecular weight

2.6.3.1 *Initiators* Typical initiators include butanol (monofunctional), ethylene or propylene glycol (difunctional) and trimethylol propane (trifunctional). Mono- and difunctional initiators both give products having linear chains but the monofunctional based products will have half of the chain ends capped by an alkyl group. Tri- (and poly-) functional initiators result in products with branched chains.

2.6.3.2 *Alkylene oxide(s)* The vast majority of commercialised polyalkylene glycols are based on ethylene oxide only, propylene oxide only, or copolymers incorporating the two. Copolymers can be synthesised as 'random' (oxides added as a mixture giving a statistical distribution throughout the chain) or 'block' (oxides added separately). Due to the more reactive nature of ethylene oxide, random copolymers will tend to preferentially incorporate propylene oxide units at the chain extremities.

Polyalkylene glycols for use in engineering and lubrication applications are usually homopolymers of propylene oxide or random copolymers. The relative proportions of ethylene oxide and propylene oxide have significant effects on several properties, e.g. pour point, water solubility and pressure–viscosity relationship.

2.6.3.3 *Molecular weight* By controlling the quantity of oxide added, polyalkylene glycols with a wide range of molecular weights (and hence viscosities) can be selected. The ability to 'engineer' polyalkylene glycol molecules in this way distinguishes them from most other products used in similar applications. Viscosities (40 °C) can be achieved from as low as 8 cSt to greater than 100 000 cSt.

2.6.4 *Key properties*

2.6.4.1 *Water solubility* As previously mentioned, polyalkylene glycols can be either water soluble or water insoluble. Water solubility is primarily governed by the ratio of ethylene oxide: propylene oxide in the polymer, with a higher proportion of ethylene oxide leading to greater solubility. Molecular weight also influences water solubility, although to a much lesser extent, with lower molecular weight polymers showing greater solubility.

Homopolymers of propylene oxide are effectively water insoluble (except for very low molecular weight, < 900). Typical oxide ratios used for water soluble copolymers are between 1:1 and 3:1 ethylene oxide: propylene oxide (by weight).

All polyalkylene glycols exhibit inverse solubility in water, i.e. water solubility decreases as temperature increases. This is explained by the loss of

Table 2.3 Comparison of pour points for two polyalkylene glycols with different proportions of propylene oxide

Product	Molecular weight	Ethylene oxide: propylene oxide	Pour point (°C)
PEG 1000	1000	1:0	+37
VG 70W	1300	1:1	−46

hydrogen bonding at elevated temperatures. The temperature of polymer/water separation is usually referred to as the cloud point and is higher for copolymers with larger proportions of ethylene oxide.

2.6.4.2 *Pour point* Polyalkylene glycols with high proportions of propylene oxide (> 50%) exhibit very low pour points (down to −50 °C). This is a result of the lateral methyl groups of the propylene oxide unit disrupting crystallisation. A typical comparison is shown in Table 2.3.

2.6.4.3 *Viscosity–pressure behaviour* The relationship between viscosity and pressure is an important parameter for lubrication performance. Better viscosity–pressure coefficients are obtained for polymers with a high degree of propylene oxide units.

2.6.4.4 *Viscosity index* Polyalkylene glycols typically exhibit excellent viscosity/temperature characteristics. For example, Emkarox VG 222 (ISO 220) has a VI of 215 which is significantly better than that for an equivalent ISO grade mineral oil (~ 100). The viscosity index tends to be better for products with a low degree of chain branching.

2.6.4.5 *Degradation* In contrast to mineral oils, polyalkylene glycols form either volatile or soluble degradation products upon oxidation, and hence do not leave unwanted solid deposits during service. Thermal stability can be significantly improved by the addition of antioxidant additives.

2.6.4.6 *Lubrication* Polyalkylene glycols show very good frictional behaviour and are inherently excellent lubricants. The high polar nature gives a strong affinity to metals, and thus lubrication films stay intact even at high surface pressures. This results in low abrasive wear.

2.6.5 *Applications*

2.6.5.1 *Introduction* Polyalkylene glycols (PAGs) were first developed in the mid 19th century but were not used commercially until the late 1930s when an application as a replacement for castor oil in automotive brake fluids was found.

The compulsory use of fire-resistant hydraulic fluids, triggered by the death of 250 miners in the Belgian pit disaster of 1956, resulted in a substantial increase in PAG consumption.

During the 1960s PAGs were used in heat treatment foundries to quench metal. A combination of inherent inverse solubility and fire-resistant properties provided an alternative to mineral oil based products.

Following strong growth during the 1970s and 1980s, PAGs are now used for many applications worldwide, and constitute the largest market share within the synthetic lubricant sector.

What of the future? There is a belief that 'current' PAGs may be near to their peak market share, but developments of new products with improved application performance may give rise to further growth.

2.6.5.2 *Brake fluids* Two predominant types of brake fluid formulations, DOT 3 and DOT 4, based on specifications issued by the US Department of Transport, are used in Western Europe. Due to increased demands placed on brake fluids, particularly with regard to water absorption, much of the polyglycol has been replaced by borate esters. However, the lubricity, limited rubber swell and low pour points of these esters means that a typical DOT 4 formulation may contain up to 10% PAG.

2.6.5.3 *Fire-resistant fluids* A major application for PAGs involves their use in fire-resistant fluids. Industrial sectors such as steel production, die casting (where leaks in high-pressure hydraulic lines can create serious fire hazards) and mining (where the consequences of fire are catastrophic) have generated a large demand foı these types of hydraulic fluids; indeed many countries now legislate their use.

Fire-resistant fluids have been classified by several international bodies:

- ISO—International Organisation for Standardisation
- CETOP—Comité Européen des Transmissions Oleohydrauliques et Pneumatiques
- Luxembourg Commission—Commission of the European Communities Safety and Health Commission for the Mining and Extractive Industries

The following designations have been used:

- HFA fluids containing more than 80% water, formulated with either mineral oil, synthetic chemical solution or synthetic emulsion
- HFB water-in-oil (invert) emulsion, water content 40–50%
- HFC water–polymer solution, minimum water content 35%
- HFD water-free chemical fluid

High molecular weight PAGs (12–30 K) have been found to be particularly suitable for the formulation of the HF-C type of fluid. PAG, water (35% minimum) and glycol, together with anticorrosion, anti-wear and antifoam additives are formulated to conform to the manufacturers' specifications for a particular system. Stringent flammability requirements, corrosion protection, low levels of wear, foam, and other demands such as viscosity and air release must all be achieved.

2.6.5.4 *Compressor lubricants* Lubrication within the pressurised area of a compressor presents many problems. A lubricant must not only seal, cool, and reduce friction and wear, but must also cope with the presence of compressed gases which may be aggressive or of an oxidising nature.

The polarity of PAGs, particularly the water-soluble type, significantly lowers the solubility with hydrocarbon gases (e.g. methane and ethylene) compared to the equivalent mineral oil derived lubricants. From a practical point of view this means that compressors used for hydrocarbon and other chemical gases can be operated without lowering the lubricant viscosity, resulting in increased efficiency. For example, in low density polyethylene compressors operating at 2000–3000 bar, the solubility of ethylene in water-soluble PAG is less than 15% of that observed with white oil. This has the added benefit of reducing lubricant contamination of the final product—an important consideration for polythene when used in food packaging.

2.6.5.5 *Metal cutting* This operation involves the removal of metal, in the form of chips, from a component. Two principal types of fluid are used to cool and lubricate:

- neat mineral oil
- soluble oils

Within the soluble oil sector, there is a small demand for fully synthetic 'true solutions', particularly for light-duty and grinding applications. These solutions may be formulated using water soluble PAGs as base stock. Their inverse solubility characteristic is advantageous in coating the metal surface with a film of polymer, thus providing lubrication and reducing tool wear. Biostability, low toxicity and little or no skin irritancy, are additional benefits.

2.6.5.6 *Gear oils and greases* The main purpose of a lubricant is to reduce friction between moving parts hence lowering temperature and improving efficiency. Increased industrial demand for high performance lubricants has resulted in a much wider acceptance of synthetic based products. PAGs are now extensively used to lubricate calender gears and bearings, within the plastics, rubber and paper industries, and are particularly suited for heavily-loaded worm gears.

The efficiency of a worm gear is related to the friction between worm drive and gear wheel. Within the contact, there is a high degree of sliding which increases operating temperatures. Consequently, a lubricant must not only provide a low coefficient of friction, but must also have a high viscosity index and possess good thermal and oxidative characteristics.

Paper and concrete mills are examples of processes where the ingress of dust and moisture can present particular problems. Gear oils based on water-soluble PAGs may be used to facilitate easier cleaning, and to extend service intervals.

Both water-soluble and insoluble types of PAGs can be used in the formulation of synthetic greases. Thickening may be achieved either by conventional soap technology, or by the addition of solids such as modified bentonite clays or fine particle silica. The clean burn-off characteristic of all PAGs makes them particularly suitable as grease bases, incorporating molybdenum disulphide or graphite. The PAGs form volatile oxidation breakdown products leaving no carbonaceous residues or sludge. A typical application is the lubrication of very high temperature chains in ovens.

2.6.5.7 *Textiles* During the processing of fibres, two important criteria must be considered (i) drag from hard surfaces causing fibre damage; and (ii) excessive lubrication between the individual filaments that may cause fibre/fibre slippage thus impairing the process. An ideal spin-finish formulation will strike a balance between fibre/fibre static friction and fibre/hard-surface dynamic friction. PAGs offer several advantages over mineral oils and simple esters:

- little or no carbonaceous residue is produced when the lubricant is volatilised during high temperature texturising and high-speed spinning
- viscosity and cloud points (inverse solubility) can be readily varied according to requirements
- scourability is good (important when polishing the fibres)
- water-solubility at ambient temperatures assists application, film coverage and removal

A typical spin-finish concentrate would consist of:

lubricant (usually blended)	75–85%
antistat	5–10%
additives (antioxidants)	5–10%

Formulations of textile machine lubricants based on water-soluble PAGs, combine gear lubricity with ease of removal from fibres during processing.

2.6.5.8 *Rubber lubrication* The negligible swelling characteristics of water-soluble and insoluble PAGs are used to effect within the rubber industry. Actual applications include antistick agent for uncured rubber, demoulding fluid in tyre production, mandrel lubricants for hoses, and lubricants for rubber packings, 'O' rings and seals. PAGs may be formulated with solvents and wetting agents and applied directly by use of a brush or spray.

2.6.5.9 *Two-stroke engine lubrication* Synthetic lubricants virtually eliminate engine problems associated with deposition and fouling, commonly seen with mineral oil lubricants. Although esters are predominant in this application, PAGs tend to have 'special' uses, for example, with model engines where a mixture of PAG/methanol provides a 'cleaner' alternative to castor oil based fuels.

2.7 Phosphate esters

2.7.1 *Introduction*

Phosphate esters have been produced commercially since the 1920s and have gained importance as plasticisers, lubricant additives, and synthetic based fluids for hydraulic and compressor oils. They are esters of phenols and alcohols with the general formula $OP(OR)_3$, where R represents aryl, alkyl or, very often, a mixture of alkyl and/or aryl components. The physical and chemical properties of phosphate esters can be varied considerably depending on the choice of substituents, and these are selected to give optimum performance for a given application. Phosphate esters are particularly used in applications that benefit from their excellent fire-resistance properties but compared to other base fluids they are fairly expensive.

2.7.2 *Manufacture*

Phosphate esters are produced by reaction of phosphoryl chloride with phenols or alcohols (or, less commonly, sodium phenoxides/alkoxides).

$$3ROH + POCl_3 \longrightarrow OP(OR)_3 + HCl$$

Early production of phosphate esters was based on the so-called 'crude cresylic acid' fraction or 'tar acids' derived by distillation of coal tar residues. This feedstock is a complex mixture of cresols, xylenols, and other materials, and includes significant quantities of ortho-cresol. The presence of ortho-cresol results in an ester that has marked neurotoxic effects, and this has led to the use of controlled coal tar fractions, in which the content of ortho-cresol and other ortho-*n*-alkylphenols is greatly reduced. Phosphate esters using coal tar fractions are generally known as 'natural', as opposed to 'synthetic' where high purity materials are used.

The vast majority of modern phosphate esters are 'synthetic', using materials derived from petrochemical sources. For example alcohols are obtained from α-olefins by the OXO process, and iso-propylated or *t*-butylated phenols are produced from phenols by reaction with propylene or butylene. The reaction of alcohol or phenol with phosphoryl chloride yields the crude product, which is generally washed, distilled, dried and decolorised to yield the finished product. Low molecular weight trialkyl esters are water soluble, requiring the use of non-aqueous techniques. When mixed alkylaryl esters are produced, the reactant phenol and alcohol are added separately, the reaction being conducted in a step-wise fashion, and the reaction temperature being kept as low as possible to avoid transesterification reactions taking place.

2.7.3 *Physical and chemical properties*

The physical properties of phosphate esters vary considerably according to the mix and type of organic substituents, the molecular weights and structural symmetry proving particularly significant (Klamann, 1984). Consequently, phosphate esters range from low viscosity, water-soluble liquids, to insoluble high-melting solids.

The use of phosphate esters arises mostly from their excellent fire resistance and superior lubricity, but is limited due to their hydrolytic and thermal stability, low temperature properties and viscosity index.

With respect to hydrolytic stability, aryl esters are superior to the alkyl esters. Increasing chain length and degree of branching of the alkyl group leads to considerable improvement in hydrolytic stabilities. However, the more sterically hindered the substituent, the more difficult it is to prepare the ester, and increasing branching leads to a progressive drop in viscosity index. Alkylaryl phosphates tend to be more susceptible to hydrolysis than the triaryl or trialkyl esters.

The thermal stability of triaryl phosphates is considerably superior to that of the trialkyl esters, which degrade thermally by a mechanism analogous to that of the carboxylic esters.

$$
\begin{array}{ccccc}
 & H\ldots\ldots O & & OR & \\
 & / & \backslash\backslash & / & \\
R{-}CH & & P & & \\
 & \backslash & / & \backslash & \\
 & CH_2{-}O & & OR &
\end{array}
\longrightarrow R{-}CH{=}CH_2 +
\begin{array}{ccc}
HO & & OR \\
 & P & \\
O & & OR
\end{array}
$$

The use of neopentyl alcohol or its homologues yields a β-hindered ester in which this decomposition mechanism is blocked, leading to much improved thermal stability. However, esterification is problematic and expensive, and the esters suffer from mediocre VIs. Consequently, β-hindered esters are not generally used in lubrication.

The low temperature properties of phosphate esters containing one or more alkyl substituent tend to be fairly good, with pour points of $-55\,°C$ being quite common. Many triaryl phosphates are fairly high melting-point solids, but an acceptable pour point can be achieved by using a mixture of aryl components. Coal tar fractions, used to make 'natural' phosphate esters, are already complex mixtures, and give esters with satisfactory pour points.

The effect of molecular mass and shape on viscosity and VI is similar to that found with other base fluid types. Thus, increasing length of straight-chain alkyl substituents increases viscosity and VI, but has a negative effect on low temperature performance with increased pour point. Increased branching of alkyl substituents of constant molecular weight results in lower viscosities and VIs, but improved pour point.

Phosphate esters are very good solvents, and are extremely aggressive towards paints and a wide range of plastics and rubbers. When selecting suitable gasket and seal materials for use with these esters, careful consideration is required. The solvency power of phosphate esters can be advantageous in that it makes them compatible with most common additives and enables them to be used as blends with other base fluids types for a number of applications.

As mentioned previously, the most important properties of phosphate esters are their fire resistance and lubricity. They have high, but not exceptional flash points and on pyrolysis generate phosphoric acid, which is a powerful flame retardant. This imparts their excellent flame-retardant properties. The extremely good lubricity might be expected from their wide use as load-carrying additives in a range of lubricants. When used as base fluids they exhibit exceptional load-carrying and anti-wear properties, well in excess of other unformulated synthetic base fluids.

2.7.4 *Applications*

Phosphate esters are particularly used for their fire resistance in applications where their moderate cost can be borne. Other requirements, such as viscosity or thermal stability, may be met by appropriate choice of substituents. Industrial fire-resistant fluids are often required to operate at sustained high temperature, and here only the triarylphosphate esters can offer satisfactory thermal stability. These are used in foundries, die casting, aluminium smelters, and mines, and in applications such as injection moulding, oven door controls, and automatic welding equipment.

Phosphate esters are also widely used as hydraulic fluids in civil aircraft. In this application, thermal stability is less important than their ability to flow at low temperature, therefore trialkyl and alkylaryl phosphates are used. When formulated with a VI improver, they give fluids with pour points of $-55\,°C$ to $-65\,°C$ and a VI of 170–300. They may also be chosen for applications in other low temperature conditions such as those found on North Sea oil rigs.

References

Antonsen, D.H., Hoffman, P.S. and Stearns, R.S. (1963) *Ind. Eng. Chem. Prod. Res. Dev.* **2** 224.
Audisio, G., Priola, A. and Rossini, A. (1988) *Makromol. Chem.* **189** 111.

Beynon, K.I., Evans, T.G., Milne, C.B. and Southern, D. (1962) *J. Appl. Chem.* **12** 33.
Beynon, K.I., Milne, C.B. and Southern, D.J. (1967) *Appl. Chem.* **17** 213.
Brennan, J.A. (1980) *Ind. Eng. Chem. Prod. Res. Dev.* **19** 2.
Cerniglia, C. E. (1984) *Petroleum Microbiology.* Atlas, R.M. (ed), Macmillan, New York.
Chaffee, A.L., Cavell, K.J., Masters, A.F. and Western, R.J. (1987) *Ind. Eng. Chem. Prod. Res. Dev.* **26** 1822.
Chen, C. and Shuihua, H. (1989) World Patent WO 89/12663.
Corno, C., Ferraris, G., Priola, A. and Cesca, S. (1979) *Macromolecules* **12** 404.
Cosmacki, E., Cottia, D., Pozzoli, L. and Leoni, R. (1988) PAH emissions of synthetic organic esters used as lubricants in two-stroke engines. *J. Syn. Lub.* **3** 251.
Evans, A.G. and Polanyi, M. (1947). *J. Chem. Soc.* 252.
Evans, A.G., Holden, D., Plesch, P.H., Polanyi, M. and Weinberger, W.A. (1946a) *Nature* (London) **157** 102.
Evans, A.G., Meadows, G.W. and Polanyi, M. (1946b) *Nature* (London) **158** 194.
Ferraris, G., Corno, C., Priola, A. and Cesca, S. (1980) *Macromolecules* **13** 1104.
Fontana, C.M. (1963) *The Chemistry of Cationic Polymerisation.* Plesch, P.H. (ed), Pergamon, Oxford, p. 209.
Garwood, W.E. (1960) US Patent 2,937,129.
Gates, D.S., Duling, I.N. and Stearns, R.S. (1969) *Ind. Eng. Chem. Prod. Res. Dev.* **8** 299.
Gunderson, R.C. and Hart, A.W. (1962) *Synthetic Lubricants,* Reinhold Publishing Corporation, London.
Higashimura, T., Miyoshi, Y. and Hasegawa, H. (1982) *J. Appl. Pol. Sci.* **27** 2593.
Isa, H. (1986) *J. Synth. Lub.* **3** 29.
Kashiwa, H. and Toyota, A. (1986) *Chem. Econ. Eng. Rev. (CEER)* **18** 14.
Klamann, D. (1984) *Lubricants and Related Products.* Verlag Chemie, Weinheim.
Krulish, J.A.C., Lowther, H.V. and Miller, B.J. (1977) An update of synthesised engine oil technology. *SAE paper No. 770634* presented at *SAE Fuels and Lubricants Meeting,* Tulsa, Oklahoma, June 1977.
Macrae, A.R. and Hammond, R.C. (1982) *Biotechnology and Genetic Engineering Review* **3**(1093) 217.
Madgavkar, A.M. (1983) US Patent 4,394,296.
Madgavkar, A.M. and Barlek, R. (1981) US Patent 4,263,467.
Madgavkar, A.M. and Swift, H.E. (1981) US Patent 4,308,414.
Madgavkar, A.M. and Swift, H.E. (1983). *Ind. Eng. Chem. Prod. Res. Dev.* **22** 675.
O'Connor, B.M. and Ross, A.R. (1989) Synthetic fluids for automotive gear oil applications: a survey of potential performance. *J. Syn. Lub.* **6**(1) 31.
Onopchenko, A., Cupples, B.L. and Kresge, A.N. (1982) *ACS Div. Petrol. Chem. Reprints* **27** 331.
Onopchenko, A., Cupples, B.L. and Kresge, A.N. (1983) *Ind. Eng. Chem. Prod. Res. Dev.* **22** 182.
Puskas, I., Banas, E.M., Nerheim, A.G. and Ray, G.J. (1979) *Macromolecules* **12** 1024.
Randles, S.J., Robertson, A.J. and Cain, R.B. (1989) *Environmentally Friendly Lubricants for the Automotive and Engineering Industries.* Presented at a Royal Society of Chemistry seminar, York, 1989.
Seger, F.M., Doherty, H.G. and Sachenen, A.N. (1950) *Ind. Eng. Chem.* **42** 2446.
Shubkin, R.L., Baylerian, M.S. and Maler, A.R. (1979) *ACS Div. Petrol. Chem. Reprints* **24** 809.
Shubkin, R.L., Baylerian, M.S. and Maler, A.R. (1980) *Ind. Eng. Chem. Prod. Res. Div.* **19** 15.
Souillard, G.J., Van Quaethoven, F. and Dyer, R.B. (1971) Polyisobutylene, a new synthetic material for lubrication. *SAE Paper 710730.*
Sugiura, K. and Kagaya, M. (1977) A study of visible smoke reduction from a small two-stroke engine using various engine lubricants. *SAE Paper 770623.*
Sullivan, F.W. and Vorhees, V. (1934) US Patent 1,955,260.
Sullivan, F.W., Vorhees, V., Neely, A.W. and Shankland, R.V. (1931) *Ind. Eng. Chem.* **23** 604.
Van der Waal, G. (1985) The relationship between chemical structure of ester base fluids and their influence on elastomer seals and wear characteristics. *J. Syn. Lub.* **1**(4) 281.
White, M.A. (1985) US Patent 4,579,991.
Whitmore, F.C. (1934). *Ind. Eng. Chem.* **26** 94.
Wits, J.J. (1989) Diester compressor lubricants in petroleum and chemical plant service. *J. Syn. Lub.* **5**(4) 321.
Wyatt, J.M. (1982) PhD Thesis, University of Kent.

3 Detergents/dispersants

C.C. COLYER and W.C. GERGEL

3.1 Introduction

A lubricant must provide the following basic functions:

- Act as a coolant by heat removal
- Provide a film between moving parts to minimize friction leading to wear

However, sophisticated complex machinery demands much more from all lubricants. Intricate engines with close tolerances between moving parts would not be possible without engine oils containing a tailored package of chemical additives. Progressive advances in additive technology not only allow today's engines to operate efficiently, but also dramatically increase the engine's useful life and reduce engine maintenance costs.

Incorporation of detergent and dispersant additives into a lubricant is essential to prevent harmful carbon and sludge deposits. Deposits in critical areas of the engine lead to engine shut-down and repair. In addition, the basic functions of a lubricant are related to proper oil flow properties which detergent and dispersant additives help to maintain by:

- Minimizing oil thickening thereby maintaining viscosity stability and flow properties.
- Containment of carbon sludge deposits that clog oil lines resulting in lack of oil flow and subsequent engine failure.

These additives are somewhat similar to the more familiar household detergents. The household detergents are water soluble while the lubricant detergent and dispersant additives are oil soluble. Detergents may also provide rust protection to engine parts, particularly valve train components.

During World War II, continued efforts were made to improve engine oils to meet the demands of engines used in combat. The petroleum additive industry responded by progressively advancing detergent additive technology to meet the challenge. After the war, continued research resulted in further detergent technological advances which were shared with the motoring public.

Dispersants began to play a major role in passenger car engine oils in the

1950s. Both detergents and dispersants have had dramatic volume growth and improved technology leading to progressively better cost/performance properties. Detergents and dispersants have a direct effect on minimizing harmful engine exhaust emissions, increasing engine life, and controlling oil consumption by maintaining clean, 'in-tune' engine operation.

Without these additives, deposits on piston ring and grooves can lead to ring sticking, resulting in loss of oil control. Loss of oil control means:

- Increased oil flow upward into the combustion chamber. When the oil burns during engine combustion, increases in engine exhaust emissions occur.
- Increased oil consumption requiring more oil added between oil drains. In addition, inadequate detergent and dispersant additive content in an engine oil can lead to valve deposits that stick valves, resulting in faulty, incomplete combustion and increased exhaust emissions.
- More contaminating combustion gases flow downward between the cylinder liner and piston rings into the oil crankcase which reduces the life of the oil, thus requiring more frequent oil drains.

Ashless dispersants are also used in gasoline and diesel fuels to provide fuel injector, carburettor, and valve cleanliness. Engine life is prolonged and undesirable engine exhaust emissions are reduced.

The terms 'detergents' and 'dispersants' are often used interchangeably because both additive types keep insoluble combustion debris and oil oxidation products dispersed within the oil. Detergents are normally utilized to minimize high-temperature engine varnish and lacquer deposits while dispersants are used to control low-temperature engine sludge deposits. Both are long chain hydrocarbons with polar ends. Detergents have a polar end containing a metal ion. Dispersants utilize oxygen and/or nitrogen for polarity and do not contain metal ions. In engine oils, the harmful products of combustion and other contaminants are rendered harmless by the polar ends. The hydrocarbon chain of these additives helps to solubilize or suspend the debris in the oil.

Detergents and dispersants are used in a wide variety of automotive and industrial lubricants. Their major applications are in engine oils. Other applications include transmission fluids, gear lubricants, and tractor hydraulic/transmission oils. Engine oil applications will be emphasized in this chapter.

3.2 Detergents

The detergent polar substrate is made up basically of two parts as illustrated in Figure 3.1. The hydrocarbon tail or the oleophilic group is the portion of the detergent polar substrate that acts as the solubilizer to enable the

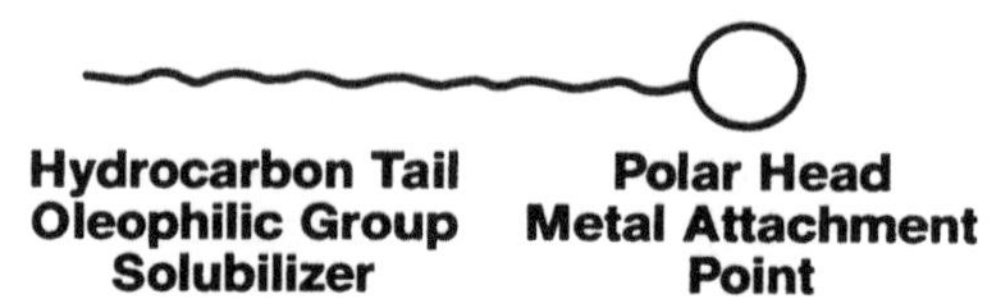

Figure 3.1 Detergent polar substrate.

detergent to be fully compatible and soluble in base oil. The other part of the detergent polar substrate is the polar head containing a metal cation. Many metals have been incorporated into detergents but currently, based on cost/performance, the three most commonly used metals are calcium, magnesium, and sodium. These metals have largely replaced the less cost/performance effective barium materials. In addition, barium compounds may have toxic properties. Some countries limit the amount of barium allowed in consumer goods like motor oils.

Detergents provide cleaning action in engine oils by neutralizing acids which arise from fuel combustion products. These acids can be of three types: sulfur acids, nitro acids, and oxy acids. In the fuel combustion process, the sulfur compounds present in the fuel oxidize to form sulfur oxides. These oxides unite with water, also a by-product of the combustion process, to form sulfur acids. Sulfur acid formation is particularly significant in diesel engine operation. Diesel fuel can contain up to 5% sulfur. The oxy acids result from the oxidation of various lubricant base oil and fuel fragments. Nitro acids are generated in the combustion process directly from air, which contains about 80% nitrogen.

Detergent polar substrate types consist of four major classes: sulfonates, phenates, salicylates, and phosphonates. Sulfonates are the most widely-used detergent additives followed by phenates, salicylates, and phosphonates. Detergents have varying capacity to provide engine rust protection. Rust protection is needed, not only in low-temperature engine operation, but also for protection when the engine is not in use. Sulfonates in particular provide excellent engine anti-rust properties. Phenates, in addition to detergency characteristics, provide oxidation inhibition properties and a somewhat lower sulfated ash content at equal alkalinity when compared with sulfonates. Some diesel engine builders include a maximum sulfated ash level in their engine oil specifications. (Elements of the petroleum and additive industries dispute that sulfated ash level *per se* is a problem.) The use of salicylates is limited while phosphonate use is minimal.

3.2.1 *Sulfonates*

Two varieties of sulfonate detergent polar substrates exist: petroleum sulfonates and synthetic sulfonates. The petroleum sulfonates are the so-called 'natural' sulfonates. These natural sulfonates occur as a by-product in the manufacture of white oil. The white oil process involves contacting of a mineral oil with sulfuric acid to form the 'white' oil plus a sulfonic acid mixture that, when treated with sodium hydroxide, yields two different types of soaps. After the separation of the sludge, the mahogany acids are generally recovered from the oil in the form of alkali salts which are usually produced by the addition of sodium carbonate or sodium hydroxide followed by the extraction of the salt with solvent, generally an alcohol such as methanol, ethanol or isopropanol. The mahogany acid soaps are named because of their reddish color. They are the sodium sulfonates or soaps which were the starting material for the original petroleum sulfonate lubricating oil additives. The mahogany acid soaps are oil soluble. The structure of these materials is not known with any degree of certainty. 'Green' acid soaps are also by-products of the white oil process. They are water soluble and normally discarded in the process.

The concentrated sodium sulfonate is then converted to other metal salts by means of a metathesis reaction with a metal chloride. As an example, a neutral calcium sulfonate can be prepared by metathesis of the sodium sulfonate with calcium chloride. An alternative route for the preparation of these metal salts is by direct neutralization of the mahogany acids with the appropriate metal hydroxide.

The source and type of base oils used in the white oil process is very important with respect to the performance characteristics of the resultant sulfonate detergent additives. Lazar and Carter (1946) determined that paraffinic base oils are greatly preferred over naphthenic base oils as the source of sulfonic acids. Sulfonates derived from the latter are reported to be more corrosive to copper-lead bearings. They do not possess the resistance to oxidation commonly exhibited by sulfonates derived from paraffinic base oils. Table 3.1 taken from the patent shows the variation in performance of an

Table 3.1 Performance of SAE 30 viscosity grade.[a]

Calcium sulfonate	Copper-lead bearing	
	Weight loss (mg)	Appearance of engine
A	1000	Clean
B	191	Very clean
C	1510	Lacquer on piston skirts

Sulfonate A—sulfonation of 1000 SUS[b] at 100 °F naphthenic base oil.

Sulfonate B—sulfonation of the base oil in run A after removal of naphthenic acids.

Sulfonate C—sulfonation of 100 SUS[b] at 100 °F naphthenic base oil.

[a] 240 hour test in 4-cycle diesel engine, SAE 30 base oil containing 3% by weight calcium sulfonate.

[b] Saybolt universal seconds—a measure of viscosity.

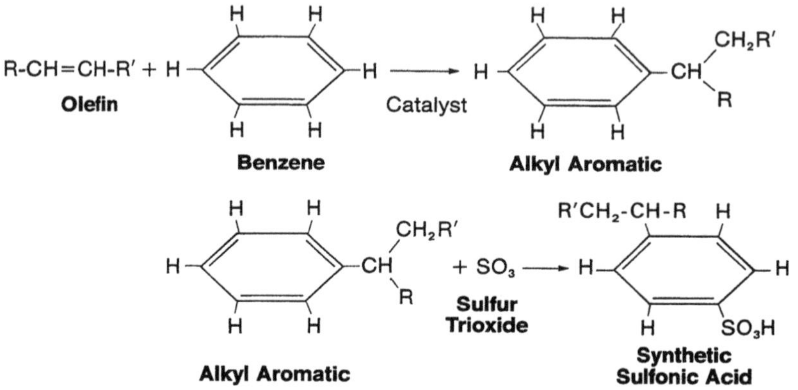

R-CH=CH-R′ + Benzene —Catalyst→ Alkyl Aromatic

Olefin Benzene Alkyl Aromatic

Alkyl Aromatic + SO_3 → Synthetic Sulfonic Acid

Alkyl Aromatic Sulfur Trioxide Synthetic Sulfonic Acid

Figure 3.2 Synthetic sulfonates.

SAE 30 viscosity grade base oil containing 3% of calcium sulfonate detergents.

Synthetic sulfonates are those specifically prepared from a synthesized alkyl aromatic substrate. They have a variety of structures. For example, if benzene is alkylated by olefin under catalytic conditions, an alkyl aromatic compound is formed. The alkyl aromatic is then contacted with sulfur trioxide to form the synthetic sulfonic acid as shown in Figure 3.2. Both the natural and synthetic sulfonates fall into two general types: neutral sulfonates and overbased sulfonates. The formation of neutral sulfonates is illustrated in Figure 3.3. A neutral sulfonate is formed by reacting the sulfonic acid with either a metal oxide or metal hydroxide to form the neutral sulfonate plus water as a by-product.

Anderson and McDonald (1960) described a procedure for producing oil soluble sulfonates from the alkylate. It involves sulfonation of the alkylate using oleum as the SO_3 source. The sulfonic acid is diluted with white oil, and the separated white oil layer containing the oil soluble sulfonic acids is blown with an inert gas to remove unreacted SO_3 and by-product SO_2. The white oil solution is then treated with activated clay and filtered. The white oil solution is extracted with ethanol to remove the oil soluble sulfonic acids. Subsequent neutralization of these acids forms the oil soluble sulfonates.

$$RSO_3H + MO \text{ or } MOH \longrightarrow RSO_3M + H_2O$$

Sulfonic Acid | Metal Oxide Metal Hydroxide | Neutral Sulfonate | Water

Figure 3.3 Sulfonate formation.

Some benzene alkylates are designed specifically for the production of oil soluble sulfonates. These materials can be converted into suitable sulfonates with a minimum of processing to remove undesirable by-products. Most are diluted with petroleum base oil or other suitable petroleum solvents prior to neutralization of the sulfonic acids.

Alternative procedures for converting the benzene alkylates to sulfonic acid may include the use of gaseous SO_3 to contact a thin film of the alkylate. This process avoids the handling and subsequent disposal of large quantities of sulfuric acid.

Sulfonate additives which contain metal in excess of the stoichiometric amount required to produce a neutral sulfonate can be prepared. In the instance of a divalent metal, the neutral sulfonate formula is represented as $(RSO_3)_2M$, where R is the alkylated benzene or hydrocarbon and M is the divalent metal.

One advantage of the basic sulfonates is their greater ability to neutralize acidic bodies. Among the earlier workers in the art who recognized this factor were Bergstrom (1942) and Van Ess and Bergstrom (1945).

When the superiority of basic soaps over the normal soaps was recognized, attempts were made to increase the basicity of the soaps, i.e. to increase the amount of metal base held in stable form as an oil soluble complex. One of the first ways employed to produce a metal salt having a large excess of metal in combination was to use an unusually large excess of neutralizing agent. Griesinger and Engelking (1946) suggested the use of a neutralizing agent up to 220% of the theoretical amount, carrying out the process in the presence of steam in order to facilitate the formation of the product. Campbell and Dellinger (1949) disclosed the use of minor amounts of an alkaline earth metal hydroxide or carbonate which is peptized, or held in a colloidal suspension in oil by means of an oil soluble mahogany sulfonate.

Overbased sulfonates are formed, as shown in Figure 3.4, by a complex reaction between a neutral metal sulfonate and a metal hydroxide. This complex reaction takes place using carbon dioxide in the presence of a promoter, which is generally an alcoholic-type material. The promoter dissolves a small amount of metal hydroxide, which is subsequently reacted with carbon dioxide to form a metal carbonate.

The amount of metal carbonate incorporated into overbased sulfonates can vary depending upon the application in which the overbased sulfonate is

$$\underset{\textbf{Neutral Sulfonate}}{RSO_3M} + \underset{\textbf{Metal Hydroxide}}{x\,MOH} \xrightarrow[\text{CARBON DIOXIDE}]{\text{PROMOTER}} \underset{\textbf{Overbased Sulfonate}}{RSO_3M\text{—}xMCO_3} + \underset{\textbf{Water}}{H_2O}$$

Figure 3.4 Overbased sulfonates.

used. Even though overbased sulfonates have been used for about 40 years, their structures have not been elucidated.

The choice of sulfonate type is usually dictated by cost versus performance. Alkylbenzene-bottoms-derived sulfonates are generally the least costly (The term 'bottoms' refers to the residue resulting from the distillation of alkylbenzene primary product. Bottoms are a complex chemical mixture.) They provide excellent performance as overbased metal derivatives in light-duty engine service. The more expensive mono and/or dialkylbenzene primary alkylate sulfonic acid derivatives are used to advantage in more severe engine service.

3.2.2 *Phenates/salicylates/phosphonates*

The structures of the other three important detergent polar substrates—namely phenates, salicylates, and phosphonates—are shown in Figures 3.5 to 3.9. The detergents formed from these three polar substrates can also be overbased. The overbased phenates, salicylates, and phosphonates are prepared in the same general manner as the overbased sulfonates.

The broad class of metal phenates includes salts of alkylphenols, alkylphenol sulfides, and the alkylphenol–aldehyde condensation products. Satisfac-

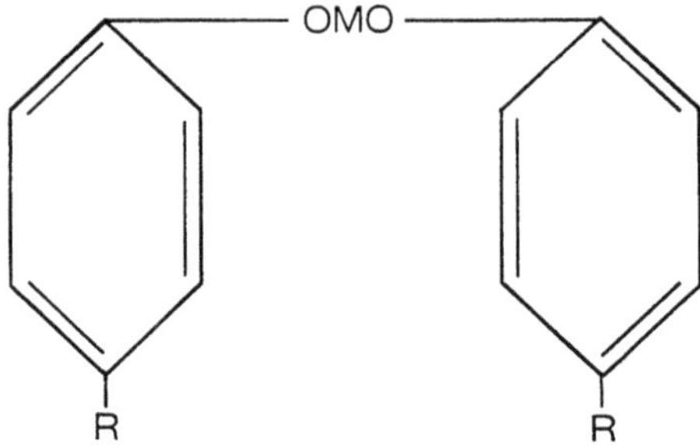

Figure 3.5 Normal phenate.

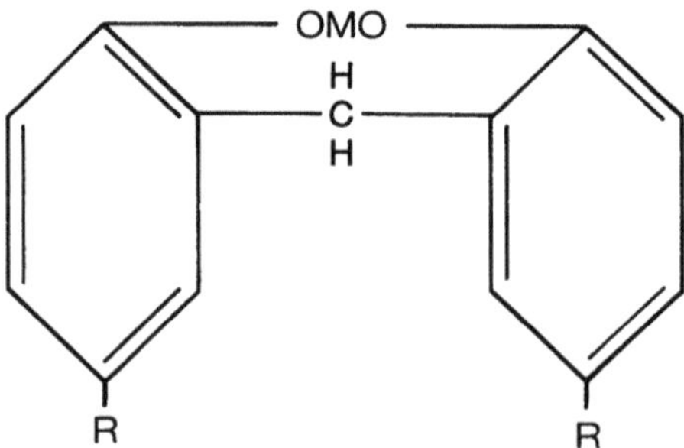

Figure 3.6 Methylene coupled phenate.

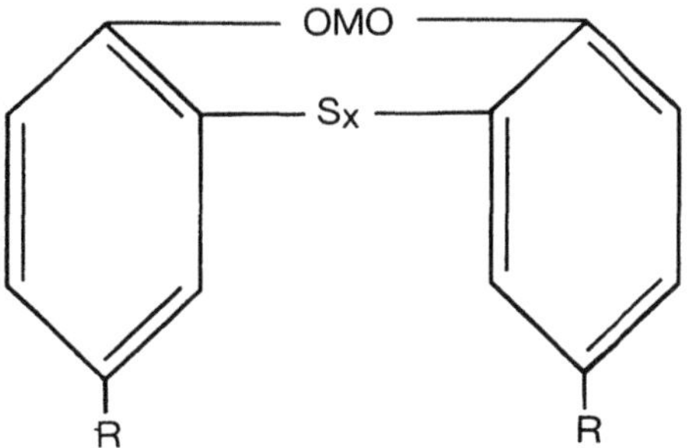

Figure 3.7 Phenate sulfide.

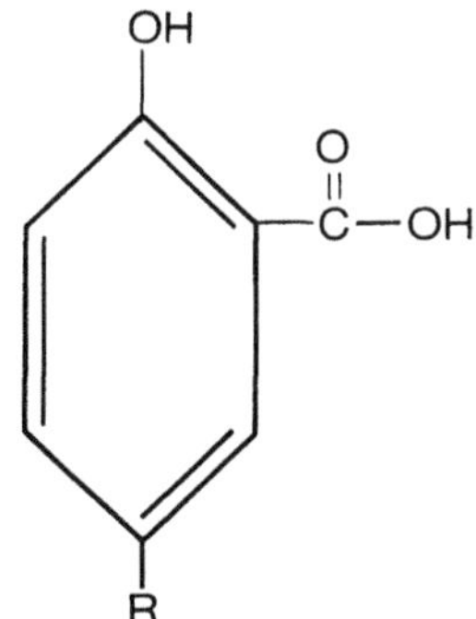

Figure 3.8 Alkyl salicylic acid.

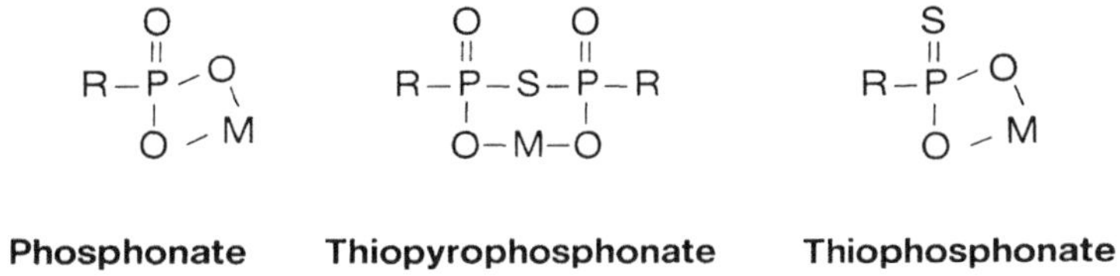

Phosphonate	Thiopyrophosphonate	Thiophosphonate
R–P(=O)(–O–M–O–)	R–P(=O)(–O–)–S–P(=O)(–O–)–R, with O–M–O	R–P(=S)(–O–M–O–)

Figure 3.9 Phosphonate family.

tory oil solubility characteristics result from the use of alkyl groups generally of eight or more carbon atoms.

The general reaction for the preparation of these materials can be described as follows:

metal carbide
or
R—PhOH + metal oxide/hydroxide ⟶ $(R—PhO)_xM$
or
metal alcoholate

where R is an alkyl group preferably of eight or more carbon atoms, M is a metallic element (i.e. Na, Ca, Ba, Mg, Al, Pb, Zn), and x can range from one to three depending on the particular metal involved. Calcium and magnesium phenates are currently employed in commercial engine lubricants.

The materials are generally prepared by carrying out the reaction in a low viscosity mineral oil at temperatures ranging up to 260 °C depending on the reactivity of the metallic base. The alkylphenol intermediates can be prepared by alkylating phenol with olefins, chlorinated paraffins, or alcohols using catalysts such as H_2SO_4 and $AlCl_3$, with the latter being employed with the chlorinated paraffin in a typical Friedel–Crafts type of alkylation.

By use of an excess of the metal base over the theoretical amounts required to form the normal phenates, it is possible to form the so-called basic phenates. Basic alkaline earth phenates containing two and three times the stoichiometric quantity of metal have been reported in the patent literature.

Since a very important function of the phenate is acid neutralization, the incorporation of excess base into these materials provides a distinct advantage over the metal-free phenates.

It is possible to introduce methylene bridging into the phenate molecule by means of an aldehyde–phenol condensation, commonly using formaldehyde as the aldehyde. It is claimed that such materials improve an oil's oxidation stability and also reduce the oil's tendency to corrode engine bearing metals.

Incorporating sulfur into the phenates is accomplished by reacting the alkylphenol with sulfur monochloride, sulfur dichloride, or elemental sulfur. Sulfurization appears to lower the corrosivity of the compound towards engine bearing metals and to improve oxidation stability.

Basic phenates can also be prepared from the phenol sulfides. This imparts the benefits of acid neutralization capacity to the phenol sulfides.

In summary, the principal functions performed by metallic phenates in lubricating oil formulations are (1) acid neutralization, (2) high temperature detergency and (3) oxidation inhibition. Phenates include the salts of alkylphenols, alkylphenol sulfides and alkylphenol aldehyde condensation products.

3.2.3 *Detergent classification*

Overbased substrates have been casually defined by the amount of total basicity contained in the product. It has become popular to label a sulfonate by its TBN (total base number), i.e. a 300 TBN synthetic sulfonate. Base number is defined in terms of the equivalent amount of potassium hydroxide contained in the material. A 300 TBN calcium sulfonate contains base equivalent to 300 milligrams of potassium hydroxide per gram or, more simply, 300 mg KOH/g. Two factors limit the degree of overbasing: oil solubility and filterability. The $Ca(OH)_2$ or CaO used to prepare calcium sulfonates should be free of inert solids.

The use of TBN alone in defining an overbased detergent is quite misleading. In order to define an overbased detergent, three of the following parameters should be known: the metal ratio (the ratio of metal base to substrate), the amount of neutral detergent ('soap' content), the molecular weight of the substrate used, the amount of metal, and/or the amount of TBN. The percentage of sulfur (if sulfonates are under consideration) is also an important parameter in detergent definition.

Using an overbased sulfonate substrate system as an example, the general formula for a calcium overbased sulfonate is:

$$(RSO_3)_aCa_b(OH)_c(CO_3)_d$$

where a, b, c and d represent moles of the various functions.

Based on this formula the following mathematical expressions can be used to define an overbased calcium sulfonate:

1. $\text{Metal ratio} = \dfrac{2b}{a}$

2. $\text{TBN (mg KOH/a)} = \dfrac{2 \times 56\,000(b-a)}{\text{Effective formula weight}}$

3. $\%\ \text{soap} = \dfrac{\text{Formula weight } (RSO_3)_2Ca \times 100}{\text{Effective formula weight}}$

where the effective formula weight is the atomic weight of $(RSO_3)_a Ca_b(OH)_c(CO_3)_d$ plus diluent.

Table 3.2 shows the variety of possible sulfonate, phenate and salicylate lubricant additives.

Although chemical descriptions of detergents are important in maximizing the performance of engine oils, it is vital to recognize that the selection of raw materials governs the performance features. Two detergents having the same total base number (TBN), molecular weight, metal ratio, etc., may have widely different performance characteristics.

Table 3.2 Range of typical detergent lubricant additives.

Parameter	Range		
	Sulfonates	Phenates	Salicylates
TBN	0–500	50–400	50–400
Metal ratio	1–30	0.8–10	1–10
Soap content, %	10–45	30–50	10–45
Metal cation	Ca, Mg, Na, Ba	Ca, Ba, Mg	Ca, Mg
Molecular weight sulfonic acid	375–700	—	—
Alkylphenol	—	160–600	—
Carboxylic acid	—	—	250–1000
Sulfur, %	0.5–4.0	0–4	—

Commercial detergents are generally available diluted about 50% in mineral oil and are used in engine oils in amounts ranging from 0.5% to 30%. Marine engine oils have TBN values as high as 100 and may contain up to 30% detergent additive. Heavy duty diesel truck oils use up to 8% detergents while gasoline engine oils may contain up to 5%.

The performance of a fully compounded oil is dependent both on the character of the base oil, and on the kind and amount of additive used. Hence, engine oil additive package formulation is enhanced by experience in choosing the proper base oil and additive combination to provide the desired performance characteristics.

3.3 Dispersants

Prior to 1955, the lubricant additives used for the purpose of keeping an engine clean were sulfonates, phenates, salicylates, and phosphonates. These materials did a good job so long as the service did not involve an excessive amount of low-temperature, short-distance, or stop-and-go type driving. Passenger car oil changes were recommended at 1000–2000 miles. If the recommendations were ignored, the oil could become so viscous that it would refuse to flow through the crankcase pan drain hole. This high viscosity was due to severe oil oxidation and the accompanying high level of insolubles in the oil. In gasoline engines the insolubles were made up of oil oxidation products, lead compounds originating from the lead tetraethyl octane improver in the gasoline, carbon and oxidation products in the combustion chamber blow-by gases, and dust contained in the combustion air. In diesel engines the major insolubles contribution was from carbon (soot) and oxidation products of the fuel and oil.

The increasing horsepower of passenger car engines in the United States from the 1950s to the 1960s led to severe sludging conditions. Although the efficiency of combustion increased, it was accompanied by lower engine temperatures and increased lubricant contamination. This was especially true when an excessive amount of stop-and-go driving conditions did not allow engine oil temperatures to rise sufficiently to vaporize the water from the oil. The condensed water mixed with the oil to promote increased sludge formation.

In the 1950s a new type of additive, a non-metallic or ashless dispersant, was introduced to help keep engines clean. This product, known as a succinimide dispersant, was a relatively high molecular weight polyisobutenyl group attached to a polar end group. The use of this type of material increased to such an extent that it became the additive component in the highest concentration in passenger car engine oils except for viscosity index improvers.

Dispersants are non-metallic or ashless cleaning agents. Figure 3.10 shows

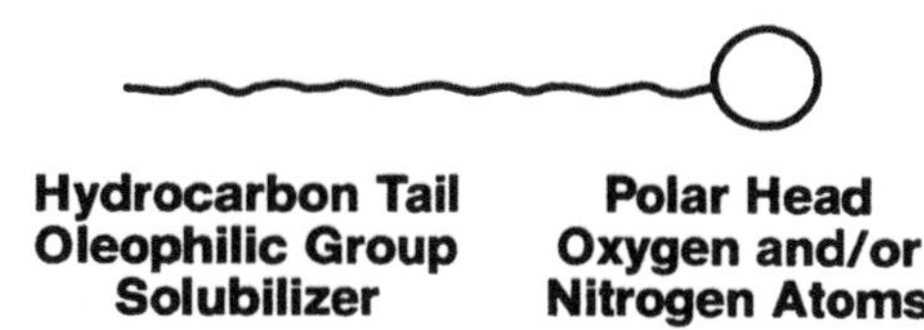

Figure 3.10 Stylized dispersant.

the stylized structure of an ashless dispersant. The structure of an ashless dispersant is similar to the structure of a detergent in that the dispersant has a hydrocarbon tail or oleophilic group which enables the dispersant to be fully soluble in the base oil used. The dispersant also has a polar head. The polarity of a dispersant is derived from the inclusion of oxygen, phosphorus, or nitrogen atoms into the molecule.

The sludge and varnish-forming precursors, resulting as a by-product of engine fuel combustion, contaminate the engine oil as they blow by the piston rings into the engine crankcase. Prior to the introduction of dispersants, these contaminants would settle out on critical parts of the engine hampering operation and eventually requiring engine overhaul. The presence of water in gasoline engine low-temperature stop-and-go operation accelerates the contaminant drop-out process. Dispersants are a vital component in gasoline engine oils and are also used to advantage in diesel engine oils to suspend harmful soot contaminants in order to provide longer engine life between overhauls. Diesel engine oil temperatures are generally sufficiently high enough to vaporize water from the oil.

Ashless dispersants are designed to have their polar chemical heads attached to rather large hydrocarbon groups. As shown in Figure 3.11 these polar heads interact with sludge. The hydrocarbon groups provide the solubilizing action which maintains the potentially harmful debris in suspension in the oil. By use of an engine oil well-fortified with a dispersant additive, as well as by practicing engine manufacturers' oil drain recommendations, virtually all of the harmful deposit-forming debris is removed from the engine when the oil is periodically drained.

There are four different types of ashless dispersants: (1) succinimides, (2) succinate esters, (3) Mannich types, and (4) phosphorus types. As with

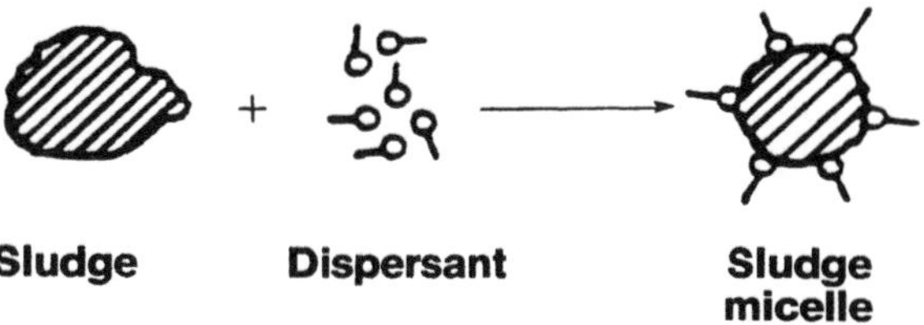

Figure 3.11 Sludge dispersion.

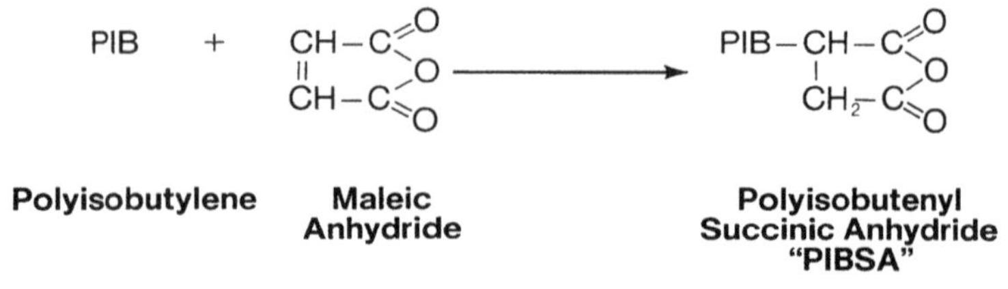

$$\text{PIB} + \begin{matrix} CH-C(=O) \\ \| \quad \;\; O \\ CH-C(=O) \end{matrix} \longrightarrow \begin{matrix} PIB-CH-C(=O) \\ | \quad \;\; O \\ CH_2-C(=O) \end{matrix}$$

Polyisobutylene **Maleic Anhydride** **Polyisobutenyl Succinic Anhydride "PIBSA"**

Figure 3.12 Succinimide/succinate ester raw materials.

detergents, dispersants are used in a variety of automotive and industrial oils, whilst combinations of dispersant types are often used in lubricant formulations. This discussion will emphasize dispersant use in engine oils.

Most dispersants currently in use are prepared from polyisobutylenes of 1000 to 10 000 molecular weight. Their polar functionality arises from amino and/or hydroxyl (alcohol) groups. The connecting groups, in most cases, are either phenols or succinic acids. The products with succinic acid groups are called alkenyl succinimides and succinate esters. The products from phenols are alkyl hydroxybenzyl polyamines (also called Mannich dispersants because of the name of the German chemist who discovered the method of preparation). These materials are generally processed as 40 to 60% concentrates in base oil.

Both the succinimides and the succinate esters are derived from the same chemical intermediate. The preparation of this intermediate is shown in Figure 3.12. Polyisobutylene is reacted with maleic anhydride to form polyisobutenyl succinic anhydride. This material is often referred to as 'PIBSA'. In the formation of succinimides, the PIBSA is reacted with a polyamine to form a structurally complex mixture which can contain imide, amide, imidazoline, diamide, and amine salt.

The polyamines generally used are those from the homologous family of ethylene diamine. Various types of succinimides can be produced, i.e. succinimides containing relatively low amounts to relatively high amounts of nitrogen. This variance in nitrogen content is accomplished by the preselection of the ratio of chemical equivalents of amino nitrogen present in the polyamine versus the number of chemical equivalents of acid in the polyisobutenyl succinic anhydride.

Figure 3.13 illustrates the formation of a simple succinate ester. The polyisobutenyl succinic anhydride is condensed with a polyol to form a succinate ester. Generally speaking, there is an excess of hydroxyl over acid moiety such that the succinate ester results in a molecule containing free (unreacted) hydroxyl groups.

Amino alcohols can also be used in place of either polyamines or polyols (Figure 3.14). If a 2,2-disubstituted-2-amino-1-alkanol is used, an oxazoline-containing dispersant can be prepared. The reaction which takes place is shown in Figure 3.15.

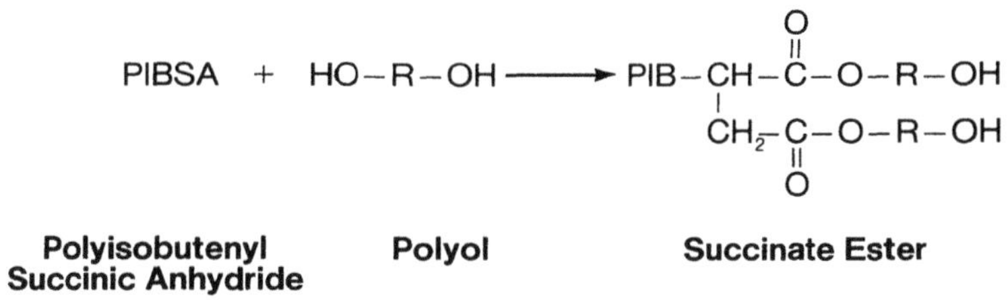

PIBSA + HO–R–OH ⟶ PIB–CH–C(=O)–O–R–OH
CH₂–C(=O)–O–R–OH

Polyisobutenyl Succinic Anhydride **Polyol** **Succinate Ester**

Figure 3.13 Simple succinate ester.

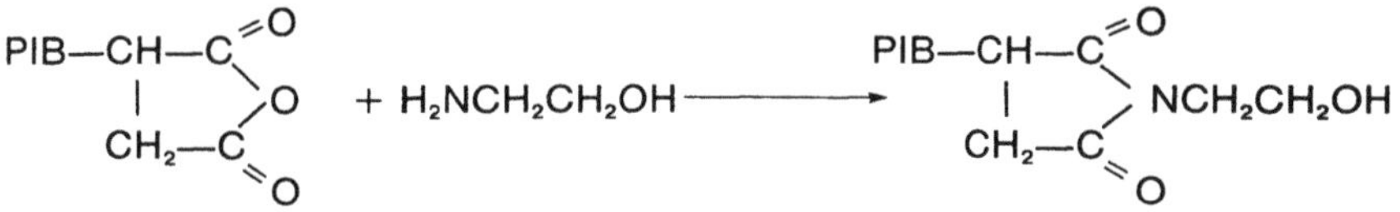

PIB—CH—C(=O)—O—C(=O)—CH₂ + $H_2NCH_2CH_2OH$ ⟶ PIB—CH—C(=O)—NCH_2CH_2OH—C(=O)—CH₂

Figure 3.14 Hydroxyethyl imide structure.

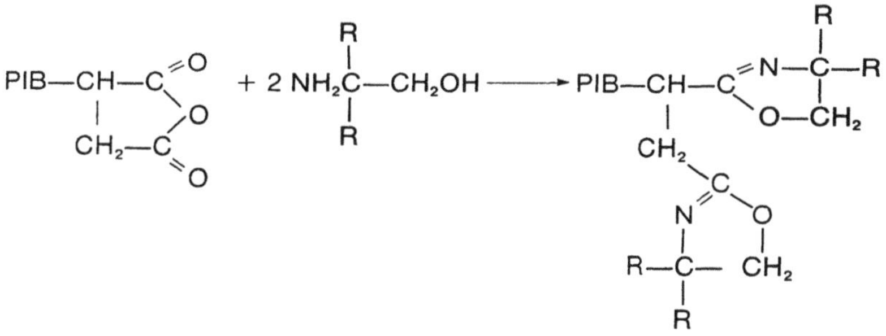

PIB—CH—C(=O)—O—C(=O)—CH₂ + 2 $NH_2C(R)_2$—CH_2OH ⟶ PIB—CH—C(=N—C(R)(R)—CH₂—O) ; CH₂—C(=N—C(R)(R)—CH₂—O)

Figure 3.15 Oxazoline structure.

The so-called Mannich dispersants, named after Mannich who discovered the reaction in 1912 (Mannich and Krosche, 1912), are the third general class of ashless dispersants for use in formulating lubricating oil additives. The Mannich reaction involves the replacement of active hydrogen atoms in organic compounds by aminomethyl or substituted aminomethyl groups by the reaction of formaldehyde with ammonia or a primary or secondary amine. The Mannich reaction as applied to lubricating oil additives is illustrated in Figure 3.16. An alkylphenol containing the active hydrogen atoms is reacted with formaldehyde and a polyamine to form the Mannich base.

The most common type of phosphorus-containing dispersants is prepared by a two-step reaction which is illustrated in Figures 3.17 and 3.18. Polyisobutylene is reacted with phosphorus pentasulfide under high-temperature conditions. The resulting reaction mixture is often treated with water to remove sulfur to form a phosphorus acid. This phosphorus acid is then

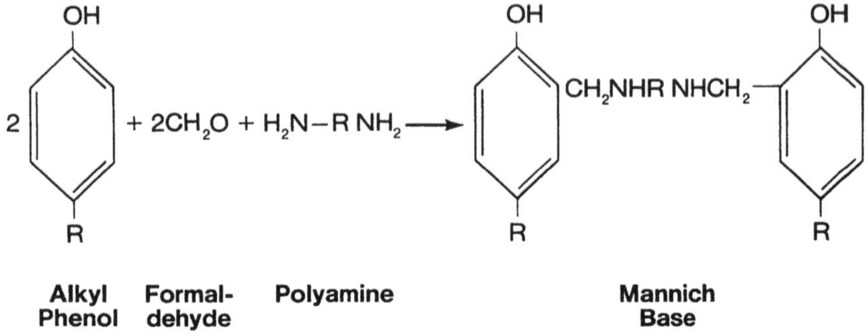

$$2\,\text{HO--C}_6\text{H}_4\text{--R} + 2CH_2O + H_2N{-}R\,NH_2 \longrightarrow \text{R--C}_6\text{H}_3(\text{OH})\text{--}CH_2NHR\,NHCH_2\text{--C}_6\text{H}_3(\text{OH})\text{--R}$$

Alkyl Phenol | Formaldehyde | Polyamine | Mannich Base

Figure 3.16 Mannich base dispersant.

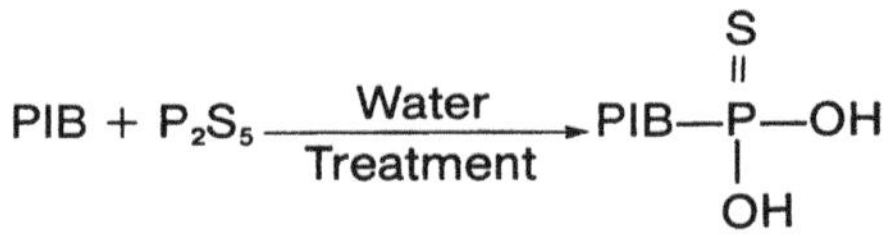

$$PIB + P_2S_5 \xrightarrow[\text{Treatment}]{\text{Water}} PIB{-}P(=S)(OH){-}OH$$

Figure 3.17 Phosphorus dispersant formation, step 1.

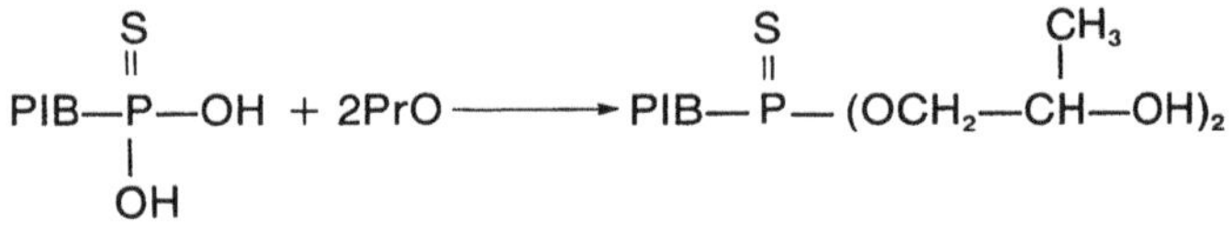

$$PIB{-}P(=S)(OH){-}OH + 2PrO \longrightarrow PIB{-}P(=S){-}(OCH_2{-}CH(CH_3){-}OH)_2$$

Figure 3.18 Phosphorus dispersant formation, step 2.

treated with propylene oxide (PrO) to form the hydroxypropyl esters of the phosphorus acid.

Phosphorus can be incorporated into dispersants using other reaction schemes. An illustrative example is the reaction of a dimeric thionophosphine sulfide and an olefin such as polyisobutylene.

Engine oil formulations may contain more than one dispersant to provide optimum performance. For example, one dispersant may excel in low-temperature gasoline engine operation, whilst another will optimize high-temperature diesel engine oil performance. It is common practice for oil companies to market 'universal' oils that can be recommended for both gasoline passenger car engines and heavy duty diesel engines. Consequently, tailor blending the oil with several dispersants would help the product to meet universal oil requirements. Different dispersants have varying effects in interacting with other additives such as oxidation, wear and rust inhibitors. In the art of engine oil formulation these factors have an impact on the choice of dispersant used to maximize performance.

Engine oil formulation combines both art and science. Formulations must be tailored to provide optimum all-around performance. It is not necessarily true that, for an engine oil to meet all the performance areas required, the more additive the better.

3.4 Other lubricants

Detergents/dispersants are incorporated into lubricants other than engine oils, for example automatic transmission fluids, gear lubricants, hydraulic oils and industrial oils, to provide significant value-added properties.

3.4.1 *Automatic transmission fluids (ATFs)*

Most US automobiles are equipped with an automatic shift transmission as opposed to a manual shift transmission. Outside the US, the percentage of cars equipped with automatic transmissions is steadily growing. ATF packages are the most complex additives in that one package may contain up to 20 separate components which are finely tuned to provide optimum performance including delicately balanced friction characteristics. Both detergents and dispersants play significant roles in ATF additive packages. As in engine oils, sulfonates neutralize oxidation and acidic products in ATFs. However, the major role of sulfonates in ATFs is to stabilize friction characteristics. An automatic transmission requires smooth shifting from gear to gear. Automatic transmission friction plate material varies from manufacturer to manufacturer and requires different fluid characteristics to provide optimum performance. Sulfonates provide durability of a given friction property in new fluid. Calcium sulfonates are generally used but the use of magnesium sulfonates in ATFs is now emerging.

Dispersants are used in ATFs to suspend thermal decomposition and oxidation products which would otherwise fill up the 'holes' in the porous surfaces of the friction material and significantly alter the shift characteristics. The dispersant must be oxidation stable and not interact unfavorably with other additive components to change the friction characteristics of the ATF.

The complexity of ATFs encourages close cooperation between transmission manufacturers and the additive company. A wide range of both metallic and non-metallic materials is used in the construction of automatic transmissions. Changes in transmission design and materials have a direct effect on the performance of an ATF. For example, in the critical area of friction plate facings, asbestos-reinforced synthetic paper has been phased out. In its place, a non-asbestos resin-impregnated synthetic paper friction material is being used. Examples of various metals used in automatic transmissions include steel, aluminum, copper, bronze, brass, tin and silver solder. Non-metal materials include plastic, paper, nylon and a variety of elastomeric sealant materials. More recently, electronic circuitry and sensors have been immersed in the ATF.

3.4.2 *Gear lubricants*

Key ingredients in gear lubricants are extreme pressure (EP) agents which protect gear teeth surfaces and provide extended gear life. Dispersants are used to provide cleanliness. They must be compatible with the EP agents. Detergents have minimal use in gear lubricants because they are liable to be incompatible with EP agents.

3.4.3 *Tractor hydraulic oils*

Tractor wet brake lubrication requires extreme pressure (EP) properties and special friction characteristics. Water compatibility is a requirement for tractor hydraulic oils, and this tends to minimize the use of dispersants. Water compatible detergents are used in tractor hydraulic oils to provide thermal stability. High alkalinity (TBN) calcium or magnesium sulfonates are preferred. They also help to provide increased dynamic friction. Like automatic transmission fluids, tractor hydraulic oil development requires close cooperation between the tractor builder, the additive manufacturer, and the oil marketer.

3.5 Performance evaluation

The principal functions performed by detergents in engine oil formulations are (1) acid neutralization, (2) high temperature detergency, (3) oxidation inhibition and (4) rust prevention. These functions provide engine cleanliness and extended trouble-free operation. Dispersants supplement these functions and, in addition, provide protection against low-temperature sludge deposits.

Engine oil formulations may contain several different natural or synthetic sulfonates comprising neutral and/or overbased detergents. Phenates may also be included with the sulfonates to fine-tune the final oil formulation to maximize performance (along with other additives which are discussed in other chapters). In addition, salicylates and phosphonates are used by some oil formulators in combination with sulfonates and phenates.

Since the introduction of chemical additives to engine oils, various organizations worldwide have established systems to define or classify engine oils by performance levels. Evaluation in actual vehicle service is time-consuming to the point that the results would be out of date when confirmed. Consequently, accelerated laboratory engine test procedures are used. The operating conditions of these laboratory engine tests are designed to provide correlation with a specific field service that the vehicle would encounter.

In worldwide use is the Engine Oil Performance and Engine Service Classification System developed by the Tripartite consisting of the American Petroleum Institute (API), the American Society for Testing and Materials (ASTM) and the Society of Automotive Engineers (SAE). Performance categories are included for both gasoline and diesel engine service. The Tripartite system is open-ended, allowing the addition of new performance classifications as engine design and service operation require.

Publication is progressively kept up to date by the SAE J183 Recommended Practice (SAE, 1988). The European engine builders' organization publishes an engine oil performance system using the US Tripartite system as the base to supplement several European laboratory engine tests. Similarly, the Japanese engine builders are developing engine test procedures based on Japanese engines to supplement the Tripartite system. The international Organization for Standardization (ISO), a non-governmental group, is attempting to develop a worldwide lubricant performance system.

The rust inhibition properties of detergents are evaluated in a laboratory Oldsmobile 5.7 l V-8 gasoline engine test designed to correlate with stop-and-go, short-trip driving under low-temperature operation. The test is currently known as the Sequence IID test and is also included in the SAE (1988) publication. Major emphasis is placed on minimizing rust deposits on the hydraulic valve lifter component.

The key engine oil evaluation tests for dispersant additive performance are two gasoline engine laboratory tests known as the Sequence III and Sequence V tests. Each of these engine tests has been progressively modified to meet the ever-increasing performance requirements of engine oils, as engine design and service conditions become more demanding on lubricants. For example, since 1970 both the Sequence III and V tests have been revised five times and are currently called Sequence IIIE and VE (previously labeled A, B, C, D and now E).

The Sequence IIIE test is conducted in a Buick 3.8 l V-6 gasoline engine designed to correlate with high-speed, high-temperature gasoline engine operation. A key performance parameter evaluated is oil viscosity increase resistance. The Sequence VE test is conducted in a Ford 2.3 l, four-cylinder gasoline engine. The test is designed to simulate suburban driving practices. Dispersant additives have a major impact in controlling sludge and varnish formation in the Sequence VE test. The sequence IIIE and VE tests are described in detail in ASTM Reports D2: 1225 (ASTM, 1989) and D2: 1002 (ASTM, 1988), respectively. Use of both tests in helping to define oil quality levels can be found in SAE J183 (SAE, 1989).

Automatic transmission fluids, gear lubricants, and tractor hydraulic oil performance evaluation tests are developed and specified by the equipment manufacturer. Various worldwide military organizations develop specifications around the manufacturers' test procedures. Oil companies endeavor to meet the manufacturers' specifications and have their product listed in a given manufacturers' approved lubricant listing.

Detergents and dispersants have a vital role in formulating engine oils. Today's transportation, construction and agricultural vehicles would not exist without oils containing tailored additive packages that include detergents and dispersants. Detergents and dispersants have many other vital applications including tractor fluids, automatic transmissions, gear lubricants and industrial oils. In all areas there is a continual need to match new engines and hardware to new lubricants for optimum vehicle performance.

References

Anderson, R.L. and McDonald, H. (1960) US Patent 2,924,618.

ASTM (1988) *Report D2: 1002—Sequence VE test.* American Society for Testing and Materials Research, Philadelphia, PA.

ASTM (1989) *Report D2: 1225—Sequence IIIE test.* American Society for Testing and Materials Research, Philadelphia, PA.

Bergstrom, R.S. (1942) US Patent 2,279,086.

Campbell, S.E. and Dellinger, I.S. (1949) US Patent 2,485,861.

Greisinger, W.K. and Engelking, E.H. (1946) US Patent 2,402,325.

Lazar, A. and Carter, J.C. (1946) US Patent 2,402,288.

Mannich, C. and Krosche, W. (1912) *Arch. Pharm.* **250** 647.

Van Ess, P.R. and Bergstrom, R.F. (1945) US Patent 2,372,411.

Further reading

Asseff, P.A. (1981) *Lubricating Oil Additives—Description and Utilization.* Lubrizol Corporation, Wickliffe, Ohio.

Asseff, P.A. (1983) *Lubrication Theory and Practice.* Publication 183-320-59, Lubrizol Corporation, Wickliffe, Ohio.

Bennett, P.A., Malone, G.K. and Murphy, C.K. (1960) *An Engine Test for Predicting the Performance of Engine Lubricants on Most Severe Passenger Car Service.* SAE Tulsa, Oklahoma Meeting, November 1960.

Borneman, W.H. and Gergel, W.C. (1980) *Lubrication of Automotive Engines With Modern Oils.* San Diego Automotive Rebuilders Association Meeting, June 1980.

Brannen, C.G. and Hunt, M.W. (1977) US Patent 4,049,560.

Bray, W.B., Dickey, C.R. and Voorhees, V. (1975) *Ind. Eng. Chem. Prod. Res. Dev.* **14** 295–298

Brois, S.J. (1978) *Olefin-Thionophosphine Sulfide Reaction Products. Their Derivatives and Use Thereof as Oil and Fuel Additives.* US Patent 4,100,187.

Brois, S.J. and Gutirrez, A. (1965) US Patent 3,219,666.

Cameron, N.A. (1976) *Basic Lubrication Theory.* Wiley & Sons, New York.

Chamberlin, W.B. (1979) *Concentrates, Lubricant Compositions and Methods for Improving Fuel Economy of Internal Combustion Engines.* UK Patent App., GB, 2023169 A (1979).

Colyer, C.C. (1983) The SAE Lubricant Review Institute—purpose and operation. *SAE Paper 830653.*

Colyer, C.C. (1967) Advances in motor oils. *Seventh World Petroleum Congress*, Mexico City, 1967.

Colyer, C.C. (1971) Gasoline engine oils—performance, evaluation and classification. *8th World Petroleum Congress*, PD 17 (2), Moscow, 1971.

Colyer, C.C. (1978) Lubrication fundamentals. *Australasia Society of Automotive Engineers*, October 26.

Colyer, C.C. (1979) Automotive lubricating oil additives past/present/future. *Chemical Marketers Research Association*, Chicago, Il, November 29.
Colyer, C.C. (1981) The ASTM test monitoring system—purpose and operation. *ASTM Standard News*, October.
Colyer, C.C. (1983) *Tailoring Oils to Engines*. SAE/I Mech.E. Exchange Lecture, June 7, 1983.
Colyer, C.C. and Tom, T.B. (1960) *Does the MS Test Predict Field Performance*. SAE Presentation, Tulsa, Oklahoma, November 1960.
Colyer, C.C. and Kollman, R.E. (1979) *Engine Oils for the 1980s*. Japan Petroleum Institute, Tokyo, October 29, 1979.
Cooke, V.B. (1982) *The Role of Additives in the Automotive Industry*. ASLE Detroit Section Oral Presentation. Publication 982-320-48, Lubrizol Corporation, Wickliffe, Ohio.
Drummond, A.Y., Anderson, R.G. and Stuart, F.A. (1962) *Lubricating Oil Compositions of Alkyldiperazine Alkenyl Succinimides*. US Patent 3,024,195 (1962).
Gagliardi, J.C., Ghannam, F.E., Gasvoda, R.F. and Potter, R.I. (1960) Engine tests for evaluating crankcase oils in stop-and-go MS service. *SAE National Fuels and Lubricants Meeting*, Tulsa, Oklahoma, November 1960.
Gergel, W.C. (1968) *Oxidized, Degraded Interpolymer of Ethylene and Propylene and Fuel Composition Containing the Same*. US Patent 3,374,073, 3/19/68.
Gergel, W.C. (1975) *Thiourea-Acylated Polyamine Reaction Product*. US Patent 3,865,813, 2/11/75.
Gergel, W.C. (1975) The facts and fiction of synthetic oils. *I.O.C.A. Meeting*, October 1975.
Gergel, W.C. (1975) Detergents—what are they. *Proceedings of the JSLE-ASLE International Lubrication Conference*, Tokyo, Japan, 1975.
Gergel, W.C. (1976) Componenti sintetici per lubrificanti. *Tribologia e Lubrificazione* **XI** (2) Giugno 1976.
Gergel, W.C. (1976) Additivi detergenti per lubrificanti, *Oleodinamica-pneumatica* **XVII** (8) Agosto 1976.
Gergel, W.C. (1977) Anti-wear protection for valve train components. Presented at *SME Clinic on the Design and Manufacturing of Cams*, St. Louis, Missouri, February 1–3, 1977.
Gergel, W.C. (1981) *The Technical Need for Higher Quality Diesel Engine Lubricants*. Presented in Cairo, Egypt, February 1981.
Gergel, W.C. (1984) Lubricant additive chemistry. *The International Symposium Technical Organic Additives and Environment*, Interlaken, Switzerland.
Gergel, W.C. (1988) New technical diesel engine oil requirements. Presented at *Sixth International Seminar on New Development in Engine and Industrial Oils, Fuels, and Additives*, Cairo, Egypt, March 7–10, 1988.
Gergel, W.C. and LaTour, G.G. (1977) Extended engine oil life through new technology. Presented at the *National Petroleum Refiners Association 1977 Annual Meeting*, March 27–29, San Francisco, California.
Gergel, W.C. and Ranstl, H.P. (1986) Valve train protection: an international challenge. Presented at the *5th International Colloquium—Additives for Lubricants and Operational Fluids*, Esslingen, Germany, January 14–16, 1986.
Gergel, W.C. and Robins, R.D. (1974) Low ash motor oils—performance in the laboratory and field. SAE 740141 presented at the *Automotive Engineering Congress*, February 25–March 1, 1974.
Gergel, W.C. and Sheahan, T.J. (1976) Maximizing petroleum utilization through extension of passenger car drain periods—what's required. Presented at *1976 API Automotive Forum*, Southfield, Michigan, January 29, 1976.
Gergel, W.C., Karn, J.L. and King, L.E. (1971) *Basic Magnesium Salts, Processing, and Lubricants and Fuels Containing the Same*. US Patent 3,629,109, 12/21/71.
Gergel, W. C., Richardson, J.P. and von Eberan-Eberhorst, C. (1989) Status of diesel engine oil development—test procedures and specification. Presented at *Technische Akademie Esslingen*, West Germany, 31.05.1988/01.06/1989.
Hutchison, D.A. and Stauffer, R.D. (1966) US Patent 3,272,746.
Inoue, K. and Watanabe, N. (1983) Interactions of Engine Oil Additives. *ASLE Trans.* **26** (2) 189–199.
Kane, J.E.J. and Jones, J.R. (1957) US Patent 2,781,403.
LeSeur, W.M. and Norman, G.R. (1965) *Reaction Product of High Molecular Weight Succinic Acids and Succinic Anhydrides With an Ethylene Polyamine*. US Patent 3,172,892.

Loane, C.M. and Gaynor, J.W. (1943) *Lubricants.* US Patents 2,316,080 to 2,316,084.
McMillan, M.L. (1977) Engine oil viscosity classifications—past, present, and future. *SAE Paper 770373.*
McNab, J.G. and Rogers, D.T. (1948) US Patent 2,451,345.
Napper, D.H. (1983) *Polymeric Stabilization of Colloidal Dispersion.* London Academic Press.
Ottewill, R.H. (1982) *Colloidal Dispersion.* London Royal Society of Chemistry, 143–163.
Piasek, E.J. and Karll, R.E. (1962) *Dihydroxybenzyl Polyamines.* US Patent 3,539,633 (1962).
Rodgers, J.J. and Kabel, R.H. (1978) The sequence IID engine oil test. *SAE International Fuels & Lubricants Meeting*, Royal York, Toronto, November 13–16, 1978. *SAE Paper 780931.*
Ryer, J., Zielinski, J., Miller, H.N. and Brois, S.J. (1977) *Oxazoline Derivatives as Additives in Oleaginous Compositions.* US Patent 4,049,564.
Schilling, G.J. and Bright, G.S. (1977) *Fuels and Lubricant Additives Lubrication* (*Texaco*), **63** (2).
Shaub, H. and Pecoraro, J.M. (1966) US Patent 3,272,743.
Sieloff, F.X., Kasai, T. and Glover, I. (1981) *Global Impact of Equipment Trends on Future Lubricants.* 1981 Lubrizol Internal Report.
Sieloff, F.X. and Musser, J.L. (1982) *What Does the Engine Designer Need to Know About Engine Oils.* Detroit SAE Section.
Smalheer, C.V. and Smith, R.K. (1967) *Lubricant Additives.* Lubrizol Corporation, Lexius-Hiles Co., Cleveland, Ohio.
Tupa, R.C. and Dorer, C.J. (1984) Gasoline and diesel fuel additives for performance/distribution/quality. *SAE Paper 841211.*
Watson, R.W. and McDonnel T.F. Jr. (1984) Additives—the right stuff for automotive oils. *SAE Publication SP-603.*
Watson, R.W. (1975) The role of alkyl groups in petroleum additives. *Proceedings of the JSLE-ASLE International Lubrication Conference*, Tokyo, Japan, 1975.
Willette, G.L. and Ozimek, R.T. (1989) Additives for lubricants—a time of change. *NPRA Paper*, Houston, Texas, 1989.
Wilson, C.E. (1941) US Patent 2,250,188.
Winning, C. (1944) US Patent 2,362,291.
Wood, K.P. and van Wijngaarden, K. (1981) The future world demand and supply of lubricating oil. *2nd Symposium on Arab and International Lubricating Oils Industry.* June 27–29, 1981, p. 2.
Wright, E.P. (1971) The contribution of the oil additive industry to solution of automotive problems experienced in the field. *8th World Petroleum Congress*, PD 17 (3), Moscow, 1971.
Zalar, F.V. (1981) Diesel engine lubrication for the 80's. *Construction Equipment Forum*, Chicago, Illinois, November, 1981.
SAE (1983) Lubricant and additive effects on engine wear. *SAE SP-558.*
ASTM (1969) *Standard Handbook of Lubrication Engineering.* McGraw-Hill Book Company, Chapter 14.
SAE (1986) Worldwide lubricant trends. *SAE SP-676.*

4 Oxidative degradation and stabilisation of mineral oil based lubricants

M. RASBERGER

4.1 Introduction

The reactions of oxygen with most organic materials are important from an economical and ecological standpoint. Processes such as the biological O_2/CO_2 cycle, the targeted oxidation of defined organic molecules, or the enzymatic oxidation of waste are all useful reactions. In contrast, the reactions of atmospheric oxygen with hydrocarbon polymers and liquid hydrocarbons (lubricants), as well as with certain biological systems, under varying conditions of temperature and oxygen pressure are undesirable processes. Such reactions lead to a deterioration of these materials. All oxidative processes with oxygen have a common reaction pattern attributable to the biradical status of oxygen.

The aim of this contribution is to present the reaction mechanisms of the degradation processes of lubricants and the factors influencing them. In addition, the mechanisms by which antioxidants inhibit lubricant oxidation with respect to specific industrial and engine oil applications are suggested.

4.2 Autoxidation of hydrocarbons

4.2.1 *Oxidation of hydrocarbons at low temperature (30–120 °C)*

The self-accelerating oxidation of hydrocarbons is called autoxidation. Its initial stage is characterised by a slow reaction with oxygen followed by a phase of increased conversion until the process comes to a standstill. The degradation is driven by an autocatalytic reaction which can be described by the well-established free radical mechanism (Lazar *et al.*, 1989; Denisov and Khudyakor, 1987). It consists of four distinct stages:

- initiation of the radical chain reaction
- propagation of the radical chain reaction
- chain branching
- termination of the radical chain reaction

4.2.1.1 *Initiation of the radical chain reaction* Under normal conditions, i.e. moderate temperature and oxygen partial pressure greater than 50 torr, the first step is a process which is catalysed by traces of transition metal ions (Veprek-Siska, 1985):

$$\mathrm{R-\overset{\displaystyle CH_3}{\underset{\displaystyle H}{\overset{|}{\underset{|}{C}}}}-H} \xrightarrow[k_1]{M^{n+}/O_2} \mathrm{R-\overset{\displaystyle H}{\underset{\displaystyle CH_3}{\overset{|}{\underset{|}{C}}}}\cdot + HOO\cdot} \qquad (4.1)$$

R refers to a long chain alkyl substituent and the catalyst M^{n+} is a transition metal such as Co, Fe, V, Cr, Cu or Mn. The rate of initiation is very slow (k_1 = rate constant = 10^{-9}–$10^{-10}\,\mathrm{l\,mol^{-1}\,s^{-1}}$; Emanuel *et al.*, 1967; Sheldon and Kochi, 1973).

The site of the oxygen attack is determined by the strength of the C–H bond, and the reactivity for hydrogen abstraction increases in the following order:

$$\mathrm{RCH_2{-}H < R_2CH{-}H < R_3C{-}H < RCH{=}CH(R)HC{-}H < C_6H_5(R)HC{-}H} \qquad (4.2)$$

Hence oxidation of an *n*-paraffin normally commences by abstraction of a hydrogen at the second carbon atom (alpha-position to the CH_3 group).

4.2.1.2 *Propagation of the radical chain reaction* Once an alkyl radical has been formed this reacts irreversibly with oxygen to form an alkyl peroxy radical:

$$\mathrm{R(CH_3)CH\cdot + O_2} \xrightarrow{k_3} \mathrm{R(CH_3)CH{-}OO\cdot} \qquad (4.3)$$

Reaction (4.3) is extremely fast ($k_3 = 10^7$–$10^9\,\mathrm{l\,mol^{-1}\,s^{-1}}$) and has a very low activation energy (k_3 is independent of temperature).

The rate of reaction of carbon centered radicals with oxygen depends on the type of substituents attached to the C-atom and increases in the following order:

$$\mathrm{H_3C\cdot < C_6H_5(R)CH\cdot < RCH{=}CH(R)CH\cdot < R_2CH\cdot < R_3C\cdot} \qquad (4.4)$$

For instance, a tertiary alkyl radical reacts 10 times faster with oxygen than a methyl radical. This difference in reactivity explains why hydrocarbons with tertiary hydrogens (iso-paraffins) are oxidised faster than *n*-paraffins.

The next step in the chain propagation scheme is the hydrogen abstraction by a peroxy radical from another hydrocarbon:

$$\mathrm{R(CH_3)CHOO\cdot + RH} \xrightarrow{k_5} \mathrm{R(CH_3)CHOOH + R\cdot} \qquad (4.5)$$

This leads to a hydroperoxide and an alkyl radical which can again react with oxygen according to reaction (4.3). The rate of reaction of step (4.5) is slow ($k_5 = 10^{-1}$–$10^{-5}\,\mathrm{l\,mol^{-1}\,s^{-1}}$ at 30 °C, depending on the type of hydrocarbon) when compared with step (4.3) and is therefore the rate determining step for chain propagation. Due to their low reactivity, peroxy radicals are present in a relatively high concentration in the system when compared with other radicals (determined via electron spin resonance).

Since peroxy radicals possess a low energy status, they react selectively and abstract tertiary hydrogen atoms in preference to secondary and primary hydrogen atoms (C–H primary: C–H secondary: C–H tertiary = 1:30:300 as relative rates; Ingold, 1983). Their reactivity is also influenced by steric effects: primary and secondary peroxy radicals show a three to five times higher reactivity than tertiary peroxy radicals (Ingold, 1969).

A more favourable route of hydrogen abstraction by a peroxy radical (Mill and Montorsi, 1973; Jensen *et al.*, 1979, 1981) occurs via an intramolecular propagation outlined in (4.6), where x is equal to 1 or 2, R^1 is a terminal alkyl group and R^2 hydrogen or an alkyl group.

$$\underset{\displaystyle \mathrm{O{-}O\cdot}}{\mathrm{RHC}}\!-\!(\mathrm{CH_2})_x\!-\!\underset{\displaystyle \mathrm{H}}{\mathrm{C}}(\mathrm{R^2})\mathrm{R^1} \longleftrightarrow \underset{\displaystyle \mathrm{O{-}O{-}H}}{\mathrm{RHC}}\!-\!(\mathrm{CH_2})_x\!-\!\dot{\mathrm{C}}(\mathrm{R^2})\mathrm{R^1} \xrightarrow{\mathrm{O_2}}$$

$$\underset{\displaystyle \mathrm{O{-}OH}}{\mathrm{RHC}}\!-\!(\mathrm{CH_2})_x\!-\!\underset{\displaystyle \mathrm{O{-}O\cdot}}{\mathrm{C}}(\mathrm{R^2})\mathrm{R^1} \xrightarrow{\mathrm{RH}} \underset{\displaystyle \mathrm{O{-}OH}}{\mathrm{RHC}}\!-\!(\mathrm{CH_2})_x\!-\!\underset{\displaystyle \mathrm{O{-}OH}}{\mathrm{C}}(\mathrm{R^2})\mathrm{R^1} + \mathrm{R\cdot}$$

$$\mathrm{I} \qquad\qquad\qquad\qquad \mathrm{II} \qquad (4.6)$$

The hydroperoxide–peroxy radical (I) reacts with hydrocarbons through intermolecular hydrogen radical abstraction, resulting in an alkyl dihydroperoxide (II) and a chain-initiating alkyl radical. Radical (I) may also intramolecularly abstract a hydrogen radical.

$$\mathrm{R}\!-\!\underset{\displaystyle \mathrm{OOH}}{\overset{\displaystyle \mathrm{H}}{\mathrm{C}}}\!-\!(\mathrm{CH_2})_x\!-\!\overset{\displaystyle \mathrm{\cdot O{-}O}}{\mathrm{C}}(\mathrm{R^2})\mathrm{R^1} \longrightarrow \mathrm{R}\!-\!\underset{\displaystyle \mathrm{OOH}}{\dot{\mathrm{C}}}\!-\!(\mathrm{CH_2})_x\!-\!\overset{\displaystyle \mathrm{OOH}}{\mathrm{C}}(\mathrm{R^2})\mathrm{R^1} \qquad x = 1\text{–}2$$

$$\mathrm{I} \qquad\qquad\qquad\qquad \mathrm{III} \qquad (4.7)$$

The alkyl dihydroperoxide radical (III) may then react with oxygen, followed by additional intermolecular hydrogen abstraction to yield an alkyl trihydroperoxide (IV) and an alkyl radical.

$$\underset{\text{III}}{R{-}\dot{C}(OOH){-}(CH_2)_x{-}C(OOH)(R^2)R^1} \xrightarrow[\text{(ii) RH}]{\text{(i) } O_2} \underset{\text{IV}}{R{-}C(OOH)_2{-}(CH_2)_x{-}C(OOH)(R^2)R^1} + R\cdot \qquad (4.8)$$

The occurrence of these intramolecular reaction sequences leads to an increased rate of formation of hydroperoxide, which results in a reinforced autocatalytic degradation of the hydrocarbons.

4.2.1.3 *Chain branching* During the early stage of autoxidation various types of hydroperoxides are generated. At low concentrations they may be cleaved homolytically to yield an alkoxy and a hydroxy radical:

$$ROOH \xrightarrow{k_9} RO\cdot + HO\cdot \qquad (4.9)$$

However, this process is thwarted because of a high activation energy, hence reaction (4.9) only plays a significant role at higher temperatures or under catalysed conditions (see sections 4.2.2 and 4.2.3). Once formed, hydroxy and especially primary alkoxy radicals are so active that they abstract hydrogen atoms in non-selective reactions:

$$HO\cdot + CH_3{-}R \longrightarrow H_2O + RCH_2\cdot \qquad (4.10)$$

$$HO\cdot + R{-}CH_2{-}R^1 \longrightarrow H_2O + R(R^1)HC\cdot \qquad (4.11)$$

$$RCH_2O\cdot + CH_3{-}R \longrightarrow RCH_2OH + RH_2C\cdot \qquad (4.12)$$

Secondary and tertiary alkoxy radicals prefer to form aldehydes and ketones:

$$R{-}C(H)(R^1){-}O\cdot \longrightarrow RCHO + R^1\cdot \qquad (4.13)$$

$$R^2{-}C(R^1)(R^3){-}O\cdot \longrightarrow R^1COR^2 + R^3\cdot \qquad (4.14)$$

At high concentrations, i.e. at an advanced state of oxidation, hydroperoxides may react via a bimolecular mechanism:

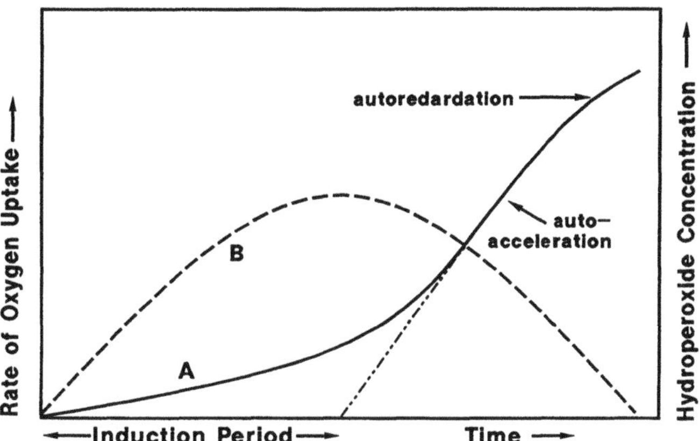

Figure 4.1 Influence of hydroperoxide concentration on the rate of oxygen uptake over time; A = rate of oxygen uptake; B = hydroperoxide concentration (schematic drawing).

$$ROOH + ROOH \rightleftharpoons \begin{matrix} R{-}O{-}O{-}H \\ \vdots \quad \vdots \\ H{-}O{-}OR \end{matrix} \longrightarrow ROO\cdot + RO\cdot + H_2O \qquad (4.15)$$

As a consequence of hydroperoxide accumulation and subsequent cleavage, the concentration of reactive free radicals initiating new chains increases. The time from the beginning of the oxidation to the autocatalytic phase of the autoxidation is called the 'induction period'.

The relationship between hydroperoxide accumulation and oxygen uptake of a hydrocarbon is schematically presented in Figure 4.1. As can be seen, during the induction period hydroperoxides are accumulated. After the induction period, the oxidation is autocatalysed.

4.2.1.4 *Termination of the radical chain reaction* As the reaction proceeds, autoxidation is followed by an autoretardation stage resulting in a standstill before the hydrocarbon is completely consumed. This autotermination is called the 'chain termination reaction' and dominates in this final phase of the oxidation process such that degradation comes to a halt. Termination may be effected by the combination of radical species (such as peroxy radicals) to yield unreactive species (such as ketones and alcohols):

$$2\,RR^1CHOO\cdot \longleftrightarrow [R(R^1)CHOOOOCH(R^1)R]$$

$$\longrightarrow R(R^1)C{=}O + O_2 + HO{-}CH(R^1)R \qquad (4.16)$$

In this example, primary and secondary peroxy radicals form intermediates which disproportionate to non-radical degradation products. In contrast, tertiary peroxy radicals may either combine to give di-tertiary alkyl peroxides or undergo a cleavage reaction leading to ketones and alkyl radicals:

$$2\,RR^1R^2C{-}OO\cdot \longleftrightarrow [RR^1R^2COOOOCRR^1R^2] \xrightarrow[-O_2]{}$$

$$2\,RR^1R^2C{-}O\cdot \begin{cases} \longrightarrow RR^1R^2COOCR^2R^1R \\ \longrightarrow 2\,RR^1C{=}O + 2\,R^{2}\cdot \end{cases} \quad (4.17)$$

Generally, the rate of termination increases across the series tertiary peroxy < secondary peroxy < primary peroxy.

If the oxygen concentration in the bulk liquid phase is limited (oxygen partial pressure below 50 torr) two additional ways of radical recombination result:

$$R\cdot + ROO\cdot \longrightarrow ROOR \quad (4.18)$$

$$R\cdot + R\cdot \longrightarrow R{-}R \quad (4.19)$$

In summary, the uncatalysed oxidation of hydrocarbons at temperatures of up to 120 °C leads to alkylhydroperoxides (ROOH), dialkylperoxides (ROOR), alcohols (ROH), aldehydes (RCHO) and ketones $R(R^1)C = O$. In addition, cleavage of a dihydroperoxide II of reaction (4.6) leads to diketones ($RCO(CH_2)_xCOR^1$), keto-aldehydes ($RCO(CH_2)_xCHO$), hydroxyketones ($RCH(OH){-}(CH_2)_xCOR^1$) and so forth.

Under metal catalysed conditions or at higher temperatures, which will be dealt with in sections 4.2.2 and 4.2.3, degradation leads to a complex mixture of final products.

4.2.2 *Oxidation of hydrocarbons at high temperature* (> *120 °C*)

Above 120 °C the degradation process can be divided into a primary and a secondary oxidation phase.

4.2.2.1 *Primary oxidation phase* Initiation and propagation of the radical chain reaction are the same as discussed under low-temperature conditions, but the selectivity is reduced and the reaction rate is increased. At high temperature the cleavage of hydroperoxides plays the most important role.

Reaction (4.9) leads to a proliferation of hydroxy radicals, which non-selectively abstract hydrogen atoms (see reactions (4.10) and (4.11)). Acids are formed by the following two reactions, which start from a hydroperoxy–peroxy radical (see reaction (4.6)) and an aldehyde (Jensen, 1979, 1981):

$$R{-}\underset{\underset{OOH}{|}}{CH}{-}CH_2{-}\overset{\overset{\cdot OO}{|}}{\underset{\underset{R^2}{|}}{C}}{-}R^1 \longleftrightarrow R\overset{\bullet}{\underset{\underset{OOH}{|}}{C}}{-}CH_2{-}\overset{\overset{OOH}{|}}{\underset{\underset{R^2}{|}}{C}}{-}R^1 \xrightarrow[-HO\cdot]{}$$

I

$$R{-}CO{-}CH_2{-}\overset{\overset{OOH}{|}}{\underset{\underset{R^2}{|}}{C}}{-}R^1 \longrightarrow R{-}COOH + R^2{-}CH_2{-}\overset{\overset{O}{\|}}{C}{-}R^1 \qquad (4.20)$$

$$R{-}CHO \xrightarrow[-ROOH]{ROO\cdot} R{-}\overset{\overset{O}{\|}}{C}\cdot \xrightarrow{O_2} R{-}\overset{\overset{O}{\|}}{C}OO\cdot \xrightarrow[-R\cdot]{RH}$$

$$R{-}\overset{\overset{O}{\|}}{C}OOH \longrightarrow R{-}\overset{\overset{O}{\|}}{C}OH + \tfrac{1}{2}O_2 \qquad (4.21)$$

In addition, when the rate of oxidation becomes limited by diffusion, ethers are formed.

$$R{-}\overset{\overset{OO\cdot}{|}}{CH}{-}CH_2{-}CH_2{-}R^1 \longrightarrow R\,\overset{\overset{OOH}{|}}{CH}{-}CH_2{-}\overset{\bullet}{C}H{-}R^1 \xrightarrow[-HO\cdot]{}$$

$$R{-}\overbrace{CH{-}CH_2{-}CH}^{O}{-}R^1 \qquad (4.22)$$

The termination reaction proceeds through primary and secondary peroxy radicals according to reaction (4.16), but at temperatures above 120 °C these peroxy radicals also interact in a non-terminating way to give primary and secondary alkoxy radicals (Ingold, 1983):

$$2\,ROO\cdot \longrightarrow [ROOOOR] \longrightarrow 2RO\cdot + O_2 \qquad (4.23)$$

These radicals again contribute to the formation of cleavage products via reactions (4.12) and (4.13).

4.2.2.2 *Secondary oxidation phase* At higher temperatures the viscosity of the bulk medium increases as a result of the polycondensation of the difunctional oxygenated products formed in the primary oxidation phase. Further polycondensation and polymerisation reactions of these high molecular weight intermediates result in products which are no longer soluble in the hydrocarbon. The resulting precipitate is called sludge. Under thin-film oxidation conditions, as in the case of a lubricant film on a metal surface, varnish-like deposits are formed (Perez *et al.*, 1987). The polycondensation reactions which lead to high molecular weight intermediates (sludge precursors) can be described as follows. In a first step, aldehydes or ketones formed in the primary oxidation phase combine via an acid- or base-catalysed aldol condensation to form alpha, beta-unsaturated aldehydes or ketones (Ali *et al.*, 1979):

$$\mathrm{R\,CO{-}(CH_2)_nCHO + CH_3{-}COR^1 \xrightarrow[\text{(or base)}]{\text{acid}} R{-}CO{-}(CH_2)_nCH{=}CHCOR^1 + H_2O}$$

$$\text{V}$$

$$\mathrm{V + R\,CO{-}(CH_2)_mCOR^2 \xrightarrow[-H_2O]{} \begin{array}{l} \mathrm{R\,CO{-}C{-}(CH_2)_{n-1}CH{=}CHCOR^1} \\ \qquad\quad \| \\ \quad \mathrm{R{-}C{-}(CH_2)_mCO{-}R^2} \end{array}} \tag{4.24}$$

$$\text{VI}$$

Further aldol condensations with species VI lead to high molecular weight but still oil-soluble polycondensation products (molecular weight about 2000).

When the reaction becomes diffusion controlled as a result of the increased viscosity of the oil, alkoxy radicals can initiate polymerisation of polycondensation products. This leads to sludge and deposit formation as well as to additional oil-soluble high molecular weight products, which contribute to the viscosity increase. This process can be described as copolymerisation of two different polycondensation species in which the alkyl groups R, R^1 and R^2 could represent oxo- or hydroxy-functionalised long hydrocarbon chains:

$$x\mathrm{R{-}CO{-}CH{=}CH{-}R^1} + y\mathrm{CH_3CO{-}CH{=}CH{-}(CH_2)_nCOOR^2}$$

$$\xrightarrow{\mathrm{RO\cdot}} \left[\underset{\mathrm{COR}}{\mathrm{CH}} {-} \underset{\mathrm{R^1}}{\mathrm{CH}} \right]_x {-} \left[\underset{\mathrm{COCH_3}}{\mathrm{CH}} {-} \underset{\mathrm{(CH_2)_n{-}COOR^2}}{\mathrm{CH}} \right]_y \tag{4.25}$$

Condensation polymerisations are accelerated in the presence of metals such as iron (Clark *et al.*, 1985), but lead suppresses polycondensation.

Under high-temperature conditions there is always the possibility of thermal cleavage of a hydrocarbon chain, especially when the availability of oxygen is limited by diffusion:

$$\mathrm{R(CH_2)_6R \longrightarrow 2\,[RCH_2CH_2{-}CH_2\cdot] \longrightarrow 2\,RCH_2CH{=}CH_2 + H_2} \tag{4.26}$$

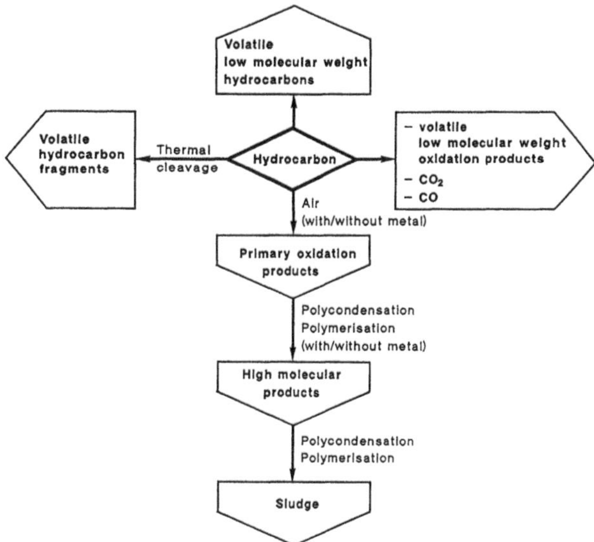

Figure 4.2 Model of lubricant degradation under high temperature conditions.

Reaction (4.26) leads to unsaturated molecules with lower molecular weight and higher volatility.

The model for high temperature oxidation (Naidu *et al.*, 1984; Hsu and Cummings, 1983; Jensen *et al.*, 1979, 1981) can be described by Figure 4.2.

4.2.3 *Metal catalysed autoxidation of hydrocarbons*

The decomposition of an alkyl hydroperoxide molecule occurs at temperatures of about 150 °C. Transition metal ions having two ionic valence states, such as $Fe^{2+/3+}$, $Pb^{2+/4+}$ and $Cu^{1+/2+}$, reduce the activation energy of this decomposition process. These ions must be present as metal soaps otherwise they are not catalytically active (Kuhn, 1973). Hence the homolytic hydroperoxide decomposition is accelerated at ambient temperatures by small concentrations (0.1–50 ppm) of these metals:

$$\mathrm{ROOH} + \mathrm{M}^{n+} \longrightarrow \mathrm{RO\cdot} + \mathrm{M}^{(n+1)+} + \mathrm{OH}^- \tag{4.27}$$

$$\mathrm{ROOH} + \mathrm{M}^{(n+1)+} \longrightarrow \mathrm{ROO\cdot} + \mathrm{M}^{n+} + \mathrm{H}^+ \tag{4.28}$$

Reactions (4.27) and (4.28) can be summarised as (4.29).

$$2\mathrm{ROOH} \xrightarrow{\mathrm{M}^{n+}/\mathrm{M}^{(n+1)+}} \mathrm{ROO\cdot} + \mathrm{RO\cdot} + \mathrm{H_2O} \tag{4.29}$$

Chain branching under the influence of a catalyst at a given temperature

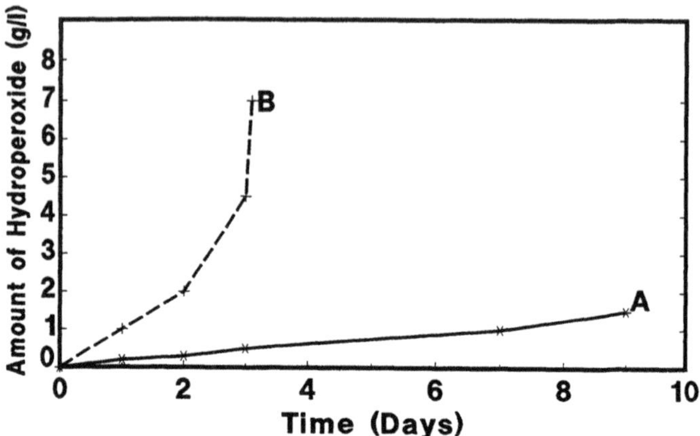

Figure 4.3 Increase of hydroperoxide concentration under the influence of Fe $(OOCC_7H_{15})_3$ catalysis: (A) pure model hydrocarbon; (B) pure model hydrocarbon plus 500 ppm Fe $(OOOC_7H_{15})_3$. Conditions: 95°C bath temperature; 300 ml oil volume; 3 l/hour oxygen flow (Payne, 1988).

proceeds faster when compared with the uncatalysed reaction. The consequence is a high rate of hydroperoxide formation (and hence oxidation) as illustrated in Figure 4.3.

The precursor steps of soap formation, shown below with Fe as the metallic surface, arise from the attack of the metal surface by alkylperoxy radicals and alkylhydroperoxide. This process may be called 'corrosive wear' (Newley *et al.*, 1980; Habeeb *et al.*, 1987):

$$ROO\cdot + Fe \longrightarrow FeO + RO\cdot \tag{4.30}$$

$$ROOH + Fe \longrightarrow FeO + ROH \tag{4.31}$$

The FeO reacts further with organic acids and forms the iron soaps:

$$2RCOOH + FeO \longrightarrow Fe(OCOR)_2 + H_2O \tag{4.32}$$

The catalytic activity of copper and iron has been discussed in the technical literature (Clark *et al.*, 1985 and quoted references; Abou El Naga and Salem, 1984; Vijh, 1985) and is summarised as follows:

- In the presence of iron, copper soaps in general retard oxidation and polycondensation/polymerisation reactions. In iron ion-free systems organocopper salts behave as pro-oxidants.
- Iron soaps accelerate oxidation and polycondensation/polymerisation reactions with increasing concentration.

An example concerning the retardation of oxidation by a copper salt (George and Robertson, 1946) during the oxidation of a model hydrocarbon (tetralin) in the presence of 0.5% iron stearate is given in Figure 4.4.

In conclusion, the oxidation of model hydrocarbons proceeds via a radical

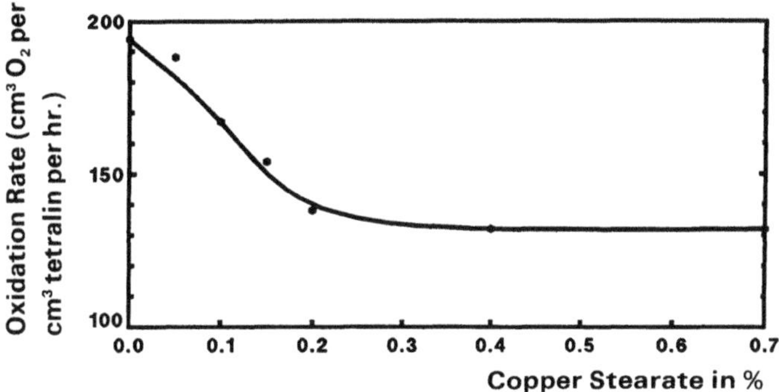

Figure 4.4 Influence of copper stearate on the oxidation rate of the iron stearate (0.5%) catalysed oxidation of tetralin.

chain mechanism and is strongly influenced by temperature and by certain transition metal ions. This results in the formation of insoluble sludge as well as a continuous increase in viscosity. The following section deals with the oxidation of base oils, which principally follows the same reaction pattern as the model hydrocarbons, but is more complex due to the presence of natural inhibitors and pro-oxidants.

4.3 Oxidation stability of base oils

The base stocks used for lubricants are a mixture of C_{20}–C_{45} hydrocarbons (see section 1.2.1), which can be subdivided into three main groups: paraffins, naphthenes and aromatics. In addition, intramolecular combinations of these three groups also comprise a part of the base stock composition. Traces of nitrogen-, sulfur- and oxygen-containing heterocycles, together with mercaptans (RSH), thioethers (R–S–R) and disulfides (R–S–S–R) are an integral part of the complex composition of lubricating base oils. Some of these compounds may act as natural antioxidants, others as pro-oxidants.

The compositions of base stocks are greatly influenced by the refining processes. Their oxidation stability shows a maximum (Fuchs and Diamond, 1942; Burn and Greig, 1972) when the aromatic content is between 10 and 20%. This so-called Fuchs concept of 'optimal aromaticity' holds only if the sulfur content of the base stock is in the range 0.005–0.070% (Burn and Greig, 1972; Korcek and Jensen, 1975).

The reason for the strong increase in oxidation stability at the optimum aromatic-sulfur ratio is the result of a 'synergistic interaction pattern' between phenols and certain thioethers (Studt, 1986). These phenols are formed by an acid-catalysed decomposition of hydroperoxides, where the peroxide group is in the side chain at the alpha-position to the aromatic nucleus. The necessary catalytic traces of acid arise from a thermal cleavage of certain oxidised sulfides (Rost, 1963) and disulfides together with further oxidation of the cleavage products to SO_3/H_2SO_4 (see also section 4.4.2.1). The rates of transformation of these aromatic hydroperoxides depend on the type of aromatic compound. Alkylated polynuclear aromatics react much faster than alkylbenzenes as they are energetically favoured to undergo acid-catalysed decomposition.

The polynuclear aromatics, together with the appropriate sulfur compounds, retard the rate of oxidation in the 'post-induction phase' when sufficient hydroperoxides have formed. However, they also contribute to sludge formation via polycondensation reactions especially when the aromatic/sulfur ratio falls outside the optimum value according to the Fuchs concept.

Benzylic radicals are more stable than alkyl radicals, therefore they react less readily with oxygen and tend to improve the oxidation stability of lubricants (Murray *et al.*, 1983). On the other hand naphthenic aromatics form unstable hydroperoxides which are very quickly degraded and serve as key sludge precursors.

Half a century ago, refined base stocks fulfilled the needs of the industry. However, with increasing demands on the lubricant it was necessary to develop stabilisers against oxidation which would extend its useful life. The types of antioxidants used and how they perform, as well as their synthesis, will be dealt with in the next section.

4.4 Inhibition of oxidative degradation of lubricants

The use of additives to control lubricant degradation requires a focus on alkyl radicals (R·), alkylperoxy radicals (ROO·) and hydroperoxides (ROOH). Primary alkoxy radicals ($RCH_2O\cdot$) and hydroxy radicals (HO·) rapidly abstract hydrogen from the substrate. It is therefore very unlikely that they can be deactivated by natural or synthetic antioxidants. In practice three additive types have proven to be successful in controlling the degradation of lubricating oils:

- radical scavengers
- hydroperoxide decomposers
- synergistic mixtures of these

4.4.1 *Radical scavengers*

The most widely used types of radical scavengers are phenolic and aminic antioxidants. Recently, organo-copper salts (Colclough *et al.*, 1981) have been introduced to control the oxidative degradation of engine oils.

4.4.1.1 *Sterically hindered phenols* Phenols which are substituted in the 2 and 6 positions with tertiary alkyl groups are called sterically hindered phenols. The most common substituent is the tertiary butyl group.

Reaction mechanism Sterically hindered phenols like (VII) compete successfully with the rate-determining steps of the propagation reaction (4.5), i.e. $k_{33} \gg k_5$ (see page 84).

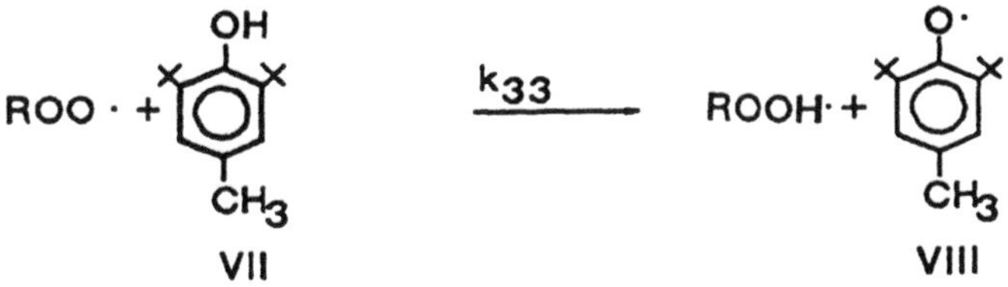

ROO · + (VII) $\xrightarrow{k_{33}}$ ROOH· + (VIII) (4.33)

The resonance-stabilised phenoxy radical (VIII) preferentially scavenges an additional peroxy radical to form a cyclohexadienone peroxide (IX).

ROO· + (VIII) ⟶ (IX) (stable < 120°C) (4.34)

Optimum protection of the substrate towards oxidation can be achieved when both 2 and 6 positions of the phenol are substituted with tertiary butyl groups and the 4 position is substituted by an *n*-alkyl group (Wasson and Smith, 1953). The replacement of a tertiary butyl group by a methyl group in the ortho position reduces the antioxidant activity considerably (Table 4.1). Steric hindrance in both ortho positions beneficially prevents hydrogen abstraction by the phenoxy radical (reaction (4.35)).

Table 4.1 Influence of ortho substitution on antioxidant activity of hindered phenols in lubricants at 0.1% and 110 °C (Wasson and Smith, 1953)

Type of phenol	Relative activity
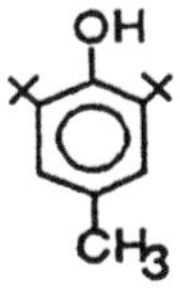	100.0
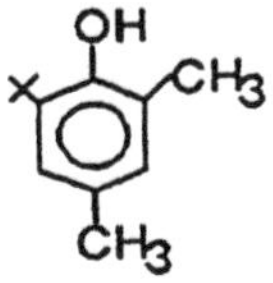	62.5

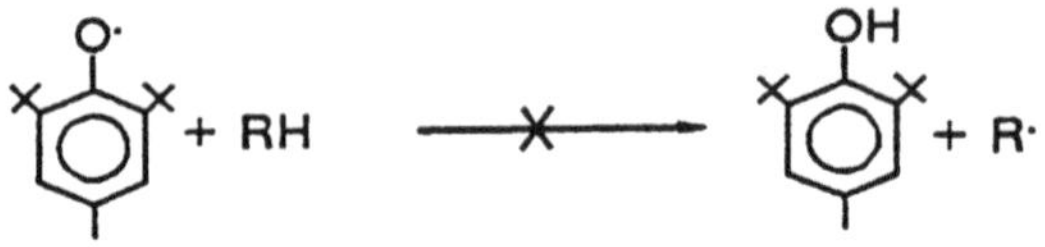

O· + RH —✕→ OH + R· (4.35)

Synthesis The worldwide annual consumption of sterically hindered phenols for industrial lubricants and engine oils amounts to 22 000–24 000 tonnes. The major structural types which are commercially available, along with their synthetic routes, are outlined in Table 4.2. (Nirula, 1983; Malec and Plonsker, 1976; Lowe and Liston, 1980).

Table 4.2 Major commercial phenolic antioxidants and their applications

Structure and synthesis	Major use
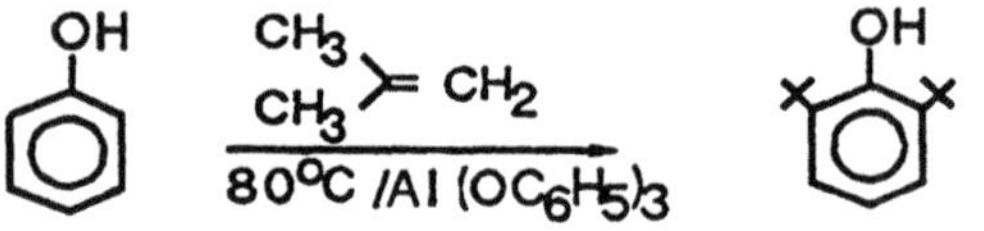 OH → (CH_3)(CH_3)C=CH_2, 80°C /Al (OC_6H_5)$_3$ → OH	Industrial oils
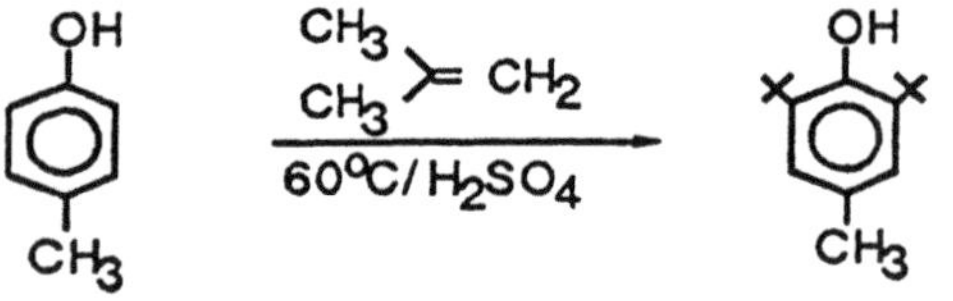 OH, CH_3 → (CH_3)(CH_3)C=CH_2, 60°C/ H_2SO_4 → OH, CH_3	Industrial oils, greases
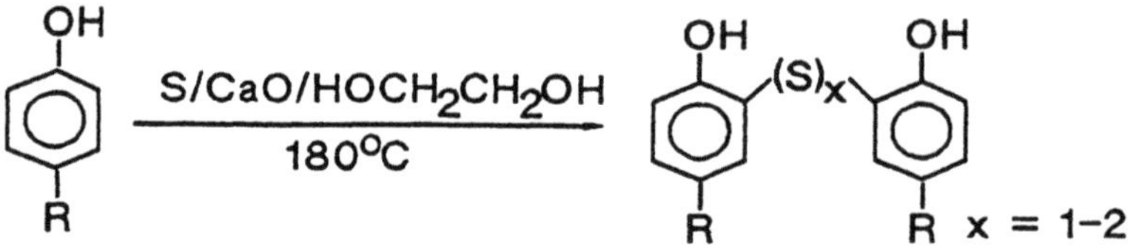 OH, R → S/CaO/$HOCH_2CH_2OH$, 180°C → OH, R –(S)$_x$– OH, R; x = 1–2	Engine oils
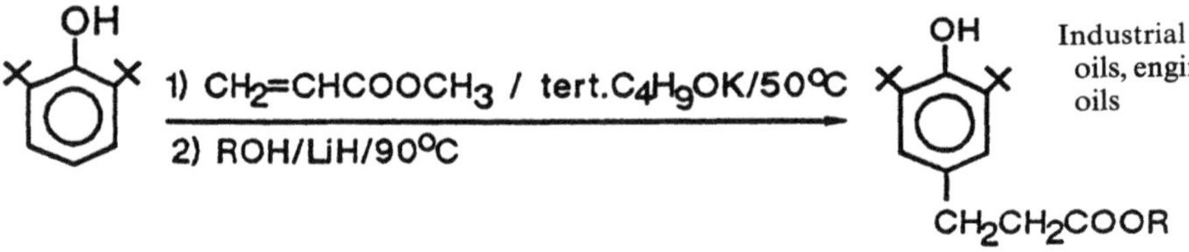 OH → 1) CH_2=$CHCOOCH_3$ / tert.C_4H_9OK/50°C, 2) ROH/LiH/90°C → OH, CH_2CH_2COOR	Industrial oils, engine oils
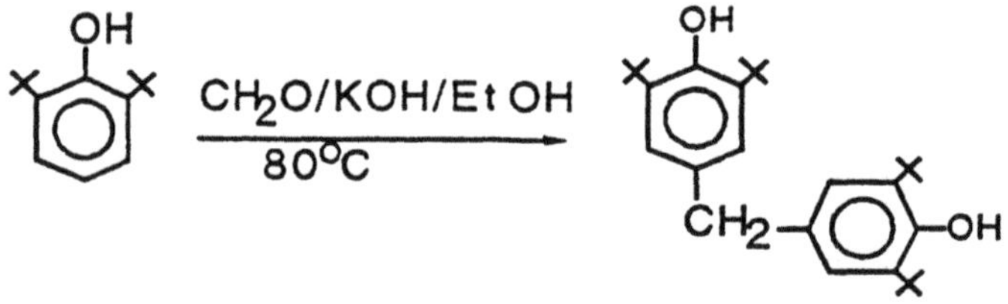 OH → CH_2O/KOH/Et OH, 80°C → OH, CH_2–(–OH)	Industrial oils, engine oils

4.4.1.2 *Aromatic amines* Secondary aromatic amines are another important class of antioxidants used in lubricants. The principal substituents of the nitrogen atom are either two aryl or one aryl and a naphthyl group.

Reaction mechanism The reaction mechanism of diphenylamines is dependent on the temperature. Under low-temperature conditions (< 120 °C) the interaction with peroxy radicals predominates (Berger *et al.*, 1983).

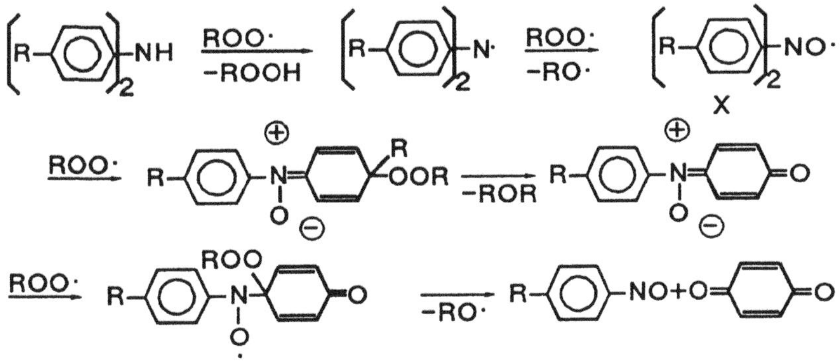

ROO· / −ROOH; ROO· / −RO·; X; ROO·; −ROR; ROO·; −RO·

(4.36)

According to the above reaction sequence, one diphenylamine molecule eliminates four peroxy radicals. By definition, the stoichiometric factor is four. In the case of a sterically hindered monophenol, this factor is two. Hence diphenylamines perform better as peroxy radical scavengers than monophenols at temperatures of < 120 °C. Under high-temperature conditions (> 120 °C), e.g. engine oils, an extended stabilisation mechanism in the form of a catalytic cycle has been suggested (Korcek *et al.*, 1988; see also Berger *et al.*, 1983).

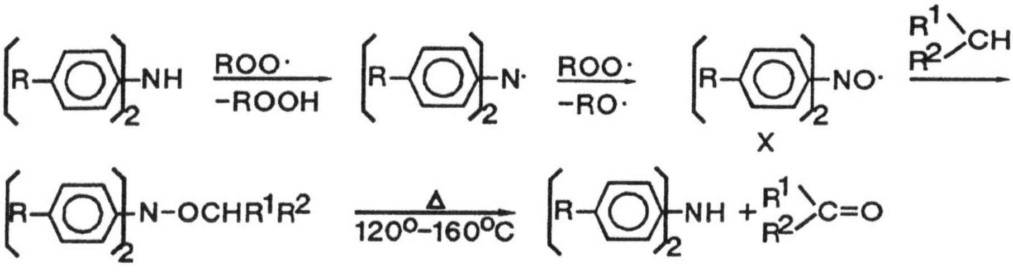

ROO· / −ROOH; ROO· / −RO·; X; $R^1R^2CH\cdot$; Δ, 120°–160°C

XI

(4.37)

It has been demonstrated that the nitroxyl radical (X) reacts with a secondary alkyl radical to form (XI) which, under high-temperature conditions (> 120 °C), regenerates the original diphenylamine molecule. In essence, this group of stabilisers acts catalytically by scavenging alternately peroxy (ROO·) and alkyl radicals (R·). As stated earlier, sterically hindered phenols deactivate only two peroxy radicals per phenol molecule. Hence, under high-temperature conditions, aromatic amines are far superior to their phenolic counter-

Table 4.3 Stoichiometric factor of different secondary aromatic amines (2×10^{-5} mol l^{-1}) in a paraffinic oil at 130 °C

Compound	Stoichiometric factor
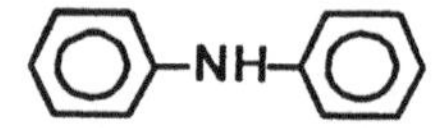 Ph–NH–Ph	41
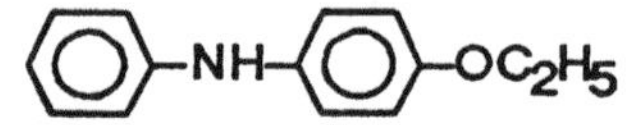 Ph–NH–C$_6$H$_4$–OC$_2$H$_5$	36
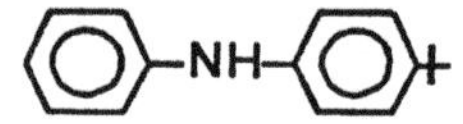 Ph–NH–C$_6$H$_4$–C(CH$_3$)$_3$	53
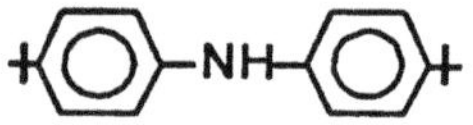 (CH$_3$)$_3$C–C$_6$H$_4$–NH–C$_6$H$_4$–C(CH$_3$)$_3$	52

parts. As shown in Table 4.3, the stoichiometric factor of the diphenylamines depends on the substituents in the para position (Berger *et al.*, 1983). The efficacy of the diphenylamine antioxidant is improved by alkylating the para positions.

The stabilisation mechanism for phenyl-alpha-naphthylamines may be described as follows (Zeman *et al.*, 1987):

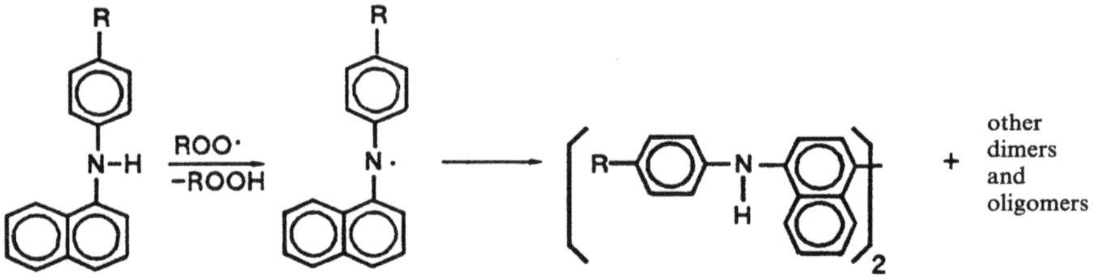

(4.38)

In contrast to the diphenylamine derivatives, a transformation of the nitrogen radical (XII) to a nitroxyl radical (N–O·) has not been observed. Due to the longer life-time of the nitrogen-centered radical (XII) by resonance stabilisation, dimerisation and oligomerisation take place whilst maintaining the –NH– function.

Synthesis The present annual consumption of this class in the lube oil industry is believed to be in the order of 17 000–19 000 tonnes. Representatives of this class of compounds as well as their syntheses are outlined in Table 4.4 (Carmichael, 1973).

Table 4.4 Major commercial aminic antioxidants and their applications

Structure and synthesis	Major use
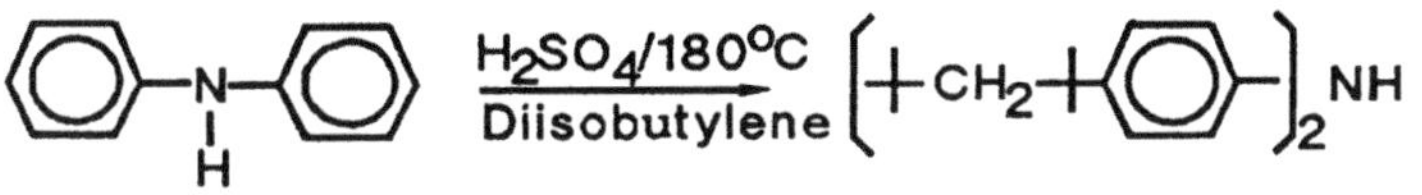 N H; H_2SO_4/180°C, Diisobutylene; $(\text{+}CH_2\text{+}\langle\bigcirc\rangle)_2NH$	Industrial lubricants, engine oils, aviation oils, automatic transmission fluids, greases
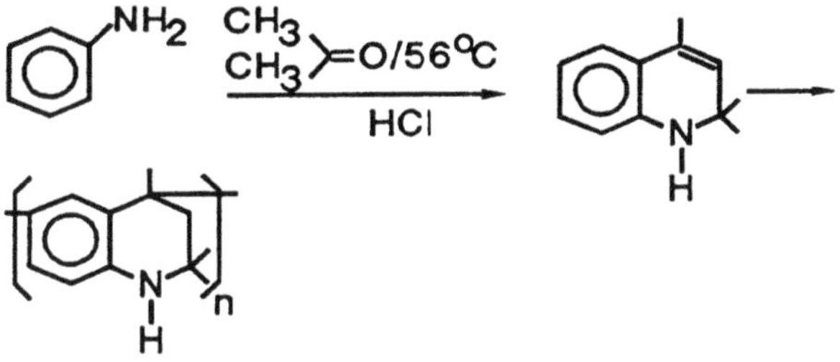 NH_2; CH_3, CH_3 >=O/56°C, HCl; N H; N H, n	Greases
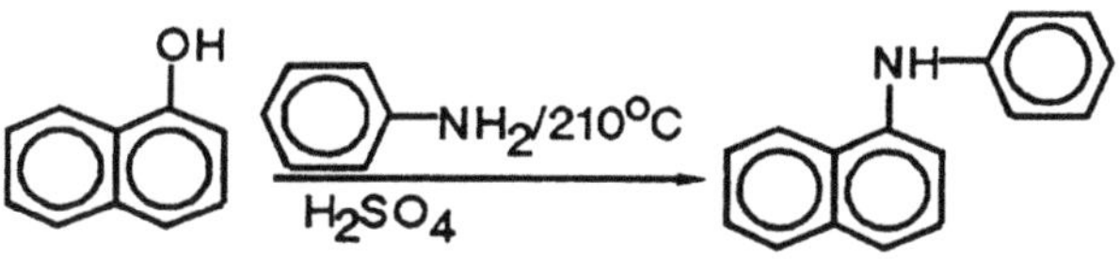 OH; NH_2/210°C, H_2SO_4; NH	Industrial lubricants, greases
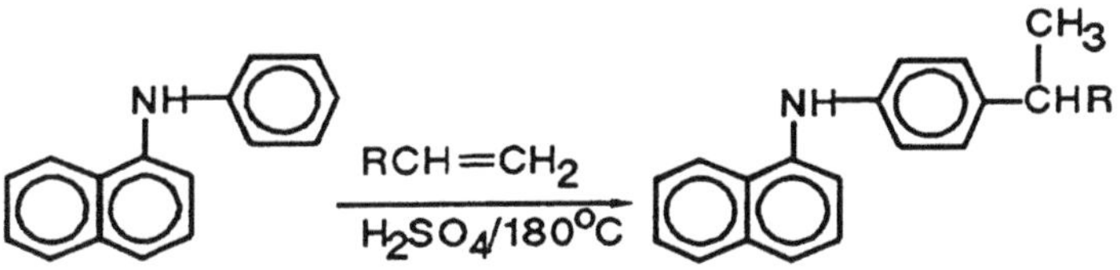 NH; $RCH{=}CH_2$, H_2SO_4/180°C; NH, CH_3, CHR	Industrial lubricants, engine oils, aviation oils, greases

4.4.1.3 *Organocopper antioxidants* A breakthrough in terms of controlling the oxidation of automotive oils has been achieved with the introduction of organocopper salts (Colclough *et al.*, 1981). As outlined earlier (section 4.2.3) they can also act as pro-oxidants, and several bench oxidation tests use copper or copper salts as a catalyst in order to make test conditions more severe and so shorten test duration. The catalytic activity of copper ions dominates up to 40 ppm of copper but in the range of 100–200 ppm in combination with ZnDTP or dithio-thiadiazoles (Colclough, 1987a) organocopper salts control the oxidation of engine oils.

Reaction mechanism The way in which copper controls oxidation may be outlined as follows (Colclough *et al.*, 1991):

- Reduction in Fe^{2+} concentration results in a stabilisation step because Fe^{2+} is more active than Fe^{3+} in catalysing oxidation:

$$Cu^{2+} + Fe^{2+} \longrightarrow Cu^{+} + Fe^{3+} \quad (4.39)$$

- Copper ions eliminate catalytically active alkyl and peroxy radicals (Scott, 1984):

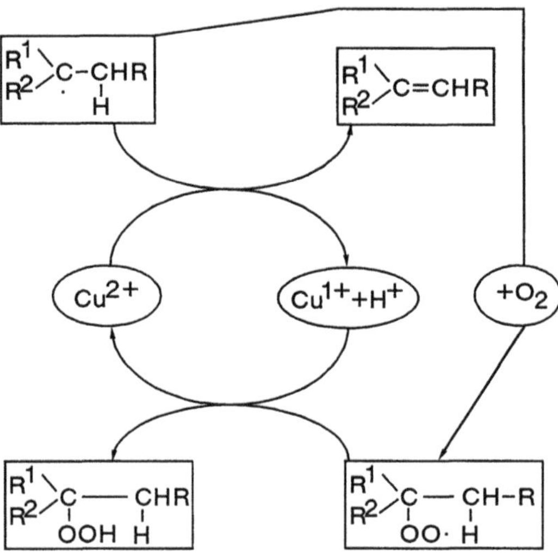

Figure 4.5 Catalytic radical scavenging mechanism of copper ions in engine oils.

Conversely, copper ions catalyse to a certain degree the decomposition of hydroperoxides:

$$ROOH \xrightarrow{Cu^{1+}/Cu^{2+}} RO\cdot + HO\cdot \quad (4.40)$$

This pro-oxidant influence of the organocopper compound has to be countered through adding to the formulation organosulfur compounds or ZnDTP as hydroperoxide decomposers.

Synthesis Cuprous or cupric metal salts are prepared according to the following reaction (Matsuura, 1965):

$$Cu^{I}Cl + RCOONa \xrightarrow[-NaCl]{} RCOOCu \quad (4.41)$$

4.4.2 *Hydroperoxide decomposers*

These compounds convert hydroperoxides into non-radical products thus preventing the chain propagation reaction. Traditionally organosulfur and organophosphorus additives have been used for this purpose.

4.4.2.1 *Organosulfur compounds*

Reaction mechanism of organosulfur compounds The most important reaction mechanism to eliminate hydroperoxides is the acid-catalysed decomposition. The catalysts are protic (RSO_2H) or Lewis (SO_2) acids (Chien and Boss, 1972; Morrison and Boyd, 1987):

$$R^1R^2CHOOH \xleftrightarrow{H^+} R^1R^2CHOO^+H_2 \xrightarrow[-H_3O^+]{} R^1R^2C=0 \qquad R^1 = \text{H or alkyl group} \tag{4.42}$$

$$R^2-\underset{CH_3}{\overset{R^1}{C}}-OOH \xleftrightarrow{H^+} R^2\underset{CH_3}{\overset{R^1}{C}}-\underset{H}{\overset{+}{O}}-OH \xrightarrow[-H_2O_2-H^+]{ROOH} R^2-\underset{CH_3}{\overset{R^1}{C}}-OO-R$$
$$\xrightarrow[-H^+]{-H_2O_2} R^1R^2C=CH_2 + \text{polymers thereof} \tag{4.43}$$

Organosulfur compounds are the main source for the formation of the acid catalyst. Compounds such as (XIII) react with hydroperoxides to yield sulfoxides (XIV) as key intermediates for the stabilisation of the lubricant (Shelton, 1981; Scott, 1981, 1983):

$$\underset{\text{XIII}}{R-S-R} \xrightarrow[-ROH]{ROOH} \underset{\text{XIV}}{R-\overset{O}{\overset{\|}{S}}-R} \tag{4.44}$$

Thermolytic cleavage of species like (XIV) leads to the formation of sulfenic acid (XV):

$$R-\overset{O}{\overset{\|}{S}}-R \xrightarrow{\Delta} \underset{\text{XV}}{RSOH} + H_2C=CR^1R^2 \qquad R = -\underset{R^2}{\overset{CH_3}{C}}-R^1 \tag{4.45}$$

The influence of the structure of the alkyl group, R, on the rate of cleavage (reaction (4.45)) has been correlated with the following order of increasing reactivity:

$$R^1CH_2- < R^1\overset{CH_3}{CH}- < R^1(R^2)\overset{CH_3}{C}- < ROOC-CH_2-CH_2- \tag{4.46}$$

Further reaction of acid XV with hydroperoxide leads to the formation of sulfinic acid:

$$R\text{—}SOH \xrightarrow[-ROH]{ROOH} R\text{—}SO_2H \qquad (4.47)$$

Sulfinic acids are the most important acid catalysts for ionic decompositions below 100 °C. At higher temperatures, SO_2 resulting from reaction (4.48) is a most efficient catalyst (Scott, 1981; 1983):

$$R\text{—}SO_2H \xrightarrow{\Delta} R\text{—}H + SO_2 \qquad (4.48)$$

Dialkyl disulfides (R–S–S–R) follow the same reaction pattern thereby finally forming a thiosulfurous acid ($RS\text{–}SO_2H$) which under higher temperature and in the presence of hydroperoxides is cleaved to give SO_2 and a sulfenic acid (RSOH).

An additional stabilisation mechanism is peroxy radical scavenging by sulf-acids (RSO_xH):

$$RSO_xH \xrightarrow{ROO\cdot} RSO_x\cdot + ROOH \qquad x = 1, 2, 3 \qquad (4.49)$$

These stabilisation mechanisms only work if a high molar $ROOH/R\text{–}S_x\text{–}R$ ratio exists. At low molar ratios sulfides (R–S–R) and disulfides (R–S–S–R) show a certain pro-oxidant effect. This behaviour is one of the reasons why the organosulfur compounds only contribute positively to oxidation stability after the phenolic antioxidants have been depleted and the molar ratio of $ROOH/R\text{–}(S)_x\text{–}R$ is high.

Synthesis The major types of organosulfur compounds along with their synthetic routes are shown in Table 4.5 (Dorinson, 1983; Fields *et al.*, 1955).

4.4.2.2 *Zinc dialkyl dithiocarbamates* These compounds are mainly used as antioxidants although, like the ZnDTPs, they also have extreme pressure activity.

Reaction mechanism The formation of sulfur acids, which serve as the catalyst for ionic hydroperoxide decomposition, arises through the oxidation of zinc dithiocarbamates by hydroperoxides (Al-Malaika *et al.*, 1987):

$$[R_2NCS_2]_2Zn \xrightarrow[-ROH]{ROOH} R_2N\text{—}\overset{\overset{\displaystyle S}{\|}}{C}\text{—}S\text{—}Zn\text{—}\overset{\overset{\displaystyle O}{\|}}{S}\text{—}\overset{\overset{\displaystyle S}{\|}}{C}\text{—}NR_2$$

$$\xrightarrow[-ROH]{ROOH} \xrightarrow[\text{steps}]{\text{several}} RN{=}C{=}S + SO_3/H_2SO_4 \qquad (4.50)$$

In addition, zinc dithiocarbamates also act as radical scavengers, whereas in the literature the main emphasis is generally put on their capability to destroy hydroperoxides. Areas of application comprise grease and engine oils.

Table 4.5 Major types of organosulfur compounds and their synthesis

Structure and synthesis	Major use
$CH_3(CH_2)_7CH{=}CH{-}(CH_2)_7COOCH_3 \xrightarrow[\Delta]{S} R{-}(S)_x{-}R + \begin{array}{c} R^1{-}CH{-}CH{-}R^2 \\ \;\;\;\; \vert \;\;\;\;\;\;\; \vert \\ \;\;\;(S)_x \;\; (S)_x \\ \;\;\;\; \vert \;\;\;\;\;\;\; \vert \\ R^1{-}CH{-}CH{-}R^2 \\ + \\ \text{further products} \end{array}$ $x = 1, 2$ $R = CH_3(CH_2)_7\underset{\vert}{C}H{-}(CH_2)_8COOCH_3$ $R^1 = CH_3(CH_2)_7{-}$ $R^2 = {-}(CH_2)_7COOCH_3$	Industrial oils, engine oils
$\begin{array}{c} N{-}N \\ \Vert \;\;\; \Vert \\ C \;\;\; C \\ HS \diagup \diagdown S \diagup \diagdown SH \end{array} \xrightarrow{Cl_2} \begin{array}{c} N{-}N \\ \Vert \;\;\; \Vert \\ C \;\;\; C \\ ClS \diagup \diagdown S \diagup \diagdown SCl \end{array} \xrightarrow[-2HCl]{CH_3{-}C(CH_3)_2{-}CH_2{-}C(CH_3)_2{-}SH/\Delta}$ $\begin{array}{c} N{-}N \\ \Vert \;\;\; \Vert \\ C \;\;\; C \\ CH_3{-}C(CH_3)_2{-}CH_2{-}C(CH_3)_2{-}S{-}S \diagup \diagdown S \diagup \diagdown S{-}S{-}C(CH_3)_2{-}CH_2{-}C(CH_3)_2{-}CH_3 \end{array}$	Industrial oils Engine oils

Synthesis The synthesis of zinc dialkyl dithiocarbamates is as follows (Polishuk and Farmer, 1979):

$$2R_2NH + 2CS_2 \xrightarrow{NaOH} 2R_2N\overset{\overset{S}{\Vert}}{C}{-}S^-Na^+ \xrightarrow[-Na_2SO_4]{ZnSO_4} (R_2NCSS)_2Zn \tag{4.51}$$

4.4.2.3 *Organophosphorus compounds* Phosphites are the main organophosphorus compounds used to control oxidative degradation of lubricants. They eliminate hydroperoxides, peroxy and alkoxy radicals, retard the darkening of lubricants over time, and also limit photodegradation. These performance characteristics may be of importance for polyalphaolefins, hydrocracked or severely hydrotreated base stocks and white oils.

Reaction mechanism The major non-radical mode of action is the reduction of hydroperoxides, especially under high-temperature conditions:

$$(RO)_3P + R^1OOH \longrightarrow (RO)_3P{=}O + R^1OH \quad (4.52)$$

Phosphites with substituted phenoxy groups also behave as peroxy and alkoxy radical scavengers forming relatively stable phenoxy radicals, which again eliminate peroxy radicals.

$$(RO)_2P\text{–}O\text{–}C_6H_2(X)(CH_3)\text{–}CH_3 \xrightarrow{R^1OO\cdot} (RO)_2P(=O)\text{–}OR^1 + CH_3\text{–}C_6H_2(X)(CH_3)\text{–}O\cdot$$

$$(RO)_2P\text{–}O\text{–}C_6H_2(X)(CH_3)\text{–}CH_3 \xrightarrow{R^1O\cdot} (RO)_2P\text{–}OR^1 + CH_3\text{–}C_6H_2(X)(CH_3)\text{–}O\cdot$$

(4.53)

When comparing the importance of these three stabilisation steps for lubricants, there appears to be a preference for the decomposition of hydroperoxide (reaction (4.52)) (Kirpitschnikow and Pobedimski, 1975).

Synthesis The following route (Russel, 1977) is widely used:

$$PCl_3 + 3\,HO\text{–}C_6H_3(R^1)\text{–}R^2 \xrightarrow[-3HCl]{\Delta} P\{O\text{–}C_6H_3(R^1)\text{–}R^2\}_3 \quad (4.54)$$

4.4.3 *Multifunctional additives*

4.4.3.1 *Zinc dithiophosphates* The dominating position of ZnDTPs as additives for lubricating oils is due to their multifunctional performance. Not only do they act as antioxidants, but they also improve the wear inhibition of the lubricant, and protect metals against corrosion. ZnDTPs are mainly used to formulate anti-wear hydraulic fluids and engine oils.

Reaction mechanism The performance of ZnDTPs is strongly influenced by the type of alcohols used for their synthesis. Table 4.6 gives an overview of the variance of performance with type of alcohol.

The way ZnDTP performs as an antioxidant is a complex interaction pattern involving hydroperoxides and peroxy radicals. The performance matrix is additionally influenced by other additives which are present in industrial or engine oil formulations.

In a model system comprising cumene hydroperoxide and diverse ZnDTPs

Table 4.6 Structure activity dependency of ZnDTPs

Structure	Function		Property	
	Oxidative inhibition	Wear protection	Thermal stability	Hydrolytic stability
Primary ZnDTP				
$Zn\left[S{-}\overset{\overset{\displaystyle S}{\|}}{P}(OCH_2R)_2\right]_2$	satisfactory	satisfactory	good	satisfactory
Secondary ZnDTP				
$Zn\left[S{-}\overset{\overset{\displaystyle S}{\|}}{P}(OCHR^1R^2)_2\right]_2$	good	good	moderate	good
Aryl ZnDTP				
$Zn\left[S{-}\overset{\overset{\displaystyle S}{\|}}{P}(O{-}C_6H_4{-}R)_2\right]_2$	moderate	bad	very good	bad

it was demonstrated that the antioxidant mechanism proceeds by an acid-catalysed ionic decomposition of the hydroperoxide (Sexton, 1984). The catalyst species is O,O′-dialkylhydrogendithiophosphate, $(RO)_2PS_2H$, derived from the ZnDTP.

$$PhC(CH_3)_2OOH \xrightarrow[110^{\circ}C]{(RO)_2PS_2H} PhOH + CH_3COCH_3 + PhC(CH_3)_2OH + PhC(CH_3){=}CH_2 + Ph\overset{\overset{\displaystyle O}{\|}}{C}CH_3 \quad (4.55)$$

Whilst the first four products are the result of an acid-catalysed cationic chain reaction, the acetophenone is formed by a free radical mechanism.

There are two inter-related mechanisms for the formation of the acid catalyst. In the first, a rapid, initial reaction of ZnDTP and hydroperoxide forms a basic ZnDTP and a disulfide (XVI):

$$4[(RO)_2PS_2]_2Zn + R^1OOH \xrightarrow[-R^1OH]{\Delta} [(RO)_2PS_2]_6Zn_4O + \underset{\text{XVI}}{[(RO)_2PS_2]_2} \quad (4.56)$$

An induction period follows where the rate of decomposition of the hydroperoxide is slow. In this reaction phase the basic ZnDTP dissociates to form ZnDTP and ZnO

$$[(RO)_2PS_2]_6Zn_4O \rightleftharpoons 3[(RO)_2PS_2]_2Zn + ZnO \qquad (4.57)$$

The ZnDTP then reacts with hydroperoxide to form additional disulfide (XVI) via the dithiophosphoryl radical (XVII):

$$[(RO)_2PS_2]_2Zn + R^1OOH \xrightarrow[-R^1O\cdot]{} [(RO)_2PS_2]ZnOH + \underset{\text{XVII}}{(RO)_2PS_2\cdot}$$

$$\underset{\text{XVII}}{2(RO)_2PS_2\cdot} \longrightarrow \underset{\text{XVI}}{[(RO)_2PS_2]_2} \qquad (4.58)$$

The kinetics of the reaction result in a final rapid decomposition of the hydroperoxide provided the concentration of the basic ZnDTP is low. Under these conditions the sulfur radical is unable to dimerise. Instead it reacts with hydroperoxide leading to the catalytically active acid:

$$(RO)_2PS_2\cdot + ROOH \xrightarrow[-ROO\cdot]{} (RO)_2PS_2H \qquad (4.59)$$

The second source of this acid arises when the ZnDTP concentration falls below a critical level. Then traces of water interact with the ZnDTP forming the acid catalyst:

$$[(RO)_2PS_2]_2Zn + H_2O \longrightarrow [(RO)_2PS_2]ZnOH + (RO)_2PS_2H \qquad (4.60)$$

It was demonstrated that at temperatures above 125 °C, the disulfide (XVI) could be an additional source of the acid catalyst via reaction (4.61) followed by reaction (4.59) (Bridgewater *et al.*, 1980):

$$\underset{\text{XVI}}{[(RO)_2PS_2]_2} \longrightarrow \underset{\text{XVII}}{2(RO)_2PS_2\cdot} \qquad (4.61)$$

$$(RO)_2PS_2\cdot + ROOH \xrightarrow[-ROO\cdot]{} (RO)_2PS_2H \qquad (4.59)$$

A general scheme for the decomposition of hydroperoxides may be:

$$ROOH \xrightarrow{H^+} [RO\overset{+}{O}H_2 \rightleftharpoons R\underset{\substack{|\\H}}{\overset{+}{O}}{-}OH] \longrightarrow \text{reaction products} \qquad (4.62)$$

The type of reaction products formed are in accordance with reactions (4.42) and (4.43).

ZnDTP may also interact with peroxy radicals according to the following mechanism (Howard and Tong, 1980):

$$[(RO)_2PS_2]_2Zn \xrightarrow{ROO\cdot} \underset{}{RO_2}^- + (RO)_2PS_2Zn^{+\cdot} + \underset{\text{XVII}}{(RO)_2PS_2\cdot} \qquad (4.63)$$

The radical (XVII) may react again with hydroperoxide according to reaction (4.59) thereby regenerating the acid $(RO)_2PS_2H$, which is a better inhibitor than the ZnDTP.

Synthesis About 150 000–160 000 tonnes per year are produced in the western world using the following synthetic route:

$$P_2S_5 + 4ROH \longrightarrow 2(RO)_2PS_2H + H_2S$$
$$2(RO)_2PS_2H + ZnO \longrightarrow [(RO)_2PS_2]_2Zn + H_2O \qquad (4.64)$$

Primary and secondary aliphatic alcohols with chain lengths from C_3 to C_{12} and alkylated phenols are used. Nowadays aryl ZnDTPs are less important than alkyl ZnDTPs. Very often mixtures of low ($< C_5$) and high molecular weight alcohols ($> C_5$) are used for the synthesis. If only one type of alcohol is used, R should contain at least five carbon atoms.

Organomolybdenum compounds Recent patent literature shows that these compounds are of general interest in the engine oil area. They are antioxidants and in addition improve the frictional and anti-wear characteristics of the lubricants. Reaction (4.65) outlines the synthesis of a typical representative of this chemistry (Sakurai *et al.*, 1978).

$$MoO_3 + NaHS + HN(isoC_8H_{17})_2 + CS_2 \xrightarrow[\text{then 4 hours at 100 °C}]{\text{pH 2.5/room temperature}}$$

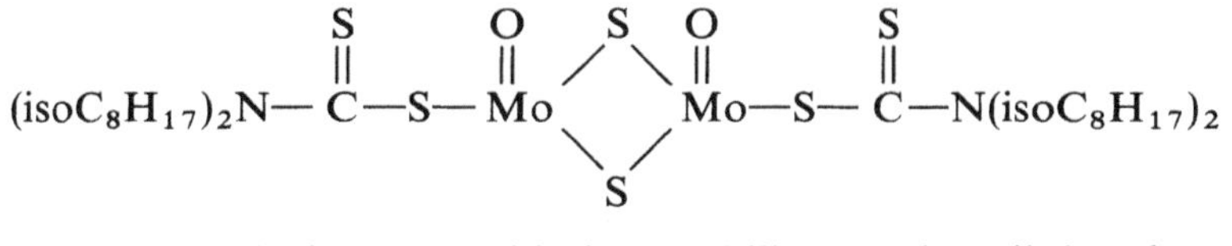

$$(isoC_8H_{17})_2N{-}\overset{S}{\overset{\|}{C}}{-}S{-}\overset{O}{\overset{\|}{Mo}}\langle{}^{S}_{S}\rangle\overset{O}{\overset{\|}{Mo}}{-}S{-}\overset{S}{\overset{\|}{C}}{-}N(isoC_8H_{17})_2 \qquad (4.65)$$

Such compounds impart oxidation stability to the oil by decomposing hydroperoxides and scavenging peroxy radicals.

Overbased phenates and salicylates Both phenates and salicylates of magnesium or calcium behave as antioxidants at high temperatures. The antioxidant activity may be related to hydroperoxide decomposition by the sulfur in structure (XVIII), and peroxy radical scavenging by the OH group in structure (XIX).

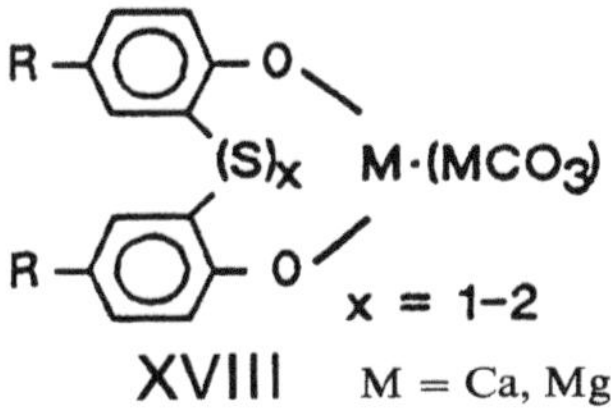

R, O, $(S)_x$, $M\cdot(MCO_3)$, R, O, x = 1–2

XVIII M = Ca, Mg

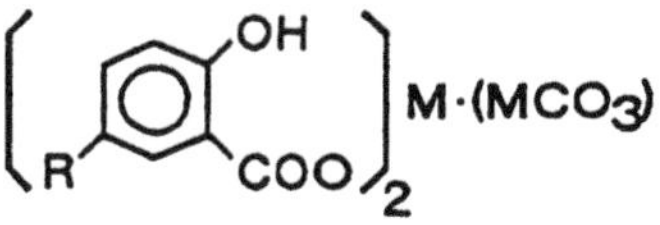

OH, $M\cdot(MCO_3)$, R, COO, 2

XIX M = Ca, Mg

Addition of MCO_3 to phenates and salicylates leads to the corresponding overbased products, which also act as acid scavengers.

Sulfur/nitrogen and sulfur/phosphorus compounds Other multifunctional sulfur/nitrogen, sulfur/phosphorus based additives have antioxidant and antiwear properties. Syntheses and generalised chemical structures are as follows (Polishuk and Farmer, 1979; Lange, 1987; Lam, 1989):

$$RR'NH \xrightarrow{CS_2/NaOH} RR'N{-}\overset{S}{\overset{\|}{C}}{-}S^{\ominus}\ Na^{+} \xrightarrow[-2NaCl]{CH_2Cl_2} \left[RR'N{-}\overset{S}{\overset{\|}{C}}{-}S\right]_2 CH_2 \qquad (4.66)$$

$$\text{(norbornene)}(COOR)_2 \xrightarrow{HS{-}\overset{S}{\overset{\|}{P}}(OR)_2} (RO)_2\overset{S}{\overset{\|}{P}}{-}S{-}\text{(norbornane)}(COOR)_2 \qquad (4.67)$$

$$\text{(norbornadiene)} \xrightarrow[70\,^{\circ}C/1\,h]{HS{-}\overset{S}{\overset{\|}{C}}NR_2} R_2N\overset{S}{\overset{\|}{C}}{-}S{-}\text{(norbornane)}{-}S{-}\overset{S}{\overset{\|}{C}}NR_2 \qquad (4.68)$$

These additives interact with peroxy radicals and hydroperoxides thus stabilising industrial lubricants and engine oils.

4.4.4 *Synergism between antioxidants*

The effectiveness of one class of antioxidant may be enhanced by combining it with another type of stabiliser. Hence, combining two different peroxy radical

Table 4.7 Oxidation stability of a lubricant stabilised with a synergistic antioxidant combination

Test fluid			
Alkylated diphenylamine	0.25%	—	0.2%
$S(CH_2CH_2{\cdot}COOR)_2$	—	0.25%	0.05%
Rust inhibitor	0.05%	0.05%	0.05%
Base stock	balance	balance	balance
Oxidation stability			
TOST (ASTM D943) (95 °C, H_2O, Fe and Cu catalyst, 3 l air/hour)			
—Time (hr) to total acid number = 2 mgKOH/g oil	2000	200	3300
—Sludge (mg) (ASTM D4310)	172	>5000	89
Base stock characteristics	ISO VG 32 CA (aromatic carbon) = 6.5% S = 0.54%		

scavengers in a lubricant gives improved oxidation stability compared with either alone. This form of additive response is called synergism or, more specifically, homosynergism because they operate by the same stabilisation mechanism.

The mechanism of this synergism can be outlined as follows (Mescina and Karpukhina, 1972):

$$\mathrm{ROO\cdot} + \mathrm{HN}\left[\!-\!\mathrm{C_6H_4}\!-\!\mathrm{R}\right]_2 \longrightarrow \mathrm{ROOH} + \underset{\mathrm{XX}}{\cdot\mathrm{N}\left[\!-\!\mathrm{C_6H_4}\!-\!\mathrm{R}\right]_2}$$

$$\mathrm{XX} + \mathrm{HO\!-\!C_6H_2X_2\!-\!R} \longrightarrow \mathrm{HN}\left[\!-\!\mathrm{C_6H_4}\!-\!\mathrm{R}\right]_2 + \mathrm{\cdot O\!-\!C_6H_2X_2\!-\!R} \qquad (4.69)$$

$$\mathrm{\cdot O\!-\!C_6H_2X_2\!-\!R} \xrightarrow{\mathrm{ROO\cdot}} \mathrm{O\!=\!C_6H_2X_2(R)(OOR)}$$

The aminic antioxidant reacts faster than the phenolic antioxidant with the peroxy radical. The less efficient sterically hindered phenol then regenerates the more effective aminic antioxidant.

An additional interesting example of how the oxidation stability of a lubricant can be optimised by using a synergistic combination of an aminic antioxidant and an organosulfur hydroperoxide decomposer, instead of the alkylated diphenylamine alone, is outlined in Table 4.7. This type of interaction is called heterosynergism because two different stabiliser mechanisms are involved.

This heterosynergism between the two antioxidants occurs due to the organosulfur compound decomposing the majority of hydroperoxides. Fewer radicals which initiate chain reactions are formed, and therefore less aminic antioxidant is consumed by the peroxy radical scavenging process.

4.5 Application of antioxidants

4.5.1 *Industrial lubricants*

Industrial lubricants in service have to prevent damage to the machinery which arises from friction between moving parts. Oxidation stability is one of the key requirements of the lubricant because the oxidation degradation products, e.g. peroxy radicals, hydroperoxides and organic acids as well as sludge and deposits, are detrimental to the equipment. The important role of antioxidants, to protect the various types of base oil under different operation conditions, is outlined in the following sections.

4.5.1.1 *Turbine oils*

Requirements The ratio between power output of turbines and oil volume has increased considerably over the years. As a consequence, operating temperatures have increased from 100 to 160°C (Sommer, 1988). Thus, protecting the lubricant against high-temperature degradation by selecting more thermally-stable and less volatile antioxidants has become important.

Antioxidant technology These more stringent requirements in terms of extended lifetime and reduced sludge can be met with new antioxidants. In Table 4.8 such a new stabiliser is compared in the TOST test against the traditionally used 2,6-ditertiarybutyl paracresol (BHT), in a hydrotreated (HT) and solvent neutral oil (SN). The antioxidant, in combination with the corrosion inhibitor, represents the most basic type of a turbine oil blend. These results can be interpreted as follows:

- The sulfur-containing, low volatility antioxidant outperforms BHT as a

Table 4.8 Comparison of oxidation stability of turbine oils formulated with two different antioxidants in a hydrotreated and solvent neutral base stock

Test fluids					
HO-(ring)-R-S-R[1] mwt = 492		0.25%		0.25%	
HO-(ring)-CH_3 BHT: mwt = 220			0.25%		0.25%
Corrosion inhibitor		0.05%	0.05%	0.05%	0.05%
Base stock		balance	balance	balance	balance
Oxidation stability					
TOST (ASTM D943)					
(95 °C, H_2O, Fe and Cu catalysts 3 l air/hour)					
—Time (hr) to total acid number = 2.0 mgKOH/g oil		4300	2400	2200	1100
—Sludge (mg) (ASTM D4310)		6	39	28	47
Base stock characteristic		Hydrotreated		Solvent neutral	
ISO VG		32		32	
CA (aromatic carbon)	(%)	NIL		6.5	
Sulfur	(%)	NIL		0.54	

result of autosynergism between the phenol and the thioether groups incorporated into the same molecule.

- The oxidation stabilities of the test fluids formulated with hydrotreated base stocks are better than the other two. In the solvent neutral base stock antioxidant efficacy is reduced because of interactions with certain polar base stock components (especially aromatics and nitrogen-containing compounds). In addition, naturally present pro-oxidants reduce base stock stability further.

Most fully formulated turbine oils make use of the synergistic interaction between phenolic and aminic antioxidants (see reaction (4.69)). Thus, depending on the performance requirement and the base stock composition, phenol/amine ratios of 1:1 to 4:1 are used.

4.5.1.2 *Anti-wear hydraulic fluids*

Requirements Increased severity of pump operations and the trend to more universally applicable hydraulic fluids leads to the following requirements:

- improved oxidation stability
- reduced sludge formation (better filterability)
- higher thermal stability
- less internal friction (energy conservation)

Two new classes of lubricant have emerged which fulfil these stringent requirements:

- premium grade ashless (ZnDTP-free) lubricants
- premium grade ZnDTP-based lubricants containing calcium detergents for improved thermal and hydrolytic stability of the ZnDTP.

For less severe operations 'normal' ashless and ZnDTP-based anti-wear hydraulic fluids are still used.

All basic hydraulic fluid formulations generally comprise antioxidant, corrosion inhibitor and anti-wear additives. The premium grades may contain up to seven further additives depending on performance requirements. For example, where extremely low sludge figures are required, the use of dispersant VI-improvers (Okada and Yamashita, 1986) may be considered. The addition of metal de-activators protects the metal surface and thus boosts the oxidation stability of the fluid by preventing dissolution of metal ions.

Overall, high-performance fluids contain 0.9–1.2% additives in order to meet the relevant Denison, Vickers, DIN and Cincinnati Milacron specifications.

Antioxidant technology In order to give an indication of the oxidation and thermal stability of a variety of ZnDTP-containing and ashless anti-wear hydraulic fluids, TOST and Cincinnati Milacron data—both relevant tests for this type of fluid—have been compiled from published (Gegner, 1982; Ellissen *et al.*, 1982) and in-house results. Table 4.9 summarises these data. These data show that the oxidation resistance and thermal stability of the premium fluids, both ashless and stabilised ZnDTP, are comparable and clearly outperform the normal grade fluids. The normal ashless grade at 0.4%–0.5% additives is superior to the normal ZnDTP-based grade, especially for thermal stability. This may be due to the use of thermally unstable secondary ZnDTPs.

In order to boost the oxidation stability of fluids formulated with stabilised ZnDTPs (in most cases zinc di-2-ethylhexyldithiophosphate) phenolic or aminic antioxidants, often in combination, are added. In a similar way to the ashless antioxidants, the response pattern of the ZnDTPs used in these fluids is strongly influenced by the degree of raffination of the base stock (Hsu *et al.*, 1982). Although the performance of ZnDTP-based fluids is acceptable, the environmental aspects may bring about a change to ashless technology.

4.5.1.3 *Air compressors*

Requirements Air compressors, the most widely used compressor type, put severe oxidation stress on the oil due to the presence of

- hot air under high pressure
- catalytically active metal salts and oxides, e.g. Fe_2O_3
- condensed water, generated by compression

These demanding conditions are exacerbated by recent trends such as continuous operations beyond 8000 hours, higher air throughput and larger power output from smaller units.

In order to meet this increased oxidative stress, which could lead to sludge, carbon deposits and varnish, lubricants are exposed to increasingly severe bench testing. For top-tier compressor oils, which should meet the requirements of all types of compressor (universal compressor oil) the following specifications have been outlined (Sugiura *et al.*, 1980; Cohen, 1987a; Matthews, 1989):

- Wolf strip test (DIN 51392): deposit below 50 mg.
- IP 48 (DIN 51351–1): viscosity increase after 24 h below 50%.
- Rotary bomb oxidation test (ASTM 2272): induction period above 1000 min.
- Field tests: running time greater than 8000 h.

In the past only ester- or PAO-based formulations could meet these performance criteria. Recently it has been demonstrated that compressor oils based

Table 4.9 Oxidation and thermal stability of various anti-wear hydraulic fluids

Tests	Performance of fully formulated anti-wear hydraulic fluids (additive blend concentration = 0.9%–1.2%)			
	Ashless premium grade	Stabilised ZnDTP (P ~ 0.035%)	Ashless* normal grade	ZnDTP normal grade (P ~ 0.1%)
Oxidation stability				
TOST (ASTM D943) (95 °C, H_2O, Fe and Cu catalysts, 3 l air/hour)				
—Time (hr) to TAN = 2.0 mgKOH/g oil	2800–3800	2500–3400	1800–2800	1500–1800
—Sludge (mg) (ASTM D4310)	35–90	45–90	140–320	130–430
Thermal stability				
Cincinnati Milacron test procedure A: 168 h at 135 °C				
—Sludge (mg/100 ml)	6–30	5–46	6–40	150–263
—Cu appearance (CM colour class)	1–3	1–3(10)†	1	10–12
—Steel appearance (CM colour class)	1–4	1–4	2	8

* Additive concentration = 0.4–0.5%
† One fluid gave a rating of 10 (black) in four different base stocks

on solvent-refined, hydrotreated or hydrocracked base stocks formulated with low volatility, thermally-stable antioxidants can also fulfil these targets.

Antioxidant technology An example of a modern compressor oil, which can be used for continuous operation in excess of 8000 h is presented in Table 4.10 (Cohen, 1987b)

Table 4.10 Analysis of used oil from a reciprocating compressor

Hours of operation		34	4800	8600
Viscosity at 40 °C	(cSt)	72	73	73
Tan	(mg KOH/g)	0.70	0.72	0.84
Solids	(%)	0.01	0.04	0.04

The patent that describes this development discloses that a mixture of a thermally-stable phenolic antioxidant and a sterically hindered phosphite was used in a hydrotreated base stock. However, the heterosynergism between the radical scavenger (phenolic antioxidant) and the hydroperoxide decomposer (phosphite) was not shown in solvent-refined base stocks. An additional benefit of this technology is energy saving, due to extremely good control of viscosity.

4.5.2 *Engine oil*

4.5.2.1 *Requirements* Lubricants for internal combustion engines are exposed to severe oxidative conditions, particularly in the upper part of the piston ring and cylinder liner zone where maximum temperatures can be 250 °C in a passenger car engine and 300 °C in a heavy duty diesel. In these areas thin oil films ($\sim 5 \times 10^{-7}$ cm) are exposed to blow-by gases during a residence time of 2–5 minutes. Blow-by gases are generated in the combustion chamber and contain the following major components:

- free radicals (mainly HO·)
- HNO_3, NO and NO_2
- O_2
- partially oxidised fuel (organic acids and olefins)
- partially oxidised lubricant components
- metallic catalysts

High temperatures and the reactive blow-by gases lead to rapid oxidation of the thin oil film, which is initiated by hydroxyl and NO_2 radicals (Korcek and Johnson, 1990). Undoubtedly, the rate of oxidation in the initial phase is dominated by NO_2 according to the following scheme (Korcek and Johnson, 1990):

$$\begin{aligned} RH + \cdot NO_2 &\longrightarrow R\cdot + HONO \\ HONO &\longrightarrow HO\cdot + \cdot NO \end{aligned} \tag{4.70}$$

The nitrogen oxides also react with fuel-derived olefins in the blow-by leading to a so-called 'sludge precursor'. Because the piston area and the crankcase are exposed to the same lubricant, all degradation intermediates such as free radicals, nitro (RNO_2) and nitroso compounds (RNO), nitrites (RONO), nitrates ($RONO_2$), acids, ketones, and oil-soluble polycondensation and polymerisation products are transferred to the sump. In addition, the blow-by gas, carrying sludge precursors, condenses in the crankcase oil. With extended oil change intervals and crankcase oil temperature up to 150°C, oil thickening and sludge formation may result.

4.5.2.2 *Antioxidant technology for passenger car engine oils*

Zinc dithiophosphates Under service conditions ZnDTPs undergo various chemical transformations and after 2000–3000 km they cannot be detected. However, 35% of the degraded ZnDTP products, containing P–O–C bonds, remain after 10 000 km and the antioxidant and antiwear performance of the lubricant is still satisfactory (Kawamura *et al.*, 1985; Killer, 1981; Gegner, 1987). Thus the antioxidant activity up to 2000–3000 km is dominated by ZnDTPs and is subsequently governed by products resulting from their thermal cleavage.

The contribution of ZnDTPs to the antioxidancy activity may be summarised as follows

- Formation of the acid catalyst $(RO)_2PSSH$ according to reactions (4.59) and (4.60) (detailed in section 4.4.3.1). The intermediate disulfide $[(RO)_2PS_2]_2$ has been observed (^{31}P NMR spectra) after engine tests (Saville *et al.*, 1988) and field tests (Coy and Jones, 1982). This disulfide has a kind of 'depot' function for the acid catalyst $(RO)_2PSSH$, such that just sufficient acid is always present. If the overall available acid had been added to the oil at once, it would have been immediately neutralised by the basic detergents (Jentsch and Okoro, 1982).
- Catalysed ionic decomposition of hydroperoxides according to reaction (4.62)

In addition, the following stabilisation steps can be anticipated:

- Peroxy radical decomposition (reaction (4.63))
- Reduction of NO_2:

$$[(RO)_2PS_2]_2Zn + \cdot NO_2 \longrightarrow NO_2^{-}(RO)_2PS_2Zn^{+} + (RO)_2PS_2\cdot \tag{4.71}$$

The thermal cleavage of the ZnDTPs with at least one secondary alkyl group, which is completed after 2000–3000 km, may be represented by:

$$[R^1\underset{\underset{\displaystyle H}{H_2C\diagdown}}{\underset{|}{H}CO}-\underset{\underset{S}{\|}}{\overset{\overset{OR}{|}}{P}}-S]_2Zn \xrightarrow{\Delta} [HO-\underset{\underset{S}{\|}}{\overset{\overset{OR}{|}}{P}}-S]_2Zn + 2R^1\underset{\underset{H}{|}}{C}=CH_2 \tag{4.72}$$

In the presence of dispersant in the engine oil the following resonance stabilised structure could be formed:

$$Zn\left\{S-\overset{\overset{RO}{|}}{P}(=S\cdots H-N(C\lessdot)_2)-O-H\cdots\right\}_2 \longleftrightarrow \left\{(\gt C)_2N\cdots H-S-\overset{\overset{OR}{|}}{P}(=O\cdots H)-S\right\}_2 Zn \tag{4.73}$$

The above complex formed between the degradation products of ZnDTP and the dispersant may be the reason for the fact that the oxidation stability and also the anti-wear performance are maintained up to 10 000 km (Kawamura *et al.*, 1985).

A possible stabilisation step under these conditions has recently been suggested (Colclough, 1987b; Colclough *et al.*, 1991) and may occur by the reaction of iron ions (Fe^{2+}/Fe^{3+}) with this complex to form the corresponding iron salts with subsequent conversion to iron polyphosphates. By this means the iron-catalysed oxidation is suppressed.

Modern engine oils with good thermo-oxidative stability are formulated with secondary or mixed primary/secondary ZnDTPs. This is because the cleavage reaction (4.72) occurs earlier for secondary ZnDTP when compared with primary ZnDTP and therefore antioxidant protection starts earlier. Recently published sequence IIIE, and VE data, on API SG oil, as well as results from the field confirm this approach (Harris *et al.*, 1989).

The influence of base oil composition on the oxidation stability of API SF/CC and SE/CC engine oils evaluated in sequence IIID and VD engine tests has recently been analysed (Roby *et al.*, 1989). The conclusions are:

- Oxidation stability of the SF/CC engine oil in the sequence IIID test is better when the base oil has a high content of sulphur and of aliphatic hydrocarbons. Olefins and nitrogen-containing organic compounds reduce this stability.
- Varnish control in the VD is achieved with the SE/CC engine oil if the

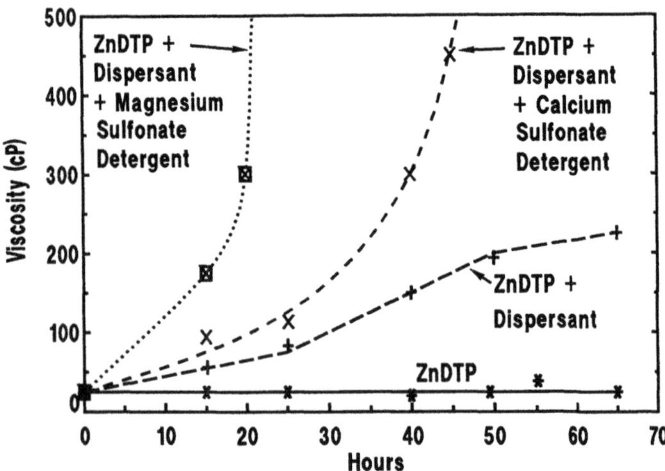

Figure 4.6 Influence of a dispersant (4.5%) and a calcium sulfonate detergent (1%) on the antioxidant performance of a ZnDTP (0.1% P). Base stock: 150 SN (S = 0.3%); temperature: 165°C; Air flow rate: 1.7 l min^{-1}; catalyst: ferric acetyl acetonate (40 ppm Fe).

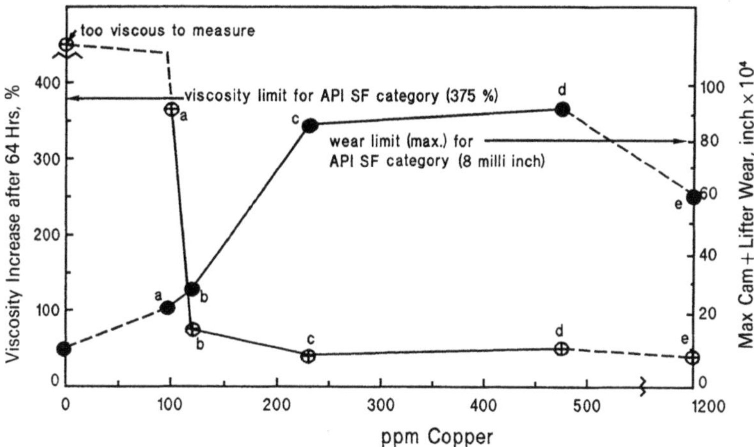

Figure 4.7 Influence of increasing copper content of different types of organocopper compounds on viscosity (⊕) and wear (●) in the sequence IIID engine test (Colclough *et al.*, 1981): a = 0.156% cupric oleate (94 ppm Cu) + 1.8% ZnDTP; b = 0.084% cuprous-di-sec.-hexyl-dithiophosphate (120 ppm Cu) + 1.7% ZnDTP; c = 0.2% cuprous-di-iso-octyl-dithiophosphate (240 ppm Cu) + 1.65% ZnDTP; d = 0.39% cuprous-di-iso-octyl-dithiophosphate (486 ppm Cu) + 1.48% ZnDTP; e = 1.5% cupric naphthenate (1200 ppm Cu) + 1.8% ZnDTP.

base oil again has a high content of aliphatic hydrocarbons, with aromatics, olefins and nitrogen being low.

Similar conclusions have been drawn by Murray *et al.* (1983). The antioxidant performance of ZnDTP is not only influenced by the base stock composition but also by the dispersants and detergents. This can be demonstrated by Figure 4.6. The deleterious effect of the dispersant on the oxidation stability of the ZnDTP-containing lubricant is clearly enhanced by calcium detergents and, to a larger extent, magnesium detergents.

In summary, the additive interaction pattern is influenced by the type of dispersant and detergent, by the base stock composition and by the concentration ratio of the additives used in a formulation (Hsu *et al.*, 1989).

Organocopper salts The successful application of organocopper salts in modern passenger car engine oils was made possible by the eventual selection of a concentration of copper in the range between 90–120 ppm. Within this range optimum control of oxidation and wear can be achieved (Colclough *et al.*, 1981). The inter-relation between oxidation and wear is outlined in Figure 4.7. The sequence IIID engine test data have been obtained in 10 W—30 fully formulated oils with different ZnDTP content to which various organocopper compounds at increasing concentrations have been added.

As has been mentioned earlier, organocopper compounds inhibit oxidation

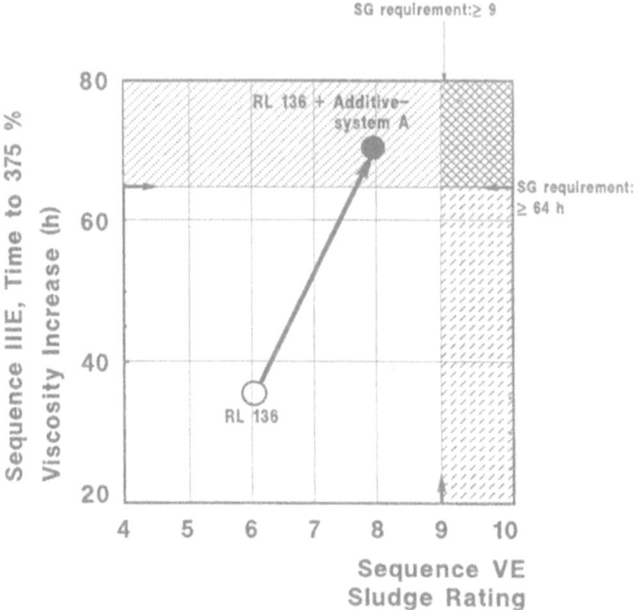

Figure 4.8 Sequence IIIE/VE results with RL 136 and RL 136+0.75% of ashless antioxidant system A.

of thin films of mineral oil (and ester) at 250 °C (Naidu *et al.*, 1984; Clark *et al.*, 1985). Under these conditions polycondensation and polymerisation reactions are suppressed such that only medium-sized oil-soluble polymers are formed. This leads to an improved control of viscosity. A further stabilisation step, mainly in the upper part of the piston, may result from an interaction between NO_x, which is a major contributor to oil degradation, and an organocopper salt:

$$Cu^{1+} + \cdot NO_x \longrightarrow Cu^{2+} + [NO_x]^- \qquad (4.74)$$

Ashless antioxidants Field problems in the United States and in Europe made it necessary to up-grade passenger car engine oils from API SF to API SG and from CCMC G3 to CCMC G4 and G5, respectively. Under these more demanding test conditions, ZnDTP as the antioxidant at a concentration corresponding to a maximum phosphorus level of 0.12%, no longer protects the oil adequately against oxidation and sludge formation. Therefore new methods for better control of oxidation are necessary.

In a recent publication (Kristen *et al.*, 1990) it has been demonstrated that a new ashless antioxidant can improve the high-temperature oxidation stability and reduce low-temperature sludge formation. Figure 4.8 shows the results of testing two lubricants. The base line is a reference oil (RL 136) with a high sludging tendency in the European PL 37 sludge test, the other lubricant is RL 136 topped up with the new antioxidant. Thus, the new antioxidant is bifunctional since it controls both viscosity increase and sludge formation in the sequence IIIE and VE engine tests.

A different approach to managing viscosity and sludge problems in engine oils has been the use of dispersant–antioxidant–olefin copolymers (Liu *et al.*, 1989). The results of IIIE and VE engine tests with an oil containing the dispersant–antioxidant–OCP compared with a dispersant OCP are clearly in favour of the former product. This new product group is synthesised by grafting the antioxidant monomer onto the dispersant-OCP. The antioxidant part could be, amongst others a 2,6-ditertiary-butyl-phenol moiety.

A new type of additive to control the formation and accumulation of sludge in gasoline engines has been introduced recently (Murphy, 1989). The structure is described as an aromatic hydrocarbon substituted with aliphatic hydrocarbons. A typical formula is:

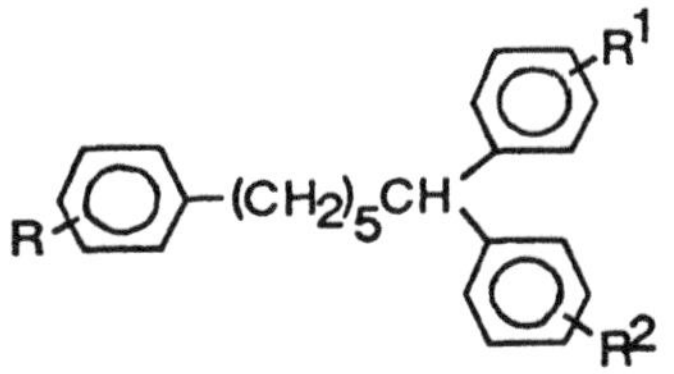

$$R-C_6H_4-(CH_2)_5CH(C_6H_4R^1)(C_6H_4R^2) \qquad (4.75)$$

An effective method of sludge reduction in a Japanese engine (Nakamura *et al.*, 1988) has been found by adding a bis-phenol type antioxidant to an engine oil containing a mixture of primary and secondary alkyl ZnDTPs.

4.5.2.3 *Antioxidant technology for heavy duty diesel engine oils* Control of deposit formation in the upper part of the piston is a major function of the diesel oil. Any failure in this respect may lead to such deleterious consequences as high oil consumption, ring sticking and bore polishing.

The deposits, which are mainly lubricant related, have been described as binders and result from thermal and oxidative degradation of the oil (Papke *et al.*, 1989; Covitch *et al.*, 1988; see also section 4.2.2.2). These binders act as absorbents for fuel combustion products, especially soot. Because high-temperature oxidation processes are the main source of deposit formation, antioxidants, in the form of the more thermally stable primary ZnDTPs, combined with sulfurised alkyl phenols and/or high molecular weight sterically hindered phenols are able to control this process. The extent to which a high molecular weight sterically hindered phenolic antioxidant improves deposit control of an API CD/SF oil in the upper part of the piston in a single cylinder diesel engine (MWM-B, test CEC L-12A–76) is demonstrated in Table 4.11.

Table 4.11 Improvement of deposit formation of a heavy-duty diesel engine oil containing a high molecular weight, sterically-hindered phenol (antioxidant B) in the MWM-B engine test (CEC L-12A-76)

Test oil Antioxidant B Base fluid*	Neat	0.6% balance
MWM-B engine test		
Piston merit rating (100 = clean)		
1st Groove (top groove)	18	40
2nd Groove average rating	84	93
Average rating	73	79

* Base fluid	SAE	:15 W–40	API	:CD/SF
	SULPH.ASH	:1.2%	VII	:8% non-dispersant OCP
	Zn	:0.11%	TBN	:8.8 mg KOH/g
	P	:0.095%	Base stock	:neat mineral oil

The large improvement in the deposit rating for the first groove from 18 to 40 may be explained in terms of the 'detergent credit' that certain antioxidants have. These types of antioxidant allow a reduction of ash content in heavy duty engine oils for those engines which are sensitive to high ash levels (see also Aggarwal and Passut, 1989).

4.6 Future antioxidant technology

ZnDTPs, sterically hindered phenols and alkylated diphenylamines are the groups of products fulfilling the present needs of the lubricant industry concerning protection of lubricants against oxidation. Will they still be the products in the year 2000? There are important driving forces leading to future antioxidants:

- Environmental concerns
- Higher performance standards
- Base stock variability

What are the effects of these factors on the selection of antioxidants for industrial and engine lubricants?

4.6.1 *Antioxidants for industrial lubricants*

Presently, there is a strong demand for 'environmentally friendly' fluids. These 'bio lubes' need to satisfy biodegradation and bioaccumulation standards, which mineral oil based fluids cannot achieve. Therefore, the use of synthetic and natural esters (Stenmark *et al.*, 1988), for many industrial applications will develop. Not only the base fluids but also the antioxidants used in them will have to fulfil certain specifications for aquatic toxicity, biodegradation and bioaccumulation. Because the antioxidant response in these new fluids is different from mineral oil based lubricants, new classes of ashless 'bio-antioxidants' may be developed.

Due to environmental concerns, the majority of ZnDTP-based anti-wear hydraulic fluids will be gradually replaced by ashless formulations using antioxidant technology which is environmentally acceptable (Grover and Perez, 1990).

There is a general trend to reduce the size of industrial equipment without reducing power output. This leads to increased thermo-oxidative stress on the lubricants, which has to be met with more effective antioxidants, i.e. substituting existing antioxidants with lower volatility, higher molecular weight products. Also, poorer quality base stocks will lead to higher treat levels of antioxidants.

4.6.2 *Antioxidants for engine oils*

Lubricants for the next generation of passenger engines will have to cope with increased power output from smaller engines, longer drain intervals and reduced oil consumption. Thus the oil will be exposed to increased temperatures, for longer times and to higher NO_x concentrations. Ceramic parts will lead to even higher temperatures. Thus, a new antioxidant type will be required, which has low volatility, high thermal stability and the ability to control the future higher NO_x levels in the blow-by.

The limitations on particulate emissions, especially in the US from 1994 onwards, and their strong impact on future engine design raises the question of how future diesel engine oils will be formulated (Fetterman and Schetelech, 1987; Cooke, 1990). Ashless antioxidants, leading to improved deposit control and, therefore via reduced oil consumption, to less particulate emission, could be an important building block. The increased operating temperatures of passenger car and heavy duty diesel engines will stimulate the use of new lubricants, which will require new, tailor-made antioxidants. This trend is expected to be reinforced by the introduction of low heat rejection diesel engines by the end of this century (Sutor *et al.*, 1990; Marolewski *et al.*, 1990).

In summary, antioxidants with improved performance will be a major design factor for future lubricants, operating under more demanding yet environmentally acceptable conditions.

Acknowledgements

I wish to thank the following colleagues in Ciba-Geigy, Basle for their helpful asssistance: Dr P.C. Hamblin, Dr D.M. Hutchings, Dr M. Ribeaud and, in particular, Dr P. Rohrbach for his critical comments, suggestions and support.

References

Abou El Naga, H.H. and Salem, A.E.M. (1984) *Wear* **96** 267.
Aggarwal, R.K. and Passut, Ch.A. (1989) *SAE Paper 892049*.
Al-Malaika, S., Marvgi, A. and Scott, G. (1987) *J. Appl. Polymer Sci.* **33** 1455.
Ali, A., Lockwood, F., Klaus, E.E., Duda, J.L. and Tewksbury, E.J. (1979) *ASLE Trans.* **22** 267.
Berger, H., Bolsmann, T.A.B. and Brouwer, D.M. (1983) In: *Developments in Polymer Stabilisation—6*. Scott, G. (ed), Elsevier Applied Science Publishers, London, pp 1–27.
Bridgewater, A.J., Dever, J.R. and Sexton, M.D. (1980) *J. Chem. Soc. Perkin II* 1006–1086.
Burn, A.J. and Greig, G. (1972) *J. Inst. Petrol.* **58** 346–350.
Carmichael, L.A. (1973) *SRI International Report No. 85*.
Chien, J.C.W. and Boss, C.R. (1972) *J. Polym. Sci.* **A-1** 1579–1600.
Clark, D.B., Klaus, E.E. and Shu, S.M. (1985) *Lub. Engng.* **41**(5) 280–287.
Cohen, S. (1987a) *ASLE Preprint No. 87-AM-2G-1*.
Cohen, S. (1987b) US Patent 4,652,385.
Colclough, T. (1987a) European Patent 318218A2.
Colclough, T. (1987b) *Ind. Eng. Chem. Res.* **26** 1888–1895.
Colclough, T., Gibson, F.A. and Marsh, J.F. (1981) UK Patent 2.056482A.
Colclough, T., Marsh, J.F. and Robson, R. (1991) *SAE Paper 910868*.
Cooke, V.B. (1990) *SAE Paper 900814*.
Covitch, M.J., Gundic, D.T. and Graf, R.T. (1988) *Lub. Engng.* **44** 128–138.
Coy, R.C. and Jones, R.B. (1982) *Inst. Mech. Eng.* **C3/82** 17–22.
Denisov, E.T. and Khudyakor, I.V. (1987) *Chem. Rev.* 1313–1357.
Dorinson, A. (1983) *Lub. Engng.* **39** 519ff.
Ellissen, W.W., Dawson, R.B. and Mattheus, B.W. (1982) *Mineralöltechnik* **27**(4).
Emanuel, N.M., Denisov, E.T. and Maizus, Z.K. (1967) *Liquid Phase Oxidation of Hydrocarbons*, Plenum, New York.
Fetterman, G.P. and Schetelech, A.A. (1987) European Patent 0311318A1.
Fields, E.K., Scanley, C.S. and Hammond, J.L. (1955) US Patent 2,719,126.
Fuchs, G.H. and Diamond, H. (1942) *Ind. Eng. Chem.* **34** 927–937.
Gegner, H. (1982) *Mineralöltechnik* **27**(11).
Gegner, H. (1987) *Mineralöltechnik* **32**(8).
George, P. and Robertson, A. (1946) *Trans. Faraday Soc.* **42** 217–228.
Grover, K.B. and Perez, R.J. (1990) *Lub. Engng* **1** 15–20.
Habeeb, J.J., Rogers, W.W. and May, C.J. (1987) *SAE Paper 872157*.
Harris, S.W., West, C.T. and Jahalka, T.L. (1989) *SAE Paper 892113*.
Howard, S.A. and Tong, S-B. (1980) *Can. J. Chem.* **58** 92–95.
Hsu, S.M. and Cummings, A.L. (1983) *SAE Paper* 831682 51–60.
Hsu, S.M., Ku, C.S. and Lin, R.S. (1982) *SAE Paper 821237*.
Hsu, S.M., Pei, P. and Ku, C.S. (1989) *Lub. Sci.* **1** 165–184.
Ingold, J.A. (1983) In: *The Chemistry of Functional Groups: Peroxides*. Patei, S. (ed), Wiley and Sons Ltd.
Ingold, K.U. (1969) *Acc. Chem. Res.* **2**(1) 1–9.
Jensen, R.K., Korcek, S., Mahoney, L.R. and Zinbo M. (1979) *J. Am. Chem. Soc.* **101** 7574–7584.
Jensen, R.K., Korcek, S., Mahoney, L.P. and Zinbo, M. (1981) *J. Am. Chem. Soc.* **103** 1742–1748.
Jentsch, C. and Okoro, E. (1982) *Erdöl und Kohle* **35**(3) 138.
Kawamura, M., Fujijata, K., Eseki, Y. and Moritani, H. (1985) *SAE Paper 852076*.
Killer, F.C.A. (1981) *Mineralöltechnik* **5**.
Kirpitschnikow, P.A. and Pobedimski, D.G. (1975) *Plask Kautchuk* **22**(5) 400–403.
Korcek, S. and Jensen, R.K. (1975) *ASLE Trans.* **19** 83–94.
Korcek, S., Jensen, R.K., Zinbo, M. and Gerlock, J.L. (1988) In: *Organic Free Radicals*. Fischer, H. and Weingarten, H. (eds), Springer-Verlag, Berlin, 95ff.
Korcek, S. and Johnson, M.D. (1990) *Automotive Lubrication*, 7th International Colloquium, Bartz, J. (ed), Technische Akademie Esslingen.
Kristen, U., Hutchings, M. and Schumacher, R. (1990) 7th International Colloquium, Bartz, J. (ed), Technische Akademie Esslingen.
Kuhn, R.R (1973) *J. Am. Chem. Soc.* (*prepr.*), *Div. Pat. Chem.* **694H.**

Lam, W.Y. (1989) US Patent 4,836,942.
Lange, R.M. (1987) US Patent 4,707,301.
Lazar, M., Rychly, J., Klimo, V., Pelikan, P. and Valko, L. (1989) *Free Radicals in Chemistry and Biology*, CRC Press, Inc., Boca Raton, Florida.
Liu, C.S., Benfaremo, N., Kapuscinski, M.M and Grina, L.D. (1989) US Patent 4,812,261.
Lowe, W. and Liston, T.V. (1980) US Patent 4,228,022.
Malec, R.E. and Plonsker, L. (1976) US Patent 3,992,308.
Marolewski, T.A., Slone, R.J. and Jung, A.K. (1990) *SAE Paper 900689*.
Matsuura, R. (1965) *Nippon Kagaku Zasshi* **86**(6) 560–572.
Matthews, P.H.D. (1989) *Syn. Lub.* **5**(4) 291–317.
Mescina, M.Y. and Karpukhina, G.V. (1972) *Zik. Maizus. Neftekhimiya* **12** 731.
Mill, T. and Montorsi, G. (1973) *Int. J. Chem. Kinet.* **5** 119.
Morrison, R.T. and Boyd, R.N. (1987) *Organic Chemistry* (5th edn) 1120.
Murphy, J.P. (1989) European Patent WO 89/00186.
Murray, D.W., MacDonald, J.M., White, A.M. and Wright, P.G. (1983) *Proc. 11th World Petroleum Congress*, Volume 4, London.
Naidu, S.K., Klaus, E.E. and Duda, J.L. (1984) *Ind. Eng. Chem. Prod. Res. Dev.* **23** 613–619.
Nakamura, K., Matsumoto, E., Kurosaka, S. and Murakami, Y. (1988) *SAE Paper 881577*.
Newley, R.A., Speiks, H.A. and Macpherson, P.B. (1980) *J. Lub. Technol.* **102** 540–544.
Nirula (1983) *SRI International Report No. 85A*.
Okada, M. and Yamashita, M. (1986) *J. Am. Soc. Lub. Eng.* **43** 459–466.
Papke, D.L., Dahlstrom, P.L. and Kreuz, K.L. (1989) *Lub. Engng* **45** 575–585.
Payne, J. (1988) *Inhouse information*.
Perez, J.M., Kelley, F.A., Klaus, E.E. and Bagrodia, V. (1987) *SAE Paper 872028*.
Polishuk, A.T. and Farmer, H.H. (1979) NLGI Spokesman 200–205.
Rost, R. and Erdöl-Kohle-Erdgas (1963) *Petrochem.* **16** 850–856.
Roby, S.H., Supp, J.A., Barner, D.E. and Hoyne, C.H. (1989) *SAE Paper 892108*.
Russel, G.E. (1977) *SRI Report No. 113*.
Sakurai, T., Nishihara, A., Handa, T., Katoh, H., Tomoda, Y., Aoki, K. and Yoto, M. (1978) US Patent 4,098,705.
Saville, S.B. Gainey, F.D., Cupples, S.D., Fox, M.F. and Picken, D.J. (1988) *SAE Paper 881586*.
Scott, G. (1981) In: *Developments in Polymer Stabilisation—4*. Scott, G. (ed), Applied Science Publishers, 1–21.
Scott, G. (1983) In: *Developments in Polymer Stabilisation—6*. Scott, G. (ed), Applied Science Publishers, 29–71.
Scott, G. (1984) In: *Developments in Polymer Stabilisation—7*. Scott, G. (ed), Elsevier Applied Science Publishers, 65–104.
Sexton, M.D. (1984) *J. Chem. Soc. Perkin Trans. II* 1771–1776.
Sheldon, R.A. and Kochi, I.K. (1973) In: *Oxidation and Combustion Reviews* 5(2). Tipper, C.F.H. (ed), 135–242.
Shelton, J.R. (1981) In: *Developments in Polymer Stabilisation—4*. Scott, G. (ed), Applied Science Publishers, 23–66.
Sommer, E. (1988) *Tribologie und Schmierungstechnik* **35**(4) 174–177.
Stenmark, G., Jokineh, K. and Kerkkonnen, H. (1988) WO 88/05808.
Studt, P. (1986) *Additives for lubricants and operational fluids*, 5th International Colloquium.
Sugiura, K., Miyagawa, T. and Nakano, A. (1980) *Lub. Engng* **38.**
Sutor, P., Bardasz, E.A. and Bryzik, W. (1990) *SAE Paper 900687*.
Veprek-Siska, J. (1985) *Oxid. Commun.* **8**(3–4) 301–307.
Vijh, A.K. (1985) *Wear* **104** 151–156.
Wasson, J.I. and Smith, W.M. (1953) *Ind. Eng. Chem.* **45** 197–200.
Zeman, A., Römer, R. and Von Roenne, V. (1987) *J. Syn. Lub.* **3** 309–324.

5 Viscosity index improvers and thickeners

R.L. STAMBAUGH

5.1 Introduction

Early in the history of the lubricants industry, the viscosity index (VI) was an important measure of quality, providing an indication of the potential of an oil for application over a wide temperature range. Pennsylvania grade oils ($\sim$ 100 VI) were the standard against which all others were measured. Hydrogenation and solvent extraction were developed to upgrade poorer quality crudes, but the refinery technology of the 1930s had a practical VI ceiling of about 110 to 115.

Early workers observed that small amounts of rubber dissolved in mineral oil could raise the VI substantially. However, high unsaturation in the polymer led to oxidation and sludge formation. Otto *et al.* (1934) discovered that this deficiency could be overcome through the use of a synthetic polymer prepared from the light ends of gasoline. Similar behavior was later described for polymethacrylates by Rohm and Haas Co. (1937a, b) and polyisobutylene by I.G. Farbenindustrie AG (1938a, b). Since these materials were initially used primarily to increase the viscosity index, they became known as viscosity index improvers.

The adoption of SAE 10W and SAE 20W requirements in the Automotive Manufacturers' Viscosity Classification in 1941 created the possibility of making an oil which would meet the requirements of more than one SAE grade. Van Horne (1949) demonstrated the formulation of 'double graded oils' with polymethacrylates, including an SAE 10W-30 oil which he suggested 'might well be called an all season motor oil.' Rapid exploitation of VI improvers for the development of multigraded engine oils followed during the 1950s.

While engine oils represent by far the largest commercial application of VI improvers, they are widely used in the lubricant field. Typical examples include automatic transmission fluids, multipurpose tractor transmission fluids, power steering fluids, shock absorber fluids, other hydraulic fluids (industrial, automotive, off-highway, aircraft), manual transmission fluids, rear axle lubricants, industrial gear oils, turbine engine oils (stationary and aircraft) and aircraft piston engine oils. Many of these have special needs as will be pointed out in following sections.

Table 5.1 Representative VI improver chemistries

Polymerized light mixed olefins
Polyisobutylene
Polymethacrylates
Polyacrylates
Poly(methacrylate-co-acrylate)
Poly(methacrylate-co-styrene)
Polybutadiene
Alkylated polystyrene
Poly(*t*-butyl styrene)
Poly(alkyl fumarate-co-vinyl acetate)
Poly(*n*-butyl vinyl ether)
Esterified poly(styrene-co-maleic anhydride)
Poly(ethylene-co-propylene)
Poly(ethylene-co-propylene diene-modified)
Poly(ethylene-co-propylene-co-olefin diene-modified)
Hydrogenated poly(styrene-co-butadiene)
Hydrogenated poly(styrene-co-isoprene)
Hydrogenated polyisoprene

5.2 Overview of VI improver chemistry

Since the early developments with polymethacrylates and polyisobutylene, a wide variety of polymers have been explored. A sampling of these is shown in Table 5.1. All of these polymers have been developed at least to the point of commercial sampling.

Five core technologies currently find important commercial usage. These include polymethacrylates (generally referred to as PMAs), poly(ethylene-co-propylene) and closely related modifications (so-called olefin copolymers or OCPs), hydrogenated poly(styrene-co-butadiene or -isoprene) and modifications (HSD or, in the case of isoprene, SIP, as well as several other abbreviations), esterified poly(styrene-co-maleic anhydride) (normally referred to by the chemically incorrect description of styrene polyester, or SPE) and a combination of the first two, concentrate-compatibilized PMA/OCP systems. Only these five will be described in any detail in the following section. Simplified chemical structures of the four basic technologies are shown in Figure 5.1.

Some VI improver compositions are chosen so as to incorporate pour point depressancy and/or dispersancy in addition to the basic viscosity control properties. The former will be mentioned as appropriate in following sections but is covered in more detail in chapter 6, section 6.2.

Dispersant VI improvers are widely used, particularly in engine oils, automatic transmission fluids and multipurpose tractor fluids. Incorporation of dispersancy into a polymer involves a carefully engineered addition of a strongly polar functional group to the base polymer backbone. The most commonly employed functional groups are amines, alcohols or amides. The

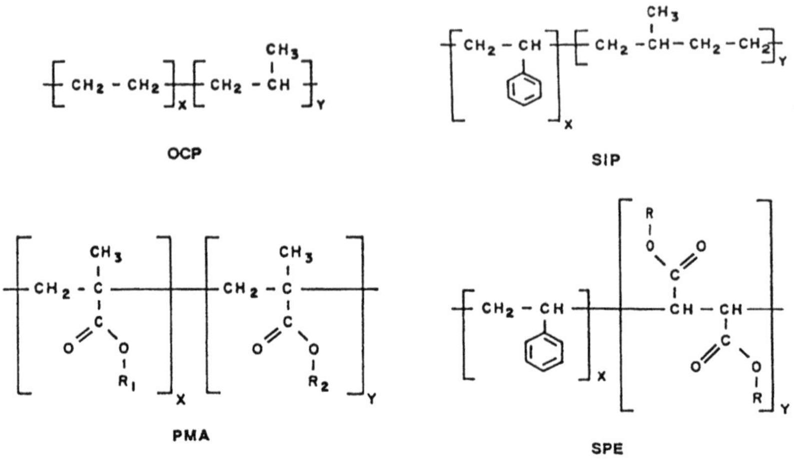

OCP

SIP

PMA

SPE

Figure 5.1 Simplified chemical structures of major VI improver classes.

mode of incorporation depends largely on the base polymer and specific details will be described in later sections.

5.3 Chemistry and manufacture of commercial VI improvers

It is interesting to note that the three most important commercial VI improver families each represent one of the most important commercial techniques for the manufacture of high molecular weight polymers: polymethacrylates by free radical chemistry, olefin copolymers by Ziegler chemistry and hydrogenated styrene–diene copolymers by anionic polymerization.

5.3.1 *Polymethacrylates*

Useful overviews of polymethacrylate VI improver chemistry are provided by Arlie *et al.* (1975a, b) and a more recent review by Neudoerfl (1986).

5.3.1.1 *Chemistry* Polymethacrylate-based VI improvers for use in engine oil are most commonly copolymers of three methacrylic acid esters, a short-, an intermediate- and a long-chain alkyl methacrylate. Both methyl and butyl (either *n*-butyl, isobutyl or mixtures of the two) methacrylates have been used for the low alkyl component, although methyl methacrylate is by far the more common. The intermediate alkyl radical can vary widely but is normally derived from a group including 2-ethylhexyl alcohol, isodecyl alcohol and alcohol mixtures which may be either C_8–C_{10}, C_{12}–C_{14} or C_{12}–C_{15}. The long chain alkyl esters are based on either C_{16}–C_{18} or C_{16}–C_{20} mixtures. Any of the alcohols in the C_{12}–C_{20} range may be derived from either natural or synthetic sources. All of the intermediate- and long-chain alkyl esters may be made by either acid-catalysed direct esterification of methacrylic acid or by transesterification of methyl methacrylate; both are practiced commercially.

The ratio of the three monomers is normally chosen so that the average alkyl side chain is about nine carbon atoms. To a first approximation, the viscosity–temperature properties of the copolymer will be the same, independent of the combination of alkyl groups chosen to reach the average. However, the preferred combination is determined by wax-interaction properties.

Linear alkyl groups of fourteen carbon atoms or more interact with wax so that polymethacrylates are frequently designed to incorporate pour point depressancy into the VI improver molecule. If the high alkyl methacrylate is used in combination with a mid-cut which has no linear C_{14} or larger, it will be used at a higher concentration than in the reverse situation where the mid-cut can carry some of the wax-interaction burden.

For non-engine oil applications, the compositions are optimized for the particular lubricant base stock as well as any specific needs of the end use. This would most often involve rebalancing polymer solubility and elimination of the high alkyl monomer where wax interaction is not an issue (synthetics or naphthenic mineral oils).

5.3.1.2 *Manufacture* Polymethacrylate VI improvers are free-radically initiated solution copolymers. A wide variety of peroxide or azo initiators may be used and several are practiced commercially. The choice is dictated largely by the half-life at reaction temperature, which in turn is driven by manufacturing convenience and kettle productivity factors. Production is a simple batch or semi-batch process at a theoretical polymer content of 50–90%, the solids level depending on both the polymer molecular weight and the mixing capabilities of the system. Reaction temperature is in the range of 100–140 °C.

The solvent for the process is most commonly mineral oil in a viscosity/volatility range that is compatible with the intended end use of the product. Volatile solvents such as toluene have been used commercially, in which case the final step of the process is a solvent-oil exchange. If a solvent is used, the primary criteria for consideration are relative freedom from significant chain transfer activity and a boiling point which is a compromise between the reaction temperature and ease of solvent transfer at the end of the reaction.

The products produced are random copolymers. No special steps are necessary to achieve this distribution since reactivity ratios of all of the alkyl methacrylate monomers are almost the same. The processes may include special conditions to help minimize molecular weight distribution.

The VI improvers for engine oils have a molecular weight (M_w) of between 250 000 and 750 000 with a distribution (M_w/M_n) of between 3 and 4. The products are sold commercially as concentrates which range from about 35–55% polymer. This is entirely a practical matter dictated by handling characteristics. The molecular weights for other applications range down to about 30 000–40 000 and product solids may range up to about 70%.

5.3.1.3 *Dispersancy* Solution copolymers are relatively easy to produce in a dispersant form since copolymerization with an appropriate polar monomer is relatively straightforward. If the polar monomer is also a methacrylate, reactivity ratios are essentially the same and no special procedures are required to produce random copolymers. Commercial examples have included dimethyl (or diethyl)aminoethyl methacrylate (E. I. du Pont de Nemours and Co., 1956), hydroxyethyl methacrylate (Shell Oil Co., 1966) and dimethylaminoethyl methacrylamide (Texaco Inc., 1977).

2-Methyl-5-vinyl pyridine (Shell Oil Co., 1960) has also been used commercially. Reactivity ratios are such that it copolymerizes slightly faster than alkyl methacrylates. Although composition drift is not severe, it could be added in a programmed fashion if uniform distribution is desired. *N*-vinyl pyrrolidinone, in contrast, copolymerizes very sluggishly with methacrylates and is best incorporated via graft reaction (Rohm and Haas Co., 1970). It is also sometimes grafted in combination with *N*-vinyl imidazole (Rohm GmbH, 1973).

Since the chemistry to produce dispersant polymethacrylates is solution chemistry, like preparation of the base polymer, only relatively simple process modifications are necessary to produce dispersants commercially.

5.3.1.4 *Utility* Because free radical solution chemistry can be used to produce a broad range of molecular weights and compositions, polymethacrylates find utility in virtually all lubricant areas where viscosity/temperature modification is desirable. This includes both mineral oil-based applications and a range of synthetic fluids.

5.3.2 *Olefin copolymers*

The chemistry of ethylene–propylene-based VI improvers has been reviewed by Spiess *et al.* (1986), Marsden (1988) and Ver Strate and Struglinski (1989).

5.3.2.1 *Chemistry* Olefin copolymer-based VI improvers are primarily derived from ethylene and propylene. However, some commercial examples are also diene-modified (so-called EPDMs) while still others contain two diolefins. Since the solution rheology is dictated almost exclusively by the ethylene and propylene, the diene modifications relate primary to bulk handling characteristics as will be noted later. The EPDMs may in some cases be somewhat more reactive towards functionalization, but by the same note they are slightly less stable towards oxidation than poly (ethylene-co-propylene). The ter- and tetrapolymers also tend to be less efficient thickeners than the copolymers.

Ethylene contents of OCP VI improvers are normally in the 40–60 wt % range (50–70 mole %). These are generally lower in ethylene content than the EP copolymers that go into most other applications. The ethylene–propylene ratio for VI improvers is a compromise between thickening efficiency and low temperature solubility. The best thickening is achieved at high ethylene content, but too much ethylene leads to polymer crystallinity and insolubility at low temperature. Copolymers with excessively high ethylene contents can interact with the wax in mineral oils at low temperatures with devastating effects on low shear flow. A further consideration in choosing the ethylene–propylene ratio is the desire to minimize the propylene content to achieve the best possible oxidative stability.

Sequence distribution of the ethylene and propylene units is also a concern. Microcrystalline regions resulting from runs of ethylene can also interact with wax at low temperatures, resulting in undesirable low temperature properties. The compromise between thickening efficiency and low temperature viscometrics has been discussed by Kapuscinski *et al.* (1989). OCP VI improvers have normally been prepared by techniques which maximize the random distribution of the two monomer units. However, newer technology has resulted in tapered compositions (controlled intramolecular monomer unit distribution) which are excellent thickeners, have improved low temperature properties and are able to avoid undesirable wax interactions.

The ethylene–propylene ratios of EPDM-based VI improvers are in the same range as above. The dienes which are used must be non-conjugated with only one of the double bonds being reactive. The most common options are 1,4-hexadiene, 5-ethylidene-2-norbornene and dicyclopentadiene. These are used at relatively low levels, i.e. about 2–5%. This is lower than found in EPDMs for most other applications, the level being driven primarily by negative effects on thickening efficiency, cost and oxidative stability.

The OCP tetrapolymers are also EPDM-based but include a second diolefin in which both of the double bonds are reactive, such as 2,5-norbornadiene (E. I. du Pont de Nemours and Co., 1974). This is used at very low levels, less than 1.0%, to introduce a small amount of branching into the polymer. The resulting polymers have better handling characteristics as solids, as will be discussed later, but this is achieved at the expense of reduced thickening efficiency.

The molecular weight (M_w) of OCP VI improvers is in the range of 50 000 to 200 000 with molecular weight distributions (M_w/M_n) of about 2.0–2.5. The newer tapered compositions tend to be slightly higher in molecular weight with narrower molecular weight distributions (less than 2.0).

5.3.2.2 *Manufacture* Olefin copolymer VI improvers are produced by solution polymerization of ethylene and propylene catalysed by soluble Ziegler–Natta catalysts. Most commonly used are an aluminum alkyl halide with a soluble vanadium compound. Heterogeneous catalysis could also be used, but the properties of the polymers are less desirable for use as VI improvers.

The polymerization is carried out in relatively dilute solution in hexane at low temperature. Excess ethylene and propylene are removed and the solutions are washed with water to coagulate the polymer and to remove catalyst residues. The polymer is then dried and baled in blocks of convenient size for further handling.

Since the polymers for VI improver use are lower in both molecular weight and ethylene content than similar polymers for other applications, the bulk polymers are extremely tacky, making handling an issue at this stage. Several options are practiced commercially to minimize the tack/melt flow problem. Most widely practiced is to produce the polymer at higher molecular weight than required for VI improver usage and to degrade it mechanically and/or oxidatively as part of the dissolution step. This can be carried out by mastication in a variety of intensive mixers such as Brabender mixers or extruders.

A second option is to produce either a diene-modified polymer or a tetrapolymer. Performance and cost limitations mentioned earlier limit this technique. Incorporation of the second diolefin is particularly effective for improving the melt flow behavior, but with significant loss in thickening efficiency.

The other obvious alternative is to simply do a solvent transfer to oil during the isolation step. This eliminates handling the solid at all so that melt flow is not an issue. The polymer is made at the correct molecular weight and further processing is avoided. However, these apparent advantages must be balanced against the potential advantage of shipping solid polymer over long distances.

OCP VI improvers are sold both as solids (typically bales of about 20–35 kg wrapped in polyethylene) or more commonly as concentrated solutions in mineral oil. These polymers are excellent thickeners in the bulk phase and polymer content is limited by handling considerations to about 6–15%. OCP VI improvers are not themselves pour point depressants, but pour-depressed versions are often sold. These are merely physical mixtures of OCP VI improver with 2–3% of a conventional pour point depressant included in the concentrate.

5.3.2.3 *Dispersancy* Incorporation of dispersancy into OCP VI improvers is considerably more difficult than is the case with free radical solution chemistry, as described above for PMAs. Direct copolymerization of the preferred N- or O-containing monomers is not practical since these Lewis bases complex and thus poison the acidic Ziegler–Natta catalysts. The only option identified so far is to use an amount of catalyst in excess of that complexed by the polar monomer as described by Entreprise de Recherches et d'Activités Petrolières (1978) for *N*-vinylimidazole or Société Nationale Elf Aquitaine (1979) for *N*-vinyl succinimide.

In spite of this there is no shortage of technologies for synthesizing dispersant OCPs. All of the useful techniques focus on reactions on the polymer involving attack at the most reactive positions, either the propylene tertiary hydrogen atom in the case of OCP copolymers or that plus the unsaturation and/or allylic hydrogens in EPDMs. The easiest attack on these sites is simply oxidative degradation. This reaction produces a complex mixture of aldehydes, ketones and acids which then can serve as reactive handles for other reactions. This mixture can be reacted with amines directly, a technique which has not proved to be very useful, or can be employed in a Mannich condensation (Standard Oil Co., 1975).

The most successful route, and most widely practiced commercially, is graft polymerization. The grafts are grown following hydrogen abstraction, presumably again at the propylene site. Grafting at the position allylic to the unsaturation in an EPDM would appear to be an attractive option, but evidence is lacking that this is the preferred site, perhaps simply because of mass effects; that is, the greater reactivity is overshadowed by the very low concentration. Also, for practical reasons, the solvent of choice for the graft process is mineral oil. Mixing considerations normally dictate that the system should contain about 60–80% oil. Hence direct hydrogen abstraction or chain transfer to oil are important side reactions limiting the yield of the desired polymer graft. The chemistry works extremely well in a non-chain-transfer solvent such as chloro- or dichlorobenzene (Rohm and Haas Co., 1979), but this introduces solvent transfer and recovery problems. A process has also been described by Exxon Chemical Patents Inc. (1988) for grafting in the absence of solvent.

In spite of the problems, graft processes have been practiced commercially both in solvent and oil. Successful examples have included both 2-vinylpyridine and *N*-vinyl pyrrolidinone grafts by Rohm and Haas Co. (1979). Maleic anhydride has also been grafted, but this route requires an additional step to further functionalize the anhydride. Amines, as for example, *N*-(3-aminopropyl)morpholine, described by Exxon Research and Engineering Co. (1979b), are most commonly used to form the corresponding imide. Exxon has also described the use of polyamines, such as diethylenetriamine, but these tend to crosslink and the terminal amine must be capped as amide (Exxon Research and Engineering Co., 1979a) or imide (Exxon Research and Engineering Co., 1986). The terminal amine in polybutene-based succinimide ashless dispersants has also been reacted successfully with the anhydride by Exxon Research and Engineering Co. (1985).

The patent literature contains numerous other examples, most of which are at this point technical curiosities. The preceding options focused on those routes which are, or appear to have the potential to be, commercial.

5.3.2.4 *Utility* The excellent thickening ability of OCPs has led to extensive use in engine oils, both diesel and gasoline. However, the relatively poor low shear, low temperature viscosity behavior of OCPs together with difficulty of achieving the necessary lower molecular weights (better shear stability) have precluded significant use in most other specialty lubricants.

5.3.3 *Hydrogenated styrene–diene copolymers*

The chemistry and performance features of hydrogenated styrene–diene VI improvers has been reviewed by Eckert and Covey (1988).

5.3.3.1 *Chemistry* The products in this general class include several subclasses covering either butadiene or isoprene as the diene, and either random, block or star-shaped polymers. The basic member is a random copolymer of styrene and butadiene described by Phillips Petroleum Co. (1971) and BASF AG (1982). These contain about 50–60 wt% styrene and 40–50 wt% butadiene. For optimum thickening, it would be desirable for the butadiene to polymerize with a high content of 1,4-configuration. However, to prevent linear polyethylene-like regimes and problems similar to those described above for OCPs, 30–40% of the butadiene in the products is in a 1,2-configuration. The final products are obtained by hydrogenation using a technique which gives very high conversion on the butadiene-derived unsaturation while hydrogenating little or none of the styrene. Molecular weights (M_w) can be in the range of 75 000 to 200 000 with distributions (M_w/M_n) of less than 1.5.

Block copolymers of styrene and isoprene have also found commercial use. Both diblock (A–B) and triblock (A–B–A) polymers, where A represents polyisoprene and B represents polystyrene, have been described by Shell Oil Co. (1973, 1988). For maximum thickening efficiency a high yield of 1,4-isoprene configuration is desirable. Hydrogenation again is nearly complete on the isoprene-derived unsaturation and negligible on styrene. Since the polystyrene regimes are oil insoluble over most of the relevant temperature range for engine oils, these block polymers function as associative thickeners. Thus, the molecular weights of the individual molecules tend to be on the low side with the styrene block having M_ws of 30 000 to 50 000 and the isoprene blocks in the range of 50 000 to 100 000.

Several star-shaped molecules have also been described, primarily by Shell Oil Co. (1978), but the most common one involves hydrogenated isoprene arms (in the range of 5–15) on a divinylbenzene-based core. The divinyl benzene content depends on the molecular weight but is typically quite low, i.e. less than 1%. Since these structures represent a compromise between thickening ability and shear stability, their molecular weights tend to be higher than other members of this class, falling in the M_w range of 300 000 to 700 000. Molecular weight distributions are under 1.5.

There is one additional group of polymers which is derived from this chemistry, that is, hydrogenated polydienes. Hydrogenated polyisoprene, as described by Shell Oil Co. (1977), is chemically equivalent to an alternating copolymer of ethylene and propylene when polymerized in a 1,4-configuration and would thus be expected to exhibit behavior which is very similar to that of OCPs. Similar characteristics can be achieved with polybutadiene as long as the 1,2 content is above about 30% per BASF (1975). Shell Oil Co. (1977) has also patented combinations of butadiene and isoprene.

5.3.3.2 *Manufacture* Styrene–diene copolymers are produced by anionic solution polymerization. A typical system employs *s*-butyl lithium in cyclohexane at a temperature which may range from 60 to 120 °C. The system may also be promoted by small amounts of an amine, such as *N*,*N*,*N*′,*N*′-tetramethylethylene diamine, or an ether, such as tetrahydrofuran. Polymerization solids is in the 20–25% range.

In the case of random polymers all of the monomers may be charged together. However, the steps are sequenced for blocks and stars. For blocks, one of the monomers is polymerized and then the second is added to the living polymer to make an A–B block. In the case of stars, the arms are polymerized first and then divinyl benzene is added to the living polymer arms to form the basic structure.

The living polymer is then terminated by addition of an alcohol or by the subsequent hydrogenation. The polymers are hydrogenated in solution using a homogeneous catalyst such as trialkyl aluminum and an organo nickel compound, such as nickel octanoate or acetylacetonate. Finally, the polymers are washed with water to remove the salts and to coagulate the polymer. Since these polymers do not have the melt flow problems associated with OCPs, the products may be isolated either in the granular form or the granules may be compressed into bales.

Hydrogenated styrene–diene VI improvers are sold in both of these solid forms as well as in oil concentrates of about 6–20% solids content. Associative thickeners must be handled in the low solids range while the other polymers may be on the higher side. These VI improvers are also available with a conventional pour point depressant included in the concentrate.

5.3.3.3 *Dispersants* There are currently no commercial examples of dispersant hydrogenated styrene–diene VI improvers. However, several options are described in the patent literature. Any polar monomer which can polymerize readily by an anionic process, such as any of the vinyl pyridine family, may obviously be included directly in the VI improver backbone, most often as a block per Shell International Research (1985a). As a practical matter, any monomer used in this fashion must also be completely free of impurities which could serve as a source of protons, thus terminating the living polymer.

A second technique, which is unique to this family, involves metallation of the backbone by reaction with butyl lithium followed by growing a graft by an anionic mechanism as described by Shell Oil Co. (1985). This may be the preferred route since addition of vinyl pyridine to the original living polymer decreases the hydrogenation rate considerably.

The final potential routes involve free radical grafting. These completely parallel the OCP options discussed earlier and will not be repeated here.

5.3.3.4 *Utility* Hydrogenated styrene–diene VI improvers have found their widest utility in gasoline and diesel engine oils. They may also be used in other applications which have only moderate requirements for shear stability, such as tractor transmission fluids and aircraft piston engine oils. However, the shear stability demands of many of the specialty applications precludes the use of products which are currently available.

5.3.4 *Styrene polyester*

5.3.4.1 *Chemistry* This family of VI improvers, described by Lubrizol Corp. (1976), are styrene–fumarate copolymers derived from esterification of an approximately 1:1 styrene–maleic anhydride copolymer. Only dispersant versions are available. The dispersancy is provided by either an amide or, probably more likely, the imide derived from *N*-(3-aminopropyl)morpholine. Molecular weight (M_w) is in the range of 350 000 to 700 000 with a M_w/M_n of about 3–4.

5.3.4.2 *Manufacture* The copolymer of styrene-maleic anhydride is slurried in mineral oil and esterified to about 70% with a C_{8-18} alcohol mixture using an acid catalyst at 150–160 °C. Esterification is then carried to 95% using *n*-butanol. *N*-(3-aminopropyl)-morpholine is finally added to react with the remaining acidity. The final product is sold as a 35–45% polymer concentrate.

5.3.4.3 *Utility* These products are used almost exclusively in automatic transmission fluids and multipurpose tractor fluids. Minor usage may remain in engine oils, but this application is largely obsolete.

5.3.5 *Concentrate-compatible PMA/OCP blends*

5.3.5.1 *Chemistry* These products, described most extensively by Rohm GmbH (1979, 1981), are a blend of conventional PMA and OCP VI improvers. These two polymers, which are normally incompatible in concentrated form, have been compatibilized by incorporation of a small amount of a polymethacrylate–OCP graft copolymer. The final product is normally about 65–75% PMA and 25–35% OCP on a polymer solids basis.

In the concentrate, the more solvent-like polymer exists as a continuous phase while the other is in a discontinuous, micellar phase. For that reason the concentrate is sold in a mineral oil carrier to which a polar solvent such as dibutyl phthalate (at about 5%) is added to invert the two polymer phases. This results in the PMA becoming the continuous phase and the OCP the discontinuous phase. As a result of having the poorer thickener in the soluble phase and the better thickener dispersed, a higher solids content can be achieved at handling viscosities than would otherwise be the case. Solids contents are in the 40–50% range.

Molecular weight and molecular weight distributions of the two components are in the conventional ranges for their classes.

5.3.5.2 *Manufacture* A conventional methacrylate monomer mixture is polymerized by free radical initiation in an oil solution about 10–15% of an OCP VI improver at a temperature of about 120–140 °C. This portion of the reaction is carried out at a fully converted polymer solids content of 40–50%.

When methacrylate conversion is complete, additional solid OCP VI improver is added to bring the total OCP content up to the desired level. The final product is then diluted with ester solvent to invert the phases and sufficient additional mineral oil is added to reach the final product solids.

5.3.5.3 *Dispersancy* Dispersancy may be incorporated by any of the means described for PMA or OCP above. Currently only grafting with *N*-vinyl pyrrolidinone or an *N*-vinyl pyrrolidinone/*N*-vinyl imidazole mixture is practiced commercially. This is conducted after the addition of OCP in the above description.

5.3.5.4 *Utility* These products are used commercially only in gasoline and diesel engine oils. As described above, the inclusion of OCP precludes use in most other applications.

5.4 Function and properties

5.4.1 *Solution properties*

Although polymer use levels in multigraded oils fall in what some have described as the semi-dilute solution regime, dilute solution theory has been useful in describing VI improver behavior. Numerous workers have pointed out that VI improvers followed the well known dilute solution behavior described by Huggins (equation 5.1) and Kraemer (equation 5.2) reasonably well. Concentrations are normally in a range where terms higher than the second order are negligible. Neveu and Huby (1988) have published an exhaustive study of the application to PMA VI improvers, including determination of the values of k' and k'' for several systems.

$$\frac{\eta_{sp}}{c} = [\eta] + k'[\eta]^2 c + \ldots \tag{5.1}$$

$$\frac{\ln \eta_r}{c} = [\eta] - k''[\eta]^2 c + \ldots \tag{5.2}$$

Several useful relationships can be derived from these equations. Working first with the Kraemer equation, rearranging and substituting, from the Mark–Houwink equation, KM_v^a for $[\eta]$, equation 5.3 is obtained.

$$\ln \eta = KM_v^a c - k''(M_v^a)^2 c^2 + \ln \eta_0 \tag{5.3}$$

This indicates that the viscosity of an oil solution, η, should increase logarithmically with polymer concentation, c, and with viscosity average molecular weight (and therefore weight average molecular weight to a first approximation) raised to the power a. The relationship has an intercept at the log of the base oil viscosity, η_0.

From a practical point of view, the relationship permits the experimentalist to create a fair understanding of the viscosity-concentration relationship for a given polymer-oil combination with a single blend. Two or three blends to define the curvature introduced by the second order interaction provide the complete relationship. Furthermore, oils which are chemically similar enough to have the same K and a values will define a series of parallel lines with intercepts at $\ln \eta_0$. Better solvents will have steeper slopes because of the

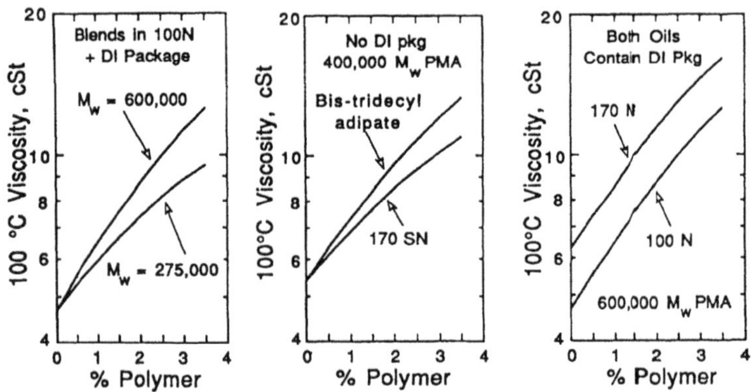

Figure 5.2 Effect of molecular weight, base stock type and base stock chemistry on VI improver thickening efficiency.

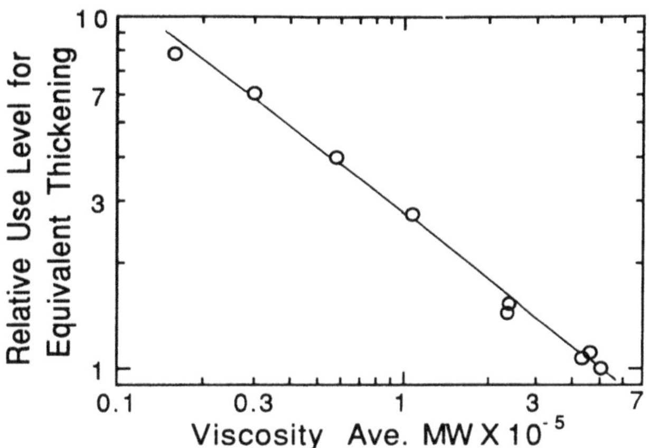

Figure 5.3 Relationship of PMA thickening efficiency to molecular weight.

higher value of *a*. The validity of all of these relationships is shown graphically in Figure 5.2.

Combining the Huggins and Kraemer equations and substituting KM_v^a for $[\eta]$, one can derive the relationship for thickening a specific base oil to a specific blend target, equation 5.4.

$$\log c = -a \log M_v + K_s \tag{5.4}$$

K_s is a very complex constant containing the constants for polymer interaction, k' and k'', from both the Huggins and Kraemer equations, K from the Mark–Houwink equation, as well as both the base oil and blend viscosities. This log–log relationship is verified for polymethacrylates in Figure 5.3.

Looking across chemistry lines, it is important to note that intrinsic viscosity, $[\eta]$, or some other measure of molecular dimensions in solution, is the driving force in thickening efficiency, not molecular weight *per se*. In other words, one should not expect two chemically different polymers of the same molecular weight to thicken the same way. Indeed, if the chemical structures differ radically, it is virtually certain that they will not.

For high molecular weight polymers, the unperturbed mean square end-to-end distance in solution is a function of only the degree of polymerization and is independent of the presence or absence of side chains. For example, addition of long alkyl side chains to the PMA is necessary to make a soluble polymer. However, they contribute nothing to thickening ability, $[\eta]$, root mean square end-to-end distance or whatever measure of size in solution one may choose. In turn, then, the relative thickening efficiency across polymer chemistry is a function of the percentage of the mass of that molecule which is in the backbone (Figure 5.4). Note that this takes the same form as the relationship shown in Figure 5.3.

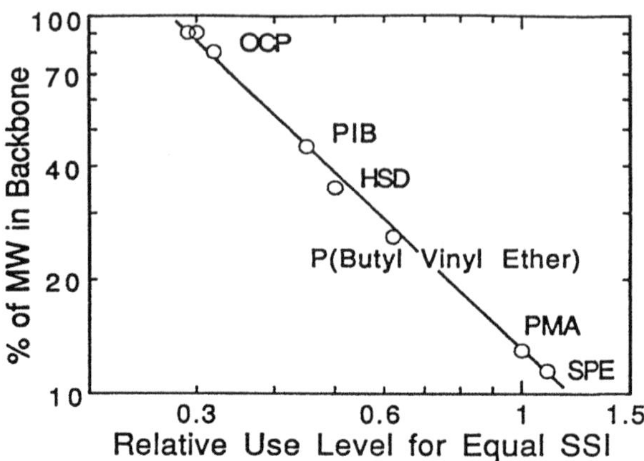

Figure 5.4 Relative thickening efficiencies of various VI improvers.

While this description is useful to understand the general concepts involved, one must recognize that it is only a first approximation to thickening. It is derived from thickening as measured by kinematic viscosity at 100 °C and makes no allowance for extensive long chain branching, dramatic differences in molecular weight distribution, associative thickening or other special effects.

5.4.2 *Mechanism of function*

The preceding discussion of thickening described some of the factors involved with thickening at 100 °C, an extremely important issue since this strongly influences use level and, therefore to a large degree, formulation economics. However, since a VI improver must work over a large temperature range, a further factor of interest is how viscosity varies with temperature.

If one looks back to dilute solution theory, one observes only that the issue falls back to the temperature dependence of k' and/or k'' in the Huggins and Kraemer equations, respectively, as well as K and a in the Mark–Houwink equation. These in turn relate in complex fashion to other basic physical chemical parameters.

Selby (1958) hypothesized that the mode of action focused on the hydrodynamic volume of the molecule as a function of temperature. In effect, he hypothesized that the molecular size in solution was highly contracted at low temperature, thus contributing little to viscosity, but was greatly expanded at high temperature, thus making a large contribution. However, Mueller (1978) and others have now reported that, of the three major VI improver classes in use today, only PMAs exhibit an intrinsic viscosity which increases uniformly with increasing temperature. The two hydrocarbon VI improvers tend to either exhibit a uniformly decreasing intrinsic viscosity with temperature or

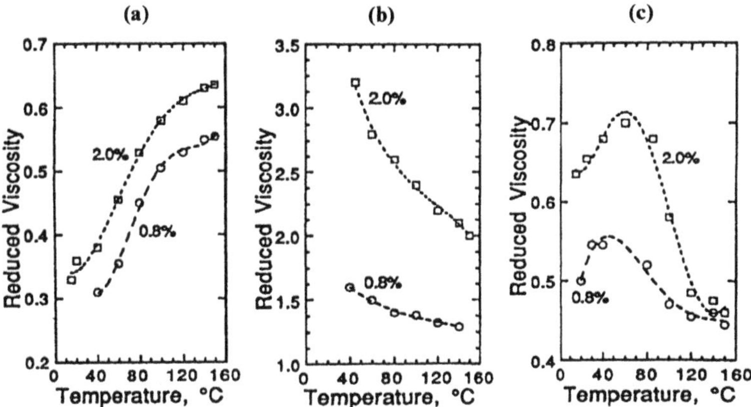

Figure 5.5 Temperature dependence of polymer coil dimensions for major VI improvers chemistries: (a) PMA, (b) OCP, (c) HSD (after Mueller, 1978).

go through a maximum somewhere in the relevant temperature range (see Figure 5.5).

Jordan *et al.* (1978) demonstrated that the effect of temperature on the flow behavior of polymethacrylate and polyacrylate blends in mineral oil was controlled to a very large extent by the entropy of activation for viscous flow. This confirmed early speculations by Bondi (1951). The increased negative entropy was presumably a result of the very sluggish translational motion of the polymer coils. On the other hand, the enthalpy of activation for viscous flow of the polymer solutions was for the most part very nearly the same as that of the oil solvent. Only the most efficient systems exhibited decreased enthalpy, suggesting that coil expansion at high temperatures may be a factor, but the effect was very small relative to the entropy effect.

Tamai *et al.* (1977) made similar observations for a polymethacrylate VI improver in both a light paraffinic and a light naphthenic oil. However, they reported that both the flow activation enthalpy and entropy increased for solutions of polyisobutylene. This apparent anomaly is explained when it is noted that the molecular weight of the PIB used in this study was only 930, well below that for a VI improver. The results for this low molecular weight PIB are in fact entirely consistent with observations reported by Jordan *et al.* (1978) for heavy base oils of high VI.

Dare-Edwards *et al.* (1988) have used NMR to study how the presence of VI improver affects the translational mobility of solvent molecules. They observed that there is a progressive reduction in translational mobility with increasing polymer concentration in both *t*-butylbenzene and *t*-butylcyclohexane. Furthermore, the self-diffusion coefficient of polymer-containing solution does not increase as fast with increasing temperature (40 to 110 °C) as it does for the solvent itself.

One must note that viscosity–temperature relationships of the type considered here barely scratch the surface. High temperature specifications are increasingly focused on high-shear-rate viscosity at temperatures of 125–175 °C, while at low temperatures, both high- (cranking) and low-shear rates (pumping) are of concern. Furthermore, the low temperature viscosity is greatly complicated by the presence of a wax dispersion and the possibility of wax-additive interactions.

5.4.3 *Shear stability*

5.4.3.1 *Permanent viscosity loss* When polymers are subjected to high shear stress in equipment, the random coil is severely distorted. In extreme cases, bond energies can be exceeded and the polymer will break, or shear, as shown schematically in Figure 5.6. This can happen in a variety of areas in equipment, for example, between the piston rings and cylinder wall or in the valve train of an engine, between the gear teeth of an axle or across the pressure relief valve of a hydraulic system.

The shearing process itself is a simple homolytic scission of a carbon–carbon bond, producing two polymeric radicals. There is no evidence that these radicals are involved in other aspects of lubricant chemistry; they are apparently rapidly terminated by either antioxidant or other species in the oil.

The backbone carbon–carbon bond strengths of all of the VI improver classes are much the same so none of the chemistries is inherently more shear stable than any other. Several workers have proposed that at least some of the shearing takes place at weak links, such as might result from severe steric hindrance. However, there is little or no evidence to support this hypothesis.

When the polymer is stretched, theory predicts that the maximum energy is concentrated in the middle of the molecule, or at least, well away from the

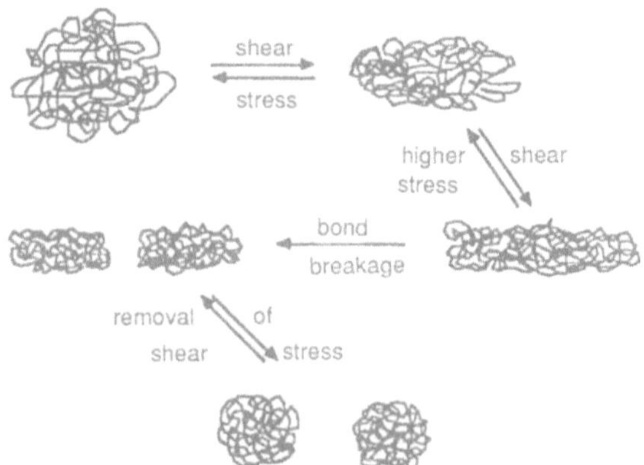

Figure 5.6 Schematic of mechanical polymer degradation.

ends. Thus, when a polymer shears mechanically, the two resulting fragments are predicted by Beuche (1960) to be about half of the molecular weight of the starting polymer. Ovenall *et al.* (1958) have hypothesized that the pieces need not be the same molecular weight, but the smaller piece would be no less than half of the limiting molecular weight, as defined below.

The higher the molecular weight, the more likely a polymer is to break. However, the mechanical degradation process is self-limiting. At some point, whatever the application, a molecular weight is reached where the energy concentration during coil distortion is insufficient to break further bonds. This is referred to as the limiting molecular weight. It should be noted, however, that molecules smaller than this will be produced since polymers only slightly larger than the limiting molecular weight still have a finite probability of breaking. Thus, the smallest molecule produced during shearing will be slightly greater than half of the limiting molecular weight. Viscosity loss during use is thus characterized by a rapid initial decrease as the bigger molecules break, followed by a slower loss and finally a plateau as the equilibrium, or terminal, molecular weight is reached.

Observations made by several workers, including Hillman *et al.* (1975) using gel permeation chromatography (GPC) are entirely consistent with theory. Figure 5.7 shows that only high molecular weight species are degraded and intermediate molecular weight polymer is produced. Note also that no low molecular weight fragments are made during mechanical shear.

The kinetics of the degradation process have been modelled by numerous workers, but most of them tend to take the general form of that described by Ovenall *et al.* (1958) as shown in equation 5.5

$$\mathrm{d}B_{\mathrm{i}}/\mathrm{d}t = k(P_{\mathrm{i}} - P_{\mathrm{e}})n_{\mathrm{i}} \tag{5.5}$$

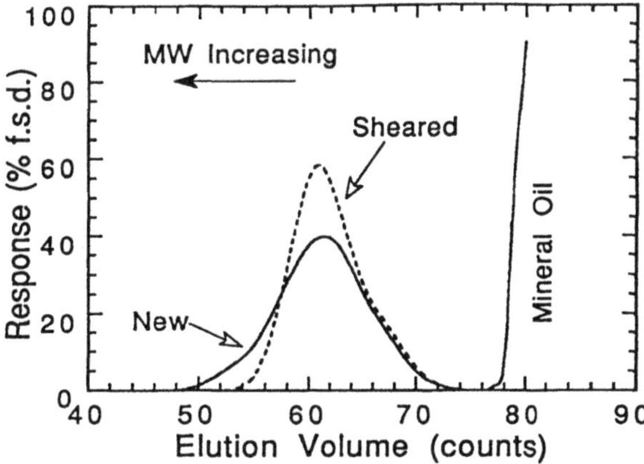

Figure 5.7 Effect of VI improver mechanical shear on molecular weight distribution (redrawn and adapted from Hillman *et al.*, 1975).

where dB_i/dt is the rate of bond breakage of polymers with a degree of polymerization P_i, P_e is the limiting degree of polymerization for bond breakage, n_i is the number of molecules of size P_i, and k is a rate constant which is largely a function of the shearing device but which may also be related to solvent and temperature. The total rate of bond breakage is summed over all of the molecular weight species that are big enough to shear degrade.

Rein *et al.* (1977) proposed that there could be more than one shear zone in a given application and that each could have a different severity. They suggested that viscosity during shear would fit an expression containing two exponential decay curves (equation 5.6), where *A*, *B*, *C* and *D* are constants for a given oil and application.

$$\eta = \eta_i - A(1-10^{-Bt}) - C(1-10^{-Dt}) \tag{5.6}$$

This was shown to fit a wide variety of applications. They also found that some applications (some hydraulic fluids and some automatic transmission fluids) appeared to be satisfied by a single term, suggesting a single shear site. However, other fluids in the same device were best fitted by two terms, suggesting that different polymers can degrade by different mechanisms.

Wright and duParquet (1983) proposed a somewhat different mechanism than earlier workers, but their model did include multiple shearing zones. They found that shear degradation in the FZG rig was best fit by a model using three shearing terms.

Klein and Mueller (1981, 1982) observed that shear degradation was a first order process, suggesting that the polymer entanglement was not a factor. They also found that the rate was proportional to the hydrodynamic volume of the VI improver and independent of the size and structure of side chain constituents. Thus, like thickening, shear stability is not a function of molecular weight across polymer classes, but rather, to a first approximation, backbone length. Arlie *et al.* (1977) had earlier made a similar observation, noting that an OCP of ~ 90 000 molecular weight degraded to a similar extent as a PMA of ~ 200 000 molecular weight.

Shear stability performance is generally expressed in terms of a scale referred to as the shear stability index, or SSI. This is defined as shown in equation 5.7

$$\mathrm{SSI} = \frac{\eta_i - \eta_f}{\eta_i - \eta_0} \times 100 \tag{5.7}$$

where η_i is the initial formulated oil viscosity, η_f is the sheared oil viscosity and η_0 is the viscosity of base oil, including all additives except VI improver. In practical terms, SSI is the percentage of the polymer-contributed viscosity which is lost as a result of shear degradation.

Care must be taken in reading the literature since SSI is sometimes misused or confused with overall viscosity loss (equation 5.8).

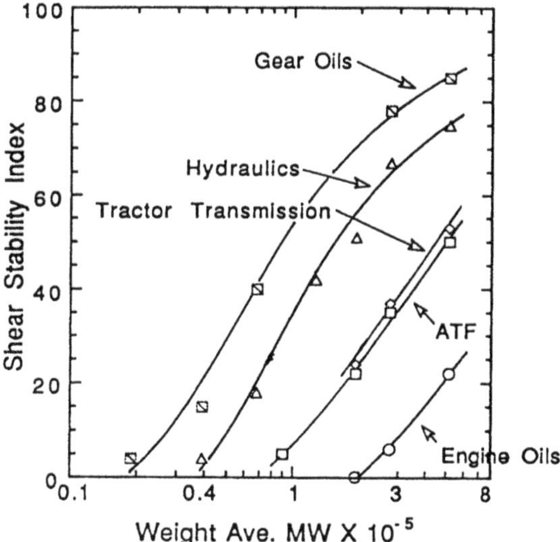

Figure 5.8 Relationship of molecular weight to SSI for commoner VI improver applications (redrawn and adapted from Kopko and Stambaugh, 1975).

$$\text{Overall viscosity loss} = \frac{\eta_i - \eta_f}{\eta_i} \times 100 \tag{5.8}$$

SSI is more useful, but it obviously requires a knowledge of the base oil viscosity, an unlikely situation when dealing with commercial lubricants.

The severity of different lubricant applications covers a wide shear stability range. Figure 5.8 from Kopko and Stambaugh (1975) shows that engine oils are the mildest application, followed by automatic transmission fluids, hydraulic fluids and rear axle lubricants. Thus, matching shear stability requirements of the application with the selection of VI improver is a key formulation consideration.

Extensive work has been done in an attempt to develop laboratory test methods which simulate viscosity loss in various applications, in particular, engine oils in field service. Virtually any device which can shear polymers has been used. Included are such things as sonic oscillators, ball mills, high speed stirring, diesel injector rigs, pumps in hydraulic circuits, gear rigs, roller bearing devices, and a variety of small engines, both motored and fired.

ASTM conducted a major program in the late 1960s with the objective of defining the ultimate test. However, none of the candidate tests was completely satisfactory and ASTM (1973, 1974) concluded that the best way to measure shear stability for engine oils was to use an engine, a technique which they regarded as impractical because of the large sample requirement and high test cost. Talbot *et al.* (1973) described the use of a small motored lawn mower engine to evaluate shear stability. The technique is elegant in its simplicity and, while it has not been widely used, it remains one of the best ways for relatively small scale evaluation.

While numerous bench tests are used in North America, diesel injector data are probably the most widely accepted. However, the 10-hour L-38 engine test is particularly important since it is part of the MIL-L-46152 and MIL-L-2104 specifications. Several bench tests are also important in defining European performance standards. European requirements are more demanding than those in North America in large part to meet the shear stability requirements of vehicles which had engines with integral gear boxes. This performance level was simulated by the Peugeot 204 viscosity stability test (CEC L-25-A-78). Since production of this vehicle configuration has been discontinued, the Peugeot test is no longer required in CCMC specifications. Nevertheless, this general performance level has been retained, and is now measured using the Kurt Orbahn diesel injector test rig. Dettman and Marsden (1981) have shown that there is a fair correlation between the two tests, but the Peugeot 204 being rather more severe.

A spur gear test rig, the FZG test, has been used for many years, and has recently been standardized by CEC (CEC L-37-T-85). This test was also studied by Dettman and Marsden (1981), who showed that it was more severe than either of the other two tests mentioned above. Laukotka (1989) made similar observations, but further reported that a tapered roller bearing rig based on a modified four-ball apparatus (P-VW 1437) was still more severe than the FZG. These two tests appear to have utility for the evaluation of gear oils and may also be of some value in severe hydraulic fluid applications.

ASTM Committee D.02, Subcommittee N is currently developing a sonic method for use specifically with hydraulic fluids. This work is on-going but Stambaugh *et al.* (1990a) have described some of the early data.

In summary, there are numerous ways to simulate shear stability performance in operating equipment. However, a common thread in the literature is that all laboratory methods tend to have some degree of VI improver chemistry bias. They are generally valuable for evaluating a given chemical type, but they are not necessarily reliable in comparing different families of VI improver. Nevertheless, these tests can be extremely useful as long as the potential for bias is recognized. It must be stated, however, that there is never a total substitute for data from real equipment.

5.4.3.2 *Temporary viscosity loss* All polymer solutions are non-Newtonian, that is, shear stress is not directly proportional to shear rate. There are several types of non-Newtonian behavior, but VI improvers at high temperature exhibit only one; pseudoplasticity, or shear-thinning.

When an oil is subjected to shear stress, polymer molecules are distorted, as described above. The size, shape and alignment with the field of flow of these elongated structures are such that the polymer contributes less to viscosity than does the spherical, random coil configuration that exists at low shear

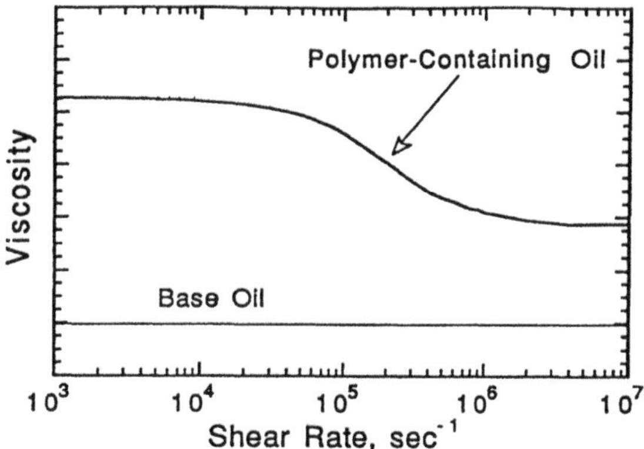

Figure 5.9 Effect of shear rate on polymer-containing lubricant.

stresses. The higher the shear stress, the lower the viscosity until a stable region, generally called the second Newtonian region is reached (Figure 5.9). As long as the shear stress does not break the molecules (see Figure 5.6), the process is completely reversible, i.e. when the shear stress is removed, the viscosity returns to its original value.

Like permanent viscosity loss, temporary viscosity loss is a function of molecular weight. The higher the molecular weight, the greater the temporary viscosity loss. It is also important to note that engine oils based on VI improvers which exhibit no permanent viscosity loss in service are still generally non-Newtonian, i.e. the molecules are distorted but there is insufficient energy to break chemical bonds.

There is no way to predict temporary viscosity loss across chemical families, or for that matter sometimes even within chemical families. Rosenberg (1975) observed that there was no correlation between permanent and temporary viscosity losses when he examined a wide variety of engine oils in a journal bearing rig. Lane *et al.* (1977) made the same observation when measuring temporary viscosity loss by the use of flow rate through a journal bearing of a running engine (Figure 5.10).

Primary polymer structure may account for some of this behavior, but VI improvers which function as associative thickeners are a major confounding factor. When the physically-associated, multi-polymer structures enter a shear field, they can dissociate into their separate molecular species. These smaller individual polymers are low enough in molecular weight that they degrade either slowly or not at all. When the molecules leave the shear field, they associate again so that there is little or no permanent loss of viscosity. However, when in the shear field, the contribution to viscosity is from the smaller, distorted individual molecules. The net result is a system which exhibits a high temporary viscosity loss relative to its low permanent viscosity loss.

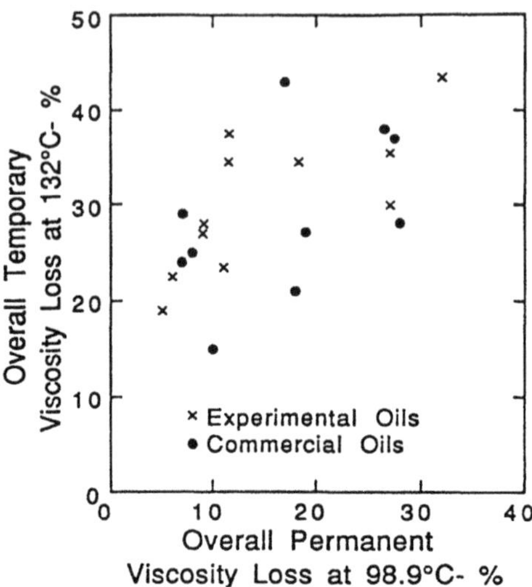

Figure 5.10 Temporary viscosity loss vs. permanent viscosity loss (after Lane *et al.*, 1977).

Coincident with the growing interest in high-shear-rate viscosity was the realization that the standard temperature for defining engine oil viscosity at high temperature (100 °C) is no longer representative of engine operation in the field. Thus, high-shear-rate viscosities and temporary viscosity losses are most often reported at 150 °C while permanent viscosity losses normally continue to be determined at 100 °C. The values at 150 °C are typically called HTHS or HTHSR (high-temperature, high-shear-rate) viscosities.

The measurement of high-shear-rate viscosity of lubricants has advanced tremendously in recent years and is now relatively routine. High pressure capillary and rotational viscometers are both commercially available and both CEC and ASTM have defined standardized test methods for their use (CEC L-36-A-87, ASTM D 4683 and ASTM D 4741 for the rotational and ASTM D 4624 for the capillary). The latter method is a generic one for a wide range of capillary viscometers. A separate method directed to a specific commercial high-pressure capillary viscometer is currently under development in ASTM Committee D.02, Subcommittee 7.

5.4.3.3 *Combined permanent and temporary viscosity loss* While much of the work that has been done in this area has focused on the isolated effects of permanent or temporary viscosity loss, equipment in the field obviously sees the combined effect. Limited work has been reported on the net effect but effort in this area appears to be increasing.

Recent literature on this subject has been reviewed by Alexander (1989). He observed that the change in high-shear-rate viscosity at 150 °C as a result of permanent viscosity loss was consistently small relative to the change in low-shear-rate viscosity at 100 °C. In general it was half or less, but this varied with the specific VI improver and DI package. However, if one considers the viscosity range of the current high temperature viscosity grades relative to a comparable range for a system based on 150 °C high-shear-rate viscosity, the HTHS change can be significant. For example, a range of 2.9–3.6 cP at 150 °C would correspond to a 100 VI extrapolation of the current SAE 30 limits at 100 °C. Alexander observed that the largest viscosity losses represented about 85% of the 100 °C SAE 30 range but still 50% of the corresponding HTHS range. Thus, the changes in high-shear-rate viscosity with permanent shear are small on an absolute basis but quite significant on a relative scale.

5.4.4 *Thermal-oxidative stability*

VI improvers can also undergo a variety of thermal and oxidative reactions. The mechanical and thermal-oxidative processes differ in several ways as shown in Table 5.2. These processes may be important for several reasons. In the first place, backbone cleavage can lead to viscosity loss, much like the mechanical process. However, since chemical degradation is a totally random process, viscosity loss will only be important if breakage is well removed from the ends of the polymer. There are also deposit implications. Rubin (1987) pointed out that oxidative breakage which produces small, reactive fragments could contribute to sludge formation. This was demonstrated in the extreme when early workers explored the use of natural rubber as a VI improver. Finally, VI improvers themselves can also become part of a deposit, the most notable example being as part of the diesel piston deposits. Degradation pathways can influence the magnitude of the VI improver contribution in such cases.

In a sense, it is unfortunate that discussions of the chemical degradation of VI improvers combine thermal and oxidative effects since the two processes are quite different. A simple thermal process is one which can take place in the absence of oxygen. This would include processes such as random thermal scission of a polymer backbone, which may be followed by depolymerization. Another possibility for PMAs or SPEs is pyrolysis of the ester side chains to form olefin and acid. The acid in turn can react with an adjacent ester to form a cyclic anhydride with elimination of alcohol. Adjacent acid groups can eliminate water to make anhydride (Mark *et al.*, 1966). There is no evidence

Table 5.2 Mechanical and thermal-oxidative processes of VI improvers

Mechanical	Thermal-oxidative
Molecular weight sensitive	Indiscriminate
Break near middle of molecule	Break at random along chain
Self-limiting	Complete degradation possible
Insensitive to chemistry	Dependent on chemistry

Table 5.3 Carbon–hydrogen bond strengths of typical organic molecules (Weast, 1989)

Carbon atom	Carbon–hydrogen bond strength D^0_{298} (kJ/mol)
Primary	419.5 ± 4.0
Secondary	401.3 ± 2.0
Tertiary	390.2 ± 2.0
Benzylic	353.1 ± 6.3
Allylic	345.2 ± 5.4

that depolymerization and ester pyrolysis are issues in the engine oil itself, but they may be a factor if the VI improver is trapped in deposits.

The oxidative process is driven either by oxygen itself or by any source of free radicals. If a polymer backbone is attacked, leading to either a polymeric carbon or oxygen radical, backbone cleavage is possible. In the case of polyethylene, polypropylene and butadiene- or isoprene-containing polymers, this may be accompanied by elimination of formaldehyde or acetaldehyde. In the case of styrene-containing polymers, formaldehyde and benzaldehyde are products of the cleavage (Mark *et al.*, 1966). Such reactions could take place either in the bulk oil phase or in deposits in which the polymer is physically trapped.

Since oxidation is a chemical process, one need look only at the carbon–hydrogen bond strengths to estimate ease of hydrogen abstraction (Weast, 1989) (see Table 5.3).

These data suggest that PMA backbones should be the most stable to oxidative attack. OCP polymers would be expected to be less stable while the styrene–diene polymers could vary depending on how much 1,2 vs 1,4 structure is present. The benzylic hydrogens from styrene are potential sites of attack and it is obviously critical that hydrogenation of the olefinic unsaturation is as complete as possible.

Wunderlich and Jost (1978) have examined thermal and oxidative stability by determining the molecular weight distributions (by GPC) of examples of major VI improvers subjected to several engine tests. Given that the chemical pathways can lead to low molecular weight fragments while mechanical shear does not, it is possible to evaluate the importance of thermal-oxidative pathways in polymer degradation. The authors observed that PMAs exhibited only mechanical instability in a 40-hour test in an Opel Kadett engine, the Peugeot 204 test and the Caterpillar 1H test while suffering some thermal degradation only in the Caterpillar 1G test. In contrast, both OCP and HSD exhibited very slight oxidative effects in the Opel Kadett and clear oxidation in all of the other tests. Furthermore, the viscosity changes observed with the block HSD copolymer were larger than would have been expected from the molecular weight change. The authors speculated that the oxidation probably changed the polymer structure in a way that interfered with the associative thickening mechanism.

Hillman *et al.* (1977) have also shown that thermal/oxidative effects contribute to viscosity instability in the FZG test. They observed that diesel injector and FZG shear data correlated reasonably well in spite of the fact that, based on GPC data, the former appeared to be a purely mechanical process while the latter had a significant thermal and/or oxidative component. This may well relate back to the observation of Rein *et al.* (1977) that a diesel injector shear process can best be simulated by assuming it involves two steps while Wright and duParquet (1983) found that the FZG appeared to involve three steps. There is also a clear chemical bias in Hillman's data which indicates that the PMA thermal/oxidative component is less than that of other VI improvers in the study.

Klein and Mueller (1979) studied oil solutions of PMA, OCP and HSD by both thermal and oxidative processes. They found that there was no viscosity change at 170 °C for any of the systems in the absence of oxygen. However, with oxygen present, the PMA lost very little viscosity while the OCP underwent severe degradation. At 260 °C all of the systems degraded significantly. In a separate study, Mueller and Leidigkeit (1979) observed that this oxidative process could be decreased but was not completely suppressed by antioxidant.

Overall, it appears that the thermal/oxidative behavior of PMA VI improvers is dominated by thermal effects while that of the other current polymers is dominated by oxidative effects. The tradeoff between the two processes does not appear to be a dominant factor in VI improver choice.

5.5 Performance

The preceding sections have attempted to summarize the physical and chemical nature of VI improvers and the lubricants containing them. This section will link those features to end use performance properties. Since engine oils are by far the largest market for VI improvers, this will be the primary focus of this section. There is no easy decision process for choosing a VI improver. Any chemical family or shear stability option represents compromises, not to mention a cost and convenience overlay on the decision process. Suffice it to say that all of the three major classes of VI improver are currently used in large quantities in both gasoline and diesel engine oils.

While it will not be discussed for the most part in the following sections, a common thread in all of the performance areas is the role of new oil vs. used oil viscosities. Most work tends to relate performance to new oil viscosities since this is how industry viscosity standards are currently set. However, it is sometimes noted that relationships can be improved if sheared oil viscosities are used.

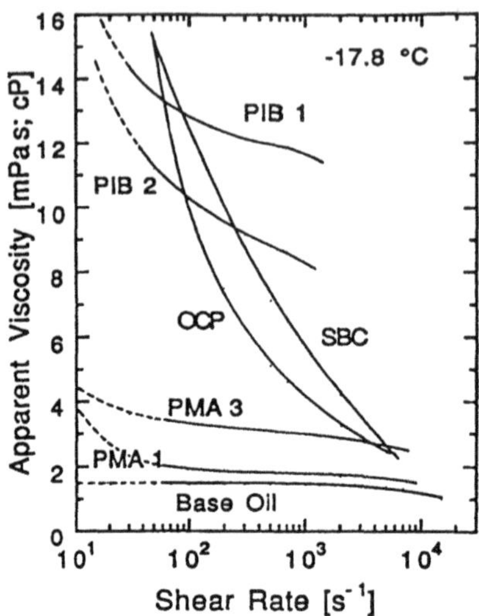

Figure 5.11 Apparent viscosity vs. shear rate for various polymer types (after von Petery *et al.*, 1978).

5.5.1 *Low temperature viscosity*

5.5.1.1 *Cranking* The benefit of low temperature cranking of polymer-modified engine oils provided one of the primary driving forces for the commercialization of multigraded engine oils. Prior to the use of VI improvers, low temperature properties were specified by extrapolation of viscosities at 210 °F (98.9 °C) and 100 °F (37.8 °C) to 0 °F (−17.8 °C). The non-Newtonian nature of polymer-containing oils completed nullified any utility of this measurement and the cold-cranking simulator (CCS) was adopted as the industry standard in SAE J300a in 1967. Developments in this area up to about 1975 have been carefully reviewed by Stewart and Selby (1977).

Bartz and Wiemann (1977) reported the low temperature properties of several VI improver systems, observing that at the high shear rates characteristic of the CCS, PMA, OCP and HSD all have comparable viscosities. In the case of the latter two, this results from substantial non-Newtonian character at low temperature (Figure 5.11). Bartz and Wiemann also reported data for polyisobutylene, which until the mid-1960s had a substantial share of the market. PIB met standards adequately when extrapolated low temperature viscosities were used. However, its measured viscosity at CCS shear rates is quite high, meaning that it could only be used with a very light base stock. This deficiency was the major factor which led to its virtual disappearance from the engine oil market.

Carlson (1983) has reported data showing that the HSD VI improvers contribute less viscosity to the CCS than PMA or OCP, permitting the use of a heavier base stock. This had a favorable effect on base oil volatility which was reflected in improved oil consumption. Both Stambaugh *et al.* (1972) and Speiss *et al.* (1986) have demonstrated that dispersant PMA or OCP VI improvers in some cases permit a significant reduction in the level of ashless dispersant in a fully formulated engine oil. This also permits the use of a heavier base stock.

The choice of shear stability level is also a major factor affecting the CCS viscosity, and therefore the base stock choice. Since shear stable (lower molecular weight) VI improvers are inherently poorer thickeners, high use levels are required with a necessary compensating reduction in base stock viscosity.

5.5.1.2 *Pumpability* As a result of abnormally high wear during the winter months in the early 1960s, low temperature pumpability requirements were introduced into SAE J300 in 1979. Unfortunately, the standards were flawed and severe field problems plagued the industry during the early 1980s. Stambaugh and O'Mara (1982) demonstrated that the severity of the problem varied greatly with engine design and that a major source of the problem could be caused by wax gelation (pour point reversion problems).

Stambaugh (1984) also showed that in the absence of wax problems, PMAs provided a superior performance to that of SIP which in turn was superior to OCP. The differences were largest in the most severe engines. Johnson (1984) confirmed this observation and noted that differences in VI improver performance were minimized in engines properly designed for low temperature flow. May and Habeeb (1989) also confirmed the VI improvers' chemistry rankings in a program which studied engine parameters further and included a broader range of SAE W-grades. Very similar observations have been made in heavy duty diesel engines by Frame *et al.* (1987) and by Stambaugh *et al.* (1990b).

While differences in VI improver chemistry do affect pumpability performance, it must be emphasized that with good base stocks and proper choice of pour point depressant, engine oils can be formulated to industry standards with any of the currently used chemistries.

5.5.2 *High temperature viscosity*

The high temperature rheology of engine oils was the subject of extensive research and controversy during the 1980s as SAE and ASTM expended considerable energy on how to deal with the high temperature portion of SAE J300. The literature has been thoroughly reviewed by an ASTM (1985) task force of Section D.02.07.0B and updated in a further review edited by Spearot (1989).

5.5.2.1 *Journal bearing oil film thickness and wear* Lubricant issues related to journal bearings have been extensively studied using everything from bearing rigs to oil film thickness measurements in running engines to catastrophic failure tests in full scale vehicles. While it remains reasonably well established from the ASTM (1985) review that high temperature, high-shear-rate viscosity is an important parameter in determining oil film thickness, debate remains as to how important a contribution can be made by pressure or viscoelastic effects.

Hutton *et al.* (1983) argued that the 150 °C high-shear-rate viscosities of polymer-thickened oils measured at high pressure can be in a different order from that observed at atmospheric pressure. Hutton *et al.* (1984), using a model lubricant based on very high molecular weight polyisobutylene, demonstrated that elasticity can make an important contribution to journal bearing load capacity.

In perhaps the most careful and thorough study of the subject, conducted by an ASTM task force and reported by Cryvoff *et al.* (1990), it was concluded that HTHS viscosity alone is sufficient to predict oil film thickness in running engines. The key to reaching this conclusion was establishing the temperature of the oil at the point of closest contact in the bearing, thus permitting the viscosity to be specified at the relevant temperature.

The only VI improver chemistry issue involved here is how associative thickeners fit into the total scheme. The issue will probably remain open for some time to come.

5.5.2.2 *Piston ring-cylinder wear* The piston ring-cylinder wall region of the engine is much more complex than the bearings in that the lubrication regime is mixed, that is, hydrodynamic during piston travel and boundary at top and bottom dead center. Furthermore, temperatures are higher than in the bearings, so much so that there may well be a volatility overlay on the rheology. The state of knowledge is included in the ASTM (1985) review and will not be covered here. Suffice it to say that viscosity and VI improver effects have been observed by several workers.

5.5.2.3 *Fuel economy* It is widely accepted that there is a viscous component of engine oil fuel economy. This was also extensively reviewed by ASTM (1985). Workers in the field seem to be in general agreement that the viscous contribution is determined by a high-shear-rate viscosity in the range of from 4×10^5 to $10^6\,s^{-1}$, measured at a temperature in the range of 100 to 150 °C, depending on the vehicle service. It is also recognized that if the viscosity is too low, boundary effects will become important and fuel economy benefits will be lost.

5.5.2.4 *Oil consumption* The role of lubricant in oil consumption control has been studied sporadically for many years. There has been relatively little debate on the subject with nearly all workers concluding that there are both viscosity and volatility components. The only issue has been the relative contributions of the two factors, which not surprisingly vary with the nature of the formulation, the design of the vehicle and the severity of the service. Since the route of oil loss is behind the piston rings and down valve guides, the viscosity component is generally accepted as being low shear viscosity. Hence, VI improver effects are correlated with kinematic viscosity (generally on the used oil) measured from 100 to 150 °C, depending on the service. Roberts (1990) has concluded that in current gasoline engine designs, volatility is the more important factor although the relative contributions of viscosity and volatility vary with engine design, including whether the engine is air- or water-cooled.

The presence of VI improver has a strong positive influence on oil consumption control in large diesel engines. The effect is not well understood, but it is clear that volatility is much less of a factor than in gasoline engines. In diesels, oil consumption is thought to result largely from oil pumped behind the rings and from a wave of oil carried upward on the top of the piston rings. When the piston reaches top dead center, the oil is thrown by inertia into the combustion chamber. McGeehan (1983) has speculated that it is the effect of polymer on the (low shear) viscosity behind the rings and in this wave which is the favorable factor.

Texaco Inc. (1988) has described the use of low levels of VI improvers in monograde oils to improve their oil consumption control properties. They note that all VI improver chemistries in current use are effective. They also argue that the effect is not the result of viscosity although they offered no alternative. It may be noteworthy that the concentration range of polymer described by Texaco as being effective is similar to that described by Wilson (1970) as being effective at controlling stray mist during mist lubrication.

There is also a long term lubricant effect on diesel engine oil consumption. Hercamp (1983) observed that oil consumption control is lost when carbon deposits build up on the top land of the piston. VI improvers may be a factor in that they contribute to piston deposits; however, the role of VI improvers in this specific phenomenon has not been demonstrated.

5.5.2.5 *High temperature pumping efficiency* High temperature pumping efficiency is influenced by the role of viscosity in control of internal pump leakage. This has been shown by Kopko and Stambaugh (1975) to be related to the high-shear-rate viscosity at test temperature. Stambaugh and Kopko (1973) have also demonstrated that high-shear viscosity directly influences pumping efficiency of hydraulic fluids as well as being a factor in gear upshifts in automatic transmissions.

5.5.3 *Deposits*

5.5.3.1 *Diesel piston deposits* Kay and O'Brien (1974) and Smith *et al.* (1976) both reported early in the development of multigraded diesel engine oils that VI improver content in the formulation was a significant factor in the contribution to piston deposits. Since it is generally accepted that all VI improvers contribute to these deposits, they argued that it was important to keep the VI improver content as low as possible, thus giving OCPs a significant advantage over other chemistries. There are also implications for shear stability level because of the higher polymer content required by a low molecular weight VI improver.

Koller *et al.* (1983), in a carefully controlled experimental design in Caterpillar 1H2 tests, observed no effect on piston top groove fill and only a very weak effect on weighted total demerits using two levels (differing by a factor of two) of dispersant OCP. Huby and Stambaugh (1986) have also reported that DI package must be customized to the VI improver chemistry to provide optimum performance.

5.5.3.2 *Dispersancy* Stambaugh *et al.* (1972) pointed out that dispersant PMAs can replace a significant portion of the ashless dispersant necessary for sludge dispersancy performance in gasoline engines using either leaded or unleaded gasoline. Significant varnish benefits were also noted from the use of dispersant PMA VI improver. Spiess *et al.* (1986) demonstrated similar dispersancy performance for dispersant OCPs. They also showed that the polymers can enhance performance in both gasoline and diesel engine oils. Dispersants can be used either to achieve higher performance levels or to displace ashless dispersant, thereby permitting the use of a heavier base stock.

Dispersant VI improvers and ashless dispersants are not necessarily a one-to-one tradeoff. Both can be extremely effective in controlling sludge deposition in gasoline engines. The very high dispersancy levels necessary for API SG engine oils can demand very high levels of ashless dispersant and corresponding light base oils. Dispersant VI improvers can be used to great advantage in such cases. However, the trade-offs in diesel engines are more complex. Ashless dispersants tend to be more effective in controlling diesel deposits, but dispersant VI improvers are better for dispersing soot in the oil. McGeehan *et al.* (1984) found that dispersant VI improvers provided the best control of soot-induced viscosity increase with an inferred favorable effect on soot deposition in the engine. He also reported improved wear control with these formulations, which he speculated resulted from a reduction in abrasive wear.

It must be noted that dispersant VI improvers cover a wide range of performance levels. These are related more to the nature of the dispersant functionality and to the way it is incorporated than to the chemical family of the backbone. Thus, the generalizations described here represent potential performance options which must be evaluated for the particular dispersant VI improver of interest.

5.6 Prognosis for future developments

It is difficult to foresee any revolutionary changes in the field of VI improvers in the future. It must be recognized that current core technologies in this area range from about 20 to 50 years old. That is not to say that the technologies in use are truly that old; all have continually undergone evolutionary change since their introduction. It is likely that this evolution will continue, driven not so much by changing polymer technology as by adapting polymer technology to meet evolving needs of the market place.

One must assume that engines will continue the trend of increasing operating temperatures. This will place continued pressure on the thermal-oxidative stability of VI improvers. It will also result in volatility being an increasingly important issue. The latter will have several spin-offs. In the first place, minimum contribution to the CCS will be a key factor to using the heaviest possible base stocks. Second, dispersancy from the VI improver will become even more important as a means of reducing ashless dispersant, thereby increasing the base stock viscosity. Finally, synthetics and part-synthetics will continue to grow, opening a market for products customized specifically for these base stocks.

The push for dispersancy will be a drive for even higher levels of dispersancy. If gasoline and diesel engine oil lines diverge, as has been speculated, a parallel series of dispersant VI improvers could emerge, one optimized for low temperature sludge dispersancy and another for high temperature deposit control and soot dispersancy.

Products will also have to be optimized for alcohol-based fuels. This evolution is likely to be one of both composition to tolerate high levels of fuel dilution, and dispersancy to handle the new types of sludge that are likely to be seen.

Low temperature pumpability requirements are likely to become more demanding. Evolution of all chemistries to improve this performance area should be expected. As the industry continues to become more aware of the low temperature properties of used engine oils, dispersant VI improvers may be designed to prevent the thickening which now occurs at extended mileages.

In summary, VI improver developments for the foreseeable future are likely to be evolutionary and market-driven.

References

Alexander, D.L. (1989) Change in high-shear-rate viscosity of engine oils during use: a review. In Spearot, J.A. (ed.) (1989), pp. 60–73.

Arlie, J.P., Denis, J. and Parc, G. (1975a) Viscosity index improvers 1. Mechanical and thermal stabilities of polymethacrylates and polyolefins. *IP Paper 75-005*, Inst. Petr., London.

Arlie, J.P., Denis, J. and Parc, G. (1975b) Viscosity index improvers 2. Relations between the structure and viscosimetric properties of polymethacrylate solutions in lube oils. *IP Paper 75-006*, Inst. Petr., London.

Arlie, J.P., Denis, J. and Parc, G. (1977) Comparative study of the shear stability of polymethacrylates and olefin copolymers. *IP Paper 77-006*, Inst. Petr., London.

ASTM (1973) Shear stability of multigrade crankcase oil, *ASTM Data Series Publication DS 49*, ASTM, Philadelphia, PA.

ASTM (1974) Shear stability of multigrade crankcase oils, *ASTM Data Series Publication DS 49 S-1*, ASTM, Philadelphia, PA.

ASTM (1985) *The relationship between high-temperature oil rheology and engine operation—a status report. ASTM DS 62*, ASTM, Philadelphia, PA.

Bartz, W.J. and Wiemann, W. (1977) Determination of the cold flow behavior of multigrade engine oils. *SAE Paper 770630.*

BASF (1975) Hydrogenated polyolefins as viscosity index improvers. *Ger. Offen. DE 2,358,764.*

BASF AG (1982) Hydrogenated styrene–butadiene copolymer VI improver for lubricating oils. *Ger. Offen. DE 3,106,959.*

Bondi, A. (1951) *Physical Chemistry of Lubricating Oils.* Reinhold, New York, 48.

Bueche, F. (1960) Mechanical degradation of high polymers. *J. Appl. Pol. Sci.* **4** (10) 101–106.

Carlson, D.C. (1983) The effect of VI improvers and resultant base oil volatility on automotive oil economy with SAE 5W-40 oils. *SAE Paper 830029.*

Cryvoff, S.A., Spearot, J.A. and Bates, T.W. (1990) Engine bearing oil film thickness measurement and oil rheology—an ASTM task force report. *SAE Paper 902064.*

Dare-Edwards, M.P., Kempsell, S.P., Barnes, J.R., Craven, C.J. and Wayne, F.D. (1988) Nuclear magnetic resonance of lubricant-related systems. *6th International Colloquium, Industrial Lubricants—Properties, Application, Disposal*, Vol. II, Bartz, W.J. (ed.), Technische Akademie Esslingen, Ostfieldern, 12.3-1 to 12.3-15.

Dettman, L.P. and Marsden, K. (1981) Shear stability testing of polymers in automotive lubricants. *CEC International Symposium on the Performance Evaluation of Automotive Fuels and Lubricants, Paper EL/4/4.*

Eckert, R.J.A. and Covey, D.F. (1988) Developments in the field of hydrogenated diene copolymers as viscosity index improvers. *Lubr. Sci.* **1** (1) 65–80.

E. I. du Pont de Nemours and Co. (1956) Lubricating Oil Compositions Containing Polymeric Additives. US Patent 2,737,496.

E. I. du Pont de Nemours and Co. (1974) Mineral Oil Composition. US Patent 3,790,480.

Entreprise de Recherches et d'Activités Petrolières (1978) Novel Lubricating Compositions Containing Nitrogen Containing Hydrocarbon Backbone Polymeric Additives. US Patent 4,092,255.

Exxon Research and Engineering Co. (1979a) Stabilized Imide Graft of Ethylene Copolymeric Additives for Lubricants. US Patent 4,137,185.

Exxon Research and Engineering Co. (1979b) Polymeric Additives for Fuels and Lubricants. US Patent 4,144,181.

Exxon Research and Engineering Co. (1985) Ethylene Copolymer Viscosity Index Improver-Dispersant Additive Useful in Oil Compositions. US Patent 4,517,104.

Exxon Research and Engineering Co. (1986) Ethylene Copolymer Viscosity Index Improver-Dispersant Additive Useful in Oil Compositions. US Patent 4,632,769.

Exxon Chemical Patents, Inc. (1988) Viscosity Index Improver-Dispersant Additive Useful in Oil Compositions. US Patent 4,780,228.

Frame, E.A., Montemayor, A.F. and Owens, E.C. (1987) *Low-Temperature Pumpability of U.S. Army Diesel Engine Oils.* Report BFLRF No. 229, US Army Belvoir Research, Development and Engineering Center, Fort Belvoir, VA.

Hercamp, R.D. (1983) Premature loss of oil consumption control in a heavy duty diesel engine. *SAE Paper 831720.*

Hillman, D.E., Lindley, H.M., Paul, J.I. and Pickles, D. (1975) Application of gel permeation chromatography to the study of shear degradation of polymeric viscosity index improvers used in automotive engine oils. *Br. Polym. J.* **7** 397–407.

Hillman, D.E., Morris, P.R., Paul, J.I. and Pickles, D. (1977) Comparison of the modes of degradation of viscosity index improvers in the Kurt Orbahn and FZG tests by gel permeation chromatography. *Materials Quality Assurance Directorate Technical Paper No. 677*, London.

Huby, F. and Stambaugh, R.L. (1986) Package optimization for diesel performance. *5th International Colloquium, Additives for Lubricants and Operational Fluids*, Vol. II, Bartz, W.J. (ed.) Technische Akademie Esslingen, Ostfieldern, 9.7-1 to 9.7-15.

Hutton, J.F., Jones, B. and Bates, T.W. (1983) Effects of isotropic pressure on the high temperature high shear viscosity of motor oils. *SAE Trans.* **92** Paper 830030.

Hutton, J.F., Jackson, K.P. and Williamson, B.P. (1984) The effects of lubricant rheology on the performance of journal bearings. *ASLE Preprint No. 84-LC-1C-1.*

I.G. Farbenindustrie AG (1938a) Hydrocarbon Lubricating Oils. US Patent 2,106,232.

I.G. Farbenindustrie AG (1938b) High Molecular Weight Iso-Olefine Polymers and Process of Producing the Same. US Patent 2,130,507.

Johnson, R.S. (1984) A laboratory engine test study of motor oil flow properties in winter service. *SAE Trans.* Paper 841387.

Jordan, E.F., Jr., Smith, S., Jr., Zabarsky, R.D., Austin, R. and Wrigley, A.N. (1978) Viscosity index. II. Correlation with rheological theories of data for blends containing *n*-octadecyl acrylate. *J. Appl. Polym. Sci.* **22** 1529–1545.

Kapuscinski, M.M., Sen, A. and Rubin, I.D. (1989) Solution studies on OCP VI improvers. *SAE Paper 892152.*

Kay, R.E. and O'Brien, J.A. (1974) New multigrade SE/CD lubricant. *SAE Paper 740523.*

Klein, J. and Mueller, H.G. (1979) Decomposition and solution behavior of polymers (in German). *Erdoel Kohle, Erdgas, Petrochem.* **32** (8) 394.

Klein, J. and Mueller, H.G. (1981) Shear stability of viscosity index improvers (in German). *Ber.-Dtsch. Ges. Mineraloelwiss. Kohlechem.*, **256**, DGMK, Hamburg.

Klein, J. and Mueller, H.G. (1982) Shear stability of viscosity index improvers (in German). *Kohle Erdgas, Petrochem.*, **35** (4) 187.

Koller, R.D., Galluccio, R.A. and Stambaugh, R.L. (1983) Deposit control in the Caterpillar 1H2 engine test—A statistical approach to identifying engine oil component effects. *SAE Paper 831723.*

Kopko, R.J. and Stambaugh, R.L. (1975) Effect of VI improver on the in-service viscosity of hydraulic fluids. *SAE Paper 750693.*

Lane, G., Roberts, D.C. and Tims, J.M. (1977) Measurement of the viscosity of multigrade oils in a running engine. *SAE Trans.* **86** Paper 770379.

Laukotka, E.M. (1989) Shear stability tests for polymer containing lubricating fluids. *CEC 3rd International Symposium of Performance Evaluation for Automotive Fuels and Lubricants, Paper 3 LT.*

Lubrizol Corp. (1976) Lubricant Containing Nitrogen-Containing Ester. US Patent 3,702,300.

Mark, H.F., Gaylord, N.G. and Bikales, N.M., eds. (1966) *Encyclopedia of Polymer Science and Technology*, Vol. 4. John Wiley & Sons, New York, pp. 661–672; 705–712.

Marsden, K. (1988) Literature review of OCP viscosity modifiers. *Lub. Sci.* **1** (3) 265–280.

May, C.J. and Habeeb, J.J. (1989) Lubricant low temperature pumpability studies—oil formulation and engine hardware effects. *SAE Paper 890037.*

McGeehan, J.A. (1983) Effect of piston deposits, fuel sulfur, and lubricant viscosity on diesel engine oil consumption and cylinder bore polishing. *SAE Trans.* **92** Paper 831721.

McGeehan, J.A., Rynbrandt, J.D. and Hansel, T.J. (1984) Effect of oil formulations in minimizing viscosity increase and sludge due to diesel engine soot. *SAE Paper 841370.*

Mueller, H.G. (1978) Mechanism of action of viscosity index improvers. *Tribol. Int.* **11** (3) 189–192.

Mueller, H.G. and Leidigkeit, G. (1979) Thermal and oxidative degradation of polymers in multiviscosity oils (in German). *Schmiertech. Tribol.* **26** (6) 201–204.

Neudoerfl, P. (1986) State of the art in the use of polymethacrylates in lubricating oils. *5th International Colloquium, Additives for Lubricants and Operational Fluids*, Vol. 11, Bartz, W.J. (ed.) Technische Akademie Esslingen, Ostfieldern, 8.2-1 to 8.2-15.

Neveu, C. and Huby, F. (1988) Solution properties of polymethacrylate VI improvers. *Lubr. Sci.* **1** (1) 27–50.

Otto, M., Miller, F.L., Blackwood, A.J. and Davis, G.H.B. (1934) Motor oils having viscosity index of 120 predicted as definite need. *Oil Gas J.* **33** (26) 98–106.

Ovenall, D.W., Hastings, G.W. and Allen, P.E. (1958) The degradation of polymer molecules in solution under the influence of ultrasonic waves. Part I. Kinetic analysis *J. Poly. Sci.* **33** 207–212.

Phillips Petroleum Co. (1971) Viscosity Index Improvers. US Patent 3,554,911.

Rein, S.W., Randall, N.P., Marshall, H.T. and Lewis, B.J. (1977) A mathematical technique for comparing shear stability in bench tests and service. *SAE Paper 770633.*

Roberts, D.C. (1990) Review of oil consumption aspects of engines. *7th International Colloquium, Automotive Lubrication*, Bartz, W.J. (ed.) Technische Akademie Esslingen, Ostfildern, 13.2-1 to 13.2-15.

Rohm and Haas Co. (1937a) Composition of Matter and Process. US Patent 2,091,627.

Rohm and Haas Co. (1937b) Process for Preparing Esters and Products. US Patent 2,100,993.

Rohm and Haas Co. (1970) Lubricating Oils and Fuels Containing Graft Copolymers. US Patent 3,506,574.

Rohm and Haas Co. (1979) Polyolefin Graft Copolymers. US Patent 4,146,489.

Rohm GmbH (1973) Graft Copolymeric Lubricating Oil Additives. US Patent 3,732,334.

Rohm GmbH (1979) Lubricating Oil Additives. US Patent 4,149,984.

Rohm GmbH (1981) Lubricating Oil Additives. US Patent 4,290,925.

Rosenberg, R.C. (1975) The influence of polymer additives on journal bearing performance. *SAE Trans.* **84** Paper 750692.

Rubin, I.D. (1987) Polymers make the grade. *Chemtech* **17** (10) 620–623.

Selby, T.W. (1958) The non-Newtonian characteristics of lubricating oils. *Trans. ASLE* **1** 68–81.

Shell International Research (1985) Hydrogenated Modified Star-shaped Polymers. UK Patent Application GB 2,144,430 A.

Shell Oil Co. (1960) Oil-Soluble Copolymers of Vinylpyridine for Use in Lubricating Oil. US Patent 2,957,854.

Shell Oil Co. (1966) Lubricating Composition Containing Non-Ash Forming Additives. US Patent 3,249,545.

Shell Oil Co. (1973) Lubricating Compositions. US Patent 3,772,196.

Shell Oil Co. (1977) Lubricating Composition Containing Hydrogenated Butadiene–Isoprene Copolymers. US Patent 4,032,459.

Shell Oil Co. (1978) Hydrogenated Star-Shaped Polymer. US Patent 4,116,917.

Shell Oil Co. (1985) Process for the Preparation of Oil-Soluble Hydrogenated Modified Star-Shaped Polymers. US Patent 4,557,849.

Shell Oil Co. (1988) Polymeric Viscosity Index Improver and Oil Composition Comprising the Same. US Patent 4,788,361.

Smith, M.F., Jr., Tunkel, N., Bachman, H.E. and Fernandez, W.J. (1976) A new look at multigrade diesel engine oils. *SAE Paper 760558.*

Société Nationale Elf Acquitaine (1979) Lubricating Oil Compositions Containing Copolymers of Olefins or of Olefins and Non-Conjugated Dienes with Unsaturated Derivatives of Cyclic Imides. US Patent 4,139,417.

Spearot, J.A., ed. (1989) High-temperature, high-shear oil viscosity—measurement and relationship to engine operation, *ASTM STP 1068*, ASTM, Philadelphia, PA.

Spiess, G.T., Johnston, J.E. and Ver Strate, G. (1986) Ethylene propylene copolymers as lube oil viscosity modifiers. *5th International Colloquium, Additives for Lubricants and Operational Fluids*, Vol. 11, Bartz, W.J. (ed.) Technische Akademie Esslingen, Ostfieldern, 8.10-1 to 8.10-11.

Stambaugh, R.L. (1984) Low temperature pumpability of engine oils. *SAE Trans.* Paper 841388.

Stambaugh, R.L. and Kopko, R.J. (1973) Behavior of non-Newtonian lubricants in high shear rate applications. *SAE Trans.* **82** Paper 730487.

Stambaugh, R.L. and O'Mara, J.H. (1982) Low temperature flow properties of engine oils. *SAE Trans.* **91** Paper 820509.

Stambaugh, R.L., Kopko, R.J. and Franklin, T.M. (1972) Effect of unleaded fuel and exhaust gas recirculation on sludge and varnish formation. *SAE Trans.* **81** Paper 720944.

Stambaugh, R.L., Kopko, R.J. and Roland, T.F. (1990a) Hydraulic pump performance—a basis for fluid viscosity classification. *SAE Paper 901633.*

Stambaugh, R.L., Machleder, W.H. and Kopko, R.J. (1990b) Heavy-duty diesel engine oil pumpability at low temperature—a study of key variables. *Proceedings of the Japan International Tribology Conference*, Nagoya, Japan, JSLE, in press.

Standard Oil Co. (1975) Oil-Soluble Lubricant Bi-Functional Additives from Mannich Condensation Products of Oxidized Olefin Copolymers, Amines and Aldehydes. US Patent 3,872,019.

Stewart, R.M. and Selby, T.W. (1977) The relationship between oil viscosity and engine performance—a literature search. *SAE Trans.* **86** Paper 770372.

Talbot, A.F., Wright, W.A. and Morris, H.I. (1973) A bench scale test for shear stability of multigrade engine oils. *SAE Paper 730485*.

Tamai, T., Toshikazu, Y. and Mogi, M. (1977) Flow activation quantities of VI-improver-blended mineral lubricating oils. *Bull. Jpn. Pet. Inst.* **19** (2) 131–134.

Texaco Inc. (1977) Multifunctional Tetrapolymer Lube Oil Additive. US Patent 4,021,357.

Texaco Inc. (1988) Diesel Lubricating Oil Consumption Control Additives. European Patent 0 302 239.

Van Horne, W.L. (1949) Polymethacrylates as viscosity index improvers and pour point depressants. *Ind. Eng. Chem.* **41** (5) 952–959.

Ver Strate, G. and Struglinski, M.J. (1989) Polymers as lubricating oil viscosity modifiers. *Polym. Mater. Sci. Eng.* **61** 252–258.

von Petery, C., Kruse, H. and Bartz, W.J. (1978) Influence of the viscosity of polymer containing engine oils on the startability of engines. *SAE Paper 780370*.

Weast, R.C., ed. (1989) *CRC Handbook of Chemistry and Physics*, 70th Edition. CRC Press, Boca Raton, FL, F-206-207.

Wilson, T.C. (1970) Oil Mist Lubrication Process and Novel Lubricating Oil Composition for Use Therein. US Patent 3,510,425.

Wright, B. and duParquet, J.P.R. (1983) Degradation of polymers in multigrade lubricants by mechanical shear. *Polym. Degrad. Stab.* **5** 425–447.

Wunderlich, W. and Jost, H. (1978) Polymer stability in engines. *SAE Paper 780372*.

6 Miscellaneous additives

J. CRAWFORD and A. PSAILA

6.1 Friction modifiers

6.1.1 *Introduction*

Friction modifiers or friction reducers have been established for several years. Originally the application was for gear oils, transmission fluids, slideway lubricants (for elimination of stick-slip from slideways) and multipurpose tractor fluids for wet brakes, etc., in fact anywhere where controlled movement is required. Following the Gulf crisis in 1978, fuel economy for all vehicles became an international issue. This led to the introduction of friction modifiers into automotive crankcase lubricants, the objective being to improve fuel efficiency via the crankcase lubricant. In the US, additional pressure was imposed on the oil and additive companies by the introduction of corporate average fuel economy (CAFE) regulations. (CAFE certification is a US Government requirement for engine manufacturers.) Reproducible measurement of fuel saving due to the lubricant was set as a target for the industry. This led to the development of the Sequence VI test for fuel consumption, which requires that the candidate crankcase lubricant results in decreased fuel consumption relative to a reference oil. To satisfy the sequence VI test the lubricant must show decreased fuel consumption relative to a reference lubricant in a General Motors 3.8 litre stationary bed engine. The EC adopted a fuels emission test for determining fuel economy. This is the EC 15 test sequence. Fuel economy testing is currently under review in both the US and Europe.

6.1.2 *Friction and lubricating regimes*

Friction is defined as 'the resistance a body meets with in moving over another body in respect of transmitting motion'. Friction coefficient is defined as

$$\frac{\text{Friction force}}{\text{Normal force}}$$

An irregular rock dragged over an unlubricated irregular surface would have a friction coefficient between 0.5 and 7. The coefficient of friction of dragging a

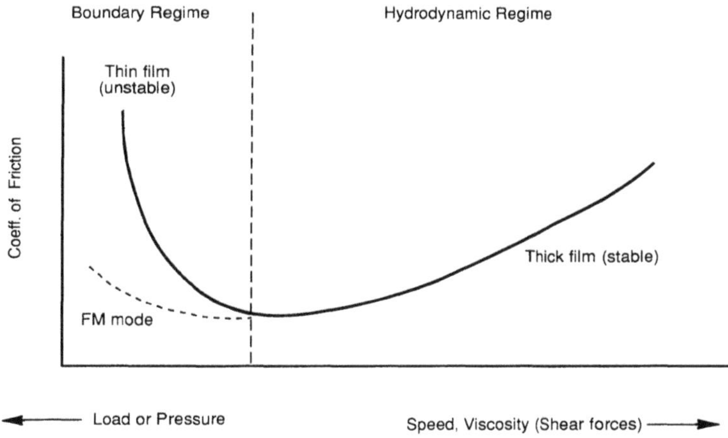

Figure 6.1 Effect of viscosity, speed and load on friction and lubrication regime.

flat stone over a flat rock is between 0.1 and 0.2. In the lubricated surface situation the coefficient of friction will be determined by the lubrication regime. In simple terms there are two lubricant regimes, hydrodynamic regime (thick lubricant film) and boundary regime (thin lubricant film) (see Figure 6.1). Friction modifiers act most successfully in the boundary regime. The friction coefficient of hydrodynamic lubrication is between 0.001 and 0.01. This is effectively hydroplaning.

Friction coefficients determined via the pin-on-disc or the modern technique of the Cameron-Plint machine depend on the friction between two smooth metal surfaces. The typical surface roughness is about 5 microinches. Such techniques investigate friction modification (FM) as a function of

pressure
temperature
surface finish
surface material identity
chemistry of the FM

Measurements from such techniques give friction coefficients of fully formulated lubricating oils of approximately 0.12 to 0.18. When friction modifiers are added to the lubricant the friction coefficient drops to between 0.08 and 0.06. The latter number is dependent on the concentration of the FM and its effectiveness.

6.1.3 *Friction modifier mechanisms*

Friction modifiers dissolved in oil are attracted to metal surfaces. The resulting adsorption forces to the metal surface are very strong, and can be as

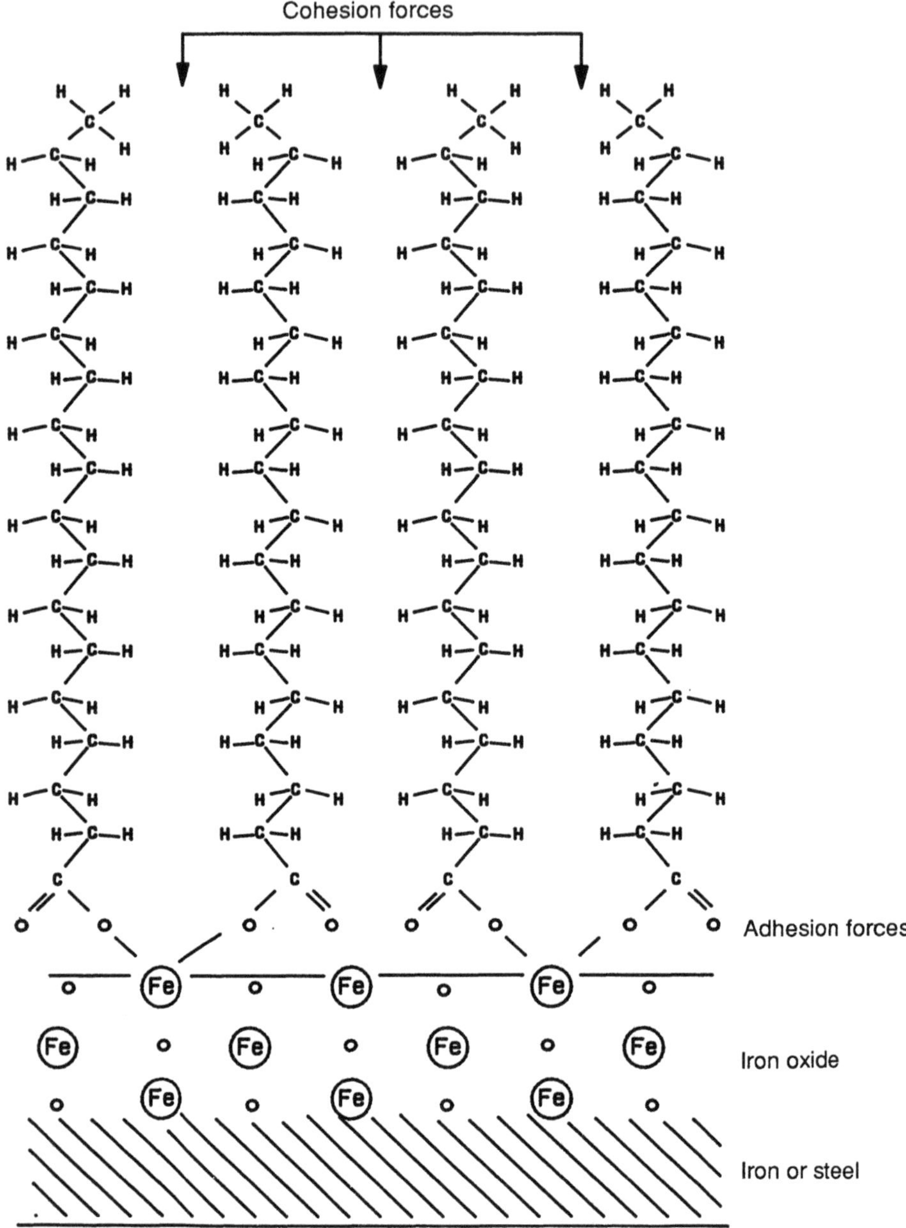

Figure 6.2 Chemisorption of stearic acid on iron–iron oxide substrate.

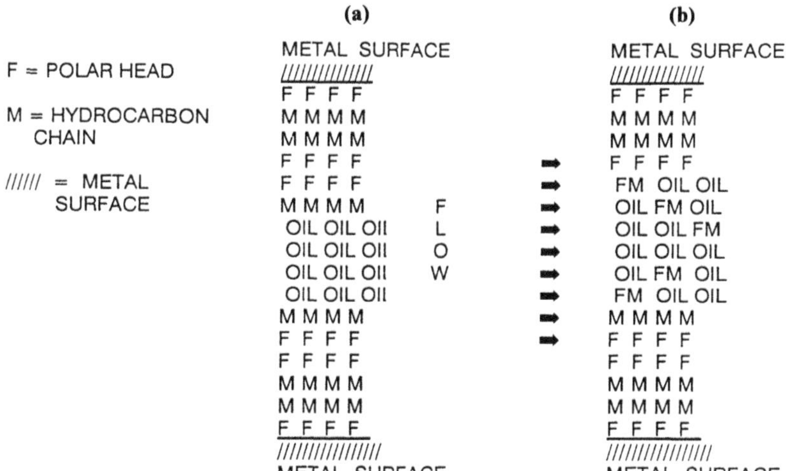

Figure 6.3 Schematic illustration of the adsorption of friction modifiers (FMs) to metal surfaces: (a) stationary situation; (b) application of force leading to shear and appearance of FMs in oil.

high as 13 kcal/mole. The polar head of the FM is attracted to the metal surface and the long hydrocarbon tail is left solubilised in the oil. The anchoring of the molecule to the metal surface results in the hydrocarbon tail being perpendicular to the surface. At normal concentrations the FM's hydrocarbon tails will line up with each other and, through hydrogen bonding and Debye orientation forces, their polar groups can be attracted with a force up to 15 kcal/mole in dimer clusters (see Figure 6.2). Van de Waals forces will cause the molecules to align themselves such that they form multimolecular clusters which are parallel to each other. This orienting of the adsorbed layer can also induce further clusters to position themselves with their respective terminal methyl groups stacking on to the methyl groups of the adsorbed molecules being formed. This is similar to iron filings in a magnetic field lining up along a particular force field. Such a layer of molecules is hard to compress but very easy to shear, thus it is easy to appreciate the slippery nature of the metal surface due to FMs.

The interactive forces between methylene carbon atoms and between methyl radicals are weak but positive (estimates of < 1 kcal/g have been recorded). Consequently, it can be seen that, under normal stress conditions, such forces will lead to simple breaking and shear. On a molecular scale it is easy to imagine that for two metal surfaces moving against each other then both surfaces would have the FM adsorbed to them.

The stationary situation is shown in Figure 6.3a. When force is applied to the adsorbed surface layers then they shear easily and appear in the oil (Figure 6.3b). A detailed structural representation is shown in Figure 6.4.

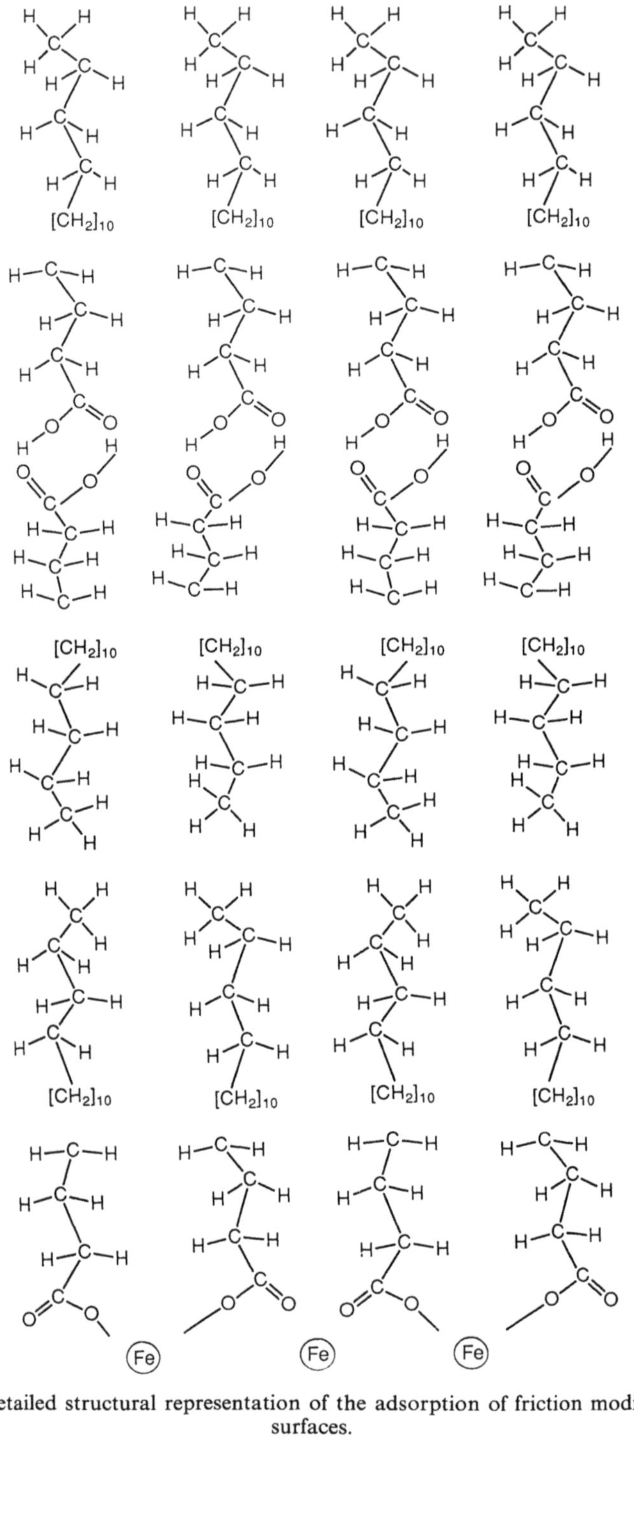

Figure 6.4 Detailed structural representation of the adsorption of friction modifiers to metal surfaces.

6.1.4 *Chemical aspects of friction modifiers*

As previously described, friction modifiers are generally long slender molecules. They normally have a straight hydrocarbon chain consisting of at least ten carbon atoms. Hydrocarbons derived from natural products are ideal for such applications. The polar head group is the dominant factor in the effectiveness of the molecule as a friction modifier. Such polar groups consist of

- Carboxylic acids or derivatives (including their salts)
- Phosphoric or phosphonic acids and their derivatives
- Amines, amides, imides, and their derivatives.

Alcohols and mercaptans polar groups are generally poor adsorbers in this situation.

6.2 Pour point depressants

The pour point of an oil is determined by the ASTM D97 pour point test. In summary the method records the temperature of the oil 3 °C above the point at which the oil will not move when tipped out of the horizontal.

The fluidity of an oil in an engine is very important under all circumstances. When starting up an engine from cold it is important that the mechanical parts move freely. Failure to do so will lead to excessive wear with the engine becoming inefficient to the point of being inoperable. The main cause is generally lack of sufficient lubrication.

6.2.1 *Low temperature operations*

When cooled to low temperatures, lubricating oil can undergo a number of changes: (i) solidification, (ii) solidification with the formation of a precipitate of macrocrystals of paraffin, and (iii) solidification with the formation of microcrystals which swell giving a crystalline structure which traps the remaining oil. The circumstances under which this occurs are dependent on the thermal history, the cooling rate, and on the composition of the lubricating oil. Naphthenic oils solidify forming an opaque solid phase. They have naturally low pour points. Paraffinic oils of high normal paraffins tend to produce relatively strong structures made up of large (macro-) crystals. Isoparaffins give smaller (micro-) crystals of a weaker structure in a solidified oil. If the cooling rate is rapid then crystal growth may be rapid and the oil will not remain fluid even on agitation. If a microcrystalline wax is formed (such deposits normally take place at about 10 °C) then the crystals can swell, behaving like a sponge with the absorption of free oil. This results in a restriction in the flow of the oil. Rapid cooling of an oil can produce a hard gel which is easily sheared. Motor oils generally use a mixture of different oils and consequently exhibit a range of behaviours. Pour point depressants are used to improve the low temperature characteristics.

6.2.2 *Mechanism of pour point depressants*

Pour point depressants act through surface adsorption on to the wax crystals. The resulting surface layer of pour point depressant inhibits the growth of the wax crystals and their capacity to adsorb oil and form gels. In the absence of long interlocking crystals or swollen particles, oil can move freely through any solid wax particles that are present.

The limitations of pour point depressants are controlled by the nature of the lubricating oil and the concentration of the pour point depressant. The effect of pour point depressant varies widely with respect to the oil but they are most effective with thinner oils such as SAE 10, SAE 20, and SAE 30 grades. With SAE 50 grade oils only small effects are seen. Different types of pour point depressant also have different efficiencies, with the maximum effect occurring at an optimum concentration level. Above this optimum level there is usually a visible effect on viscosity at higher temperatures. Typical levels of application in commercial oils are 0.1 to 1.0%.

In the winter of 1980–1981 problems with lubricating oil pumpability occurred. In the following winter it was reported that, in addition to pumpability problems, oils would not flow out of containers. The problem was suspected to be the use of olefin copolymers (OCP). The OCPs contained a small amount of crystallinity with the result that when cooled or left overnight at very low temperatures, the oil gelled. With no oil in circulation via the oil pump the camshaft sustained catastrophic wear. Special pour point depressants were developed to counteract these problems. This experience led to the development of the mini rotary viscometer (MRV) and the test method ASTM D3829. The mini rotary viscometer predicts the low temperature pumpability of an engine oil. The test is incorporated into SAE J300 (SAE, 1989). The cooling curve for the MRV simulates the cooling of the oil in the sump when the engine is shut down at the test temperature.

6.2.3 *Pour point depressant additives*

There is a range of pour point depressant additives of different chemical species.

(i) *Polymethacrylates*

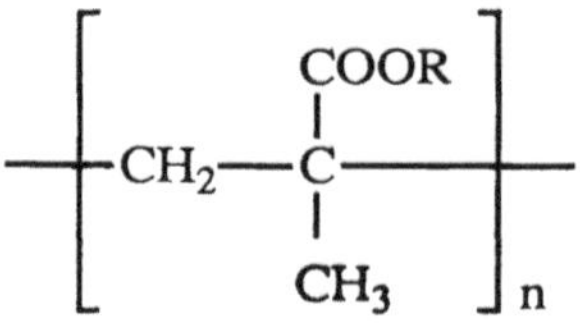

$$\left[-CH_2-\underset{\displaystyle CH_3}{\overset{\displaystyle COOR}{\underset{|}{\overset{|}{C}}}}- \right]_n$$

These are the most widely used pour point depressants. R in the ester has a major effect on the product, and is usually represented by a normal paraffinic chain of at least 12 carbon atoms. This ensures oil solubility. The molecular weight of the polymer is also very important. Typically these materials are between 7000 and 10 000 number average molecular weight. Commercial materials normally contain mixed alkyl chains which can be branched.

(ii) *Polyacrylates*

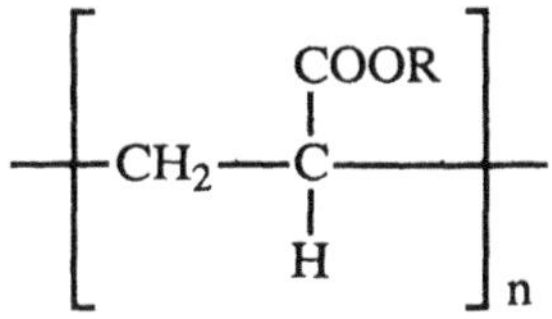

$$\left[-CH_2-\underset{H}{\overset{COOR}{\underset{|}{\overset{|}{C}}}}- \right]_n$$

These are very similar in behaviour to the polymethyacrylates.

(iii) *Di (tetra paraffin phenol) phthalate*

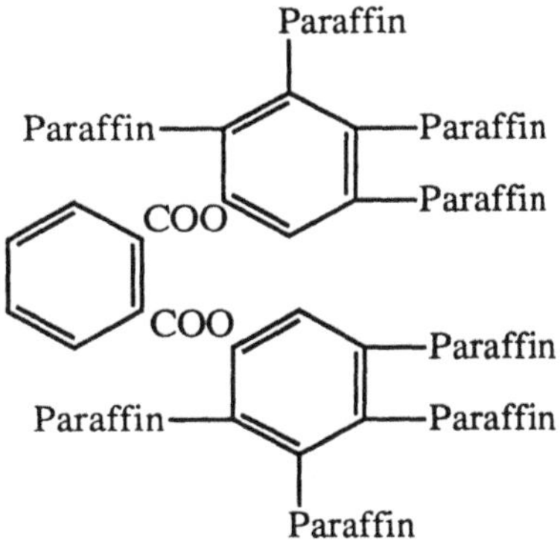

Paraffin
Paraffin
Paraffin
Paraffin
COO
COO
Paraffin
Paraffin
Paraffin
Paraffin

(iv) *Condensation products of tetra paraffin phenol*

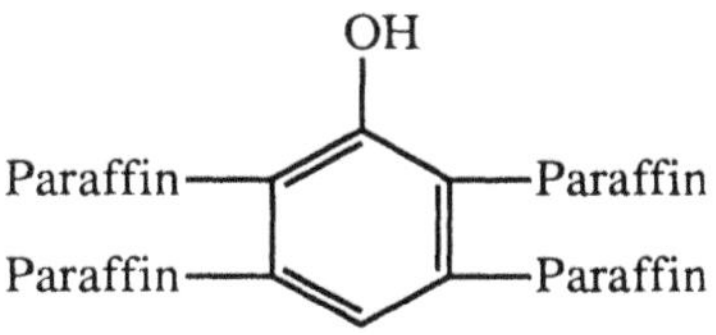

OH
Paraffin
Paraffin
Paraffin
Paraffin

(v) *Condensation product of a chlorinated paraffin wax with naphthalene*

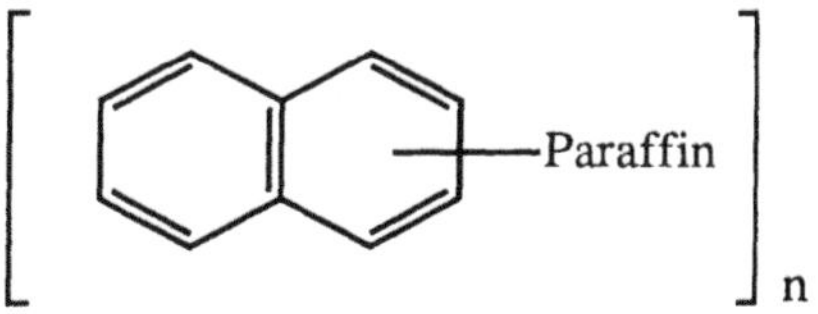

Paraffin
n

6.3 Demulsifiers and antifoams

6.3.1 *Introduction*

During normal operation internal combustion engines subject lubricants to considerable heating and mechanical shearing. In addition to the primary function of lubrication therefore, oils must also contend with various combustion products—mainly acidic gases, moisture and carbon particles—which could harm the engine. For this reason dispersant and detergent additives are incorporated into the lubricant formulation. These surfactants and other additives, such as viscosity index improvers, can have a stabilising effect on foams and emulsions that may form under the high shear regime. Emulsions and foams can seriously impair the effective lubrication of an engine by starvation of the lubricant or blockage of oil ducts.

6.3.2 *Emulsion stabilisation*

The presence of fortuitous water in the lubricant will readily give rise to the formation of water-in-oil emulsions stabilised by dispersant and detergent surfactants. Dispersants typically consist of a long hydrocarbon backbone (e.g. polyisobutene of molecular weight > 900) and a succinimide head group formed by reaction with a polyamine (e.g. tetraethylenepentaamine). Detergents, on the other hand, consist of a $CaCO_3$ core surrounded by an alkyl-aryl sulphonate or a sulphurised alkyl-phenol shell. Clearly either type of product is able to form strongly anchored interfacial layers which confer a high degree of steric stabilisation to the emulsions. Other polar materials, e.g. soot and oxidation products, tend to accumulate at the interfaces. Should the water droplets come into close proximity, they are kept apart by the repulsive interaction of the long oleophilic segment of the surfactants.

6.3.3 *Demulsification*

Satisfactory resolution of the emulsion depends on the ease with which the stabilising factors can be neutralised. A combination of mechanical and chemical means is frequently employed. Demulsification may be considered as a series of processes. In the first instance the droplets must approach one another and be allowed to form a loose aggregate. At this point it is important to 'sweep up' all the small droplets before the final stage, coalescence, progresses too far. Once coalescence occurs there is a dramatic reduction in the interfacial area resulting in the concentration of solids and other debris which have accumulated at the interfaces. This must be dealt with before the process can be completed. Once a distinct water phase is formed it can be removed by draining.

The requirements for successful demulsifiers are that they should displace the stabilising surfactants from the interface and disperse them in the bulk phase. Typical demulsifier formulations comprise more than one component. These must be attracted to the interface and therefore have limited solubility in the bulk phase. Their greater affinity to the water–oil interface ensures that the stabilising molecules are displaced. Another important feature is the reduced inter-droplet repulsion exhibited by the demulsifiers. This allows the droplets to come closer together, forming an aggregation of droplets separated only by thin layers of oil. The rheological behaviour and tension of the thin layer are critically important and determine the ease with which film rupture ensues. This stage is typified by a decrease in interfacial tension and surface area. After coalescence, the interfacial tension recovers its original value.

Mechanical aids to resolution of an emulsion are important. Although no chemical assistance is required to form a hard sediment of closely packed emulsion, stubborn emulsions do require the application of a chemical to bring about the final film-rupture stage.

6.3.4 *Demulsifiers*

It is often said that the skill of formulating a demulsifier is a 'black art'. The selection of an effective product relies on a 'trial and error' approach, supplemented by the selector's knowledge of what has worked before. The test procedure requires the production of an emulsion under reproducible conditions. The rate of water coalescence is taken as a measure of the effectiveness of the product.

A wide range of products, including every class and type of surfactant, has been found to exhibit demulsification properties. The most commonly used products in lubricant formulations contain anionic surfactants (e.g. alkylnaphthalene sulphonates). Also used are nonionic alkoxylated alkylphenol resins and block copolymers of ethylene oxide and propylene oxide.

6.3.5 *Foam stability*

The same surfactants that are responsible for the formation of stable emulsions almost certainly play an important role in the stabilisation of foams. Foams consist of a closely packed array of gaseous polyhedral cells separated by thin liquid films. The pressure differences across the films are similar; however, at the junction of the cells curved regions of low pressure (called the Plateau borders) exist. These form the causeway for the liquid drainage. Gravitational liquid drainage is the prime mechanism of film rupture. This is directly related to bulk viscosity. The liquid film can be considered to consist of a sandwich of surfactant molecules at the gas/liquid boundary separated by a layer of the bulk fluid. Clearly a point will be reached at which the monomolecular surfactant layers come close enough to create a repulsive force which will delay film rupture.

Mechanical foam breakers and chemical methods are employed to destroy unwanted foams. The strategy is similar to that discussed above for dealing with emulsions. The stabilising monomolecular surfactant must be sufficiently displaced to create a hole in the liquid layer. Anti-foams are typically low solubility surfactants.

6.3.6 *Antifoams*

The most important antifoam chemicals used to deal with non-aqueous foams are polydimethylsiloxanes. These are available in a range of molecular weights (from low viscosity fluids of 1000 cSt to several million cSt). Silicone fluids of the correct molecular weight have limited solubility in oil and a surface tension of 21 mN/m, lower than that of hydrocarbon oils. Fluorosilicones exhibit even lower surface tension but are considerably more expensive.

A variety of other materials including alkoxylated aliphatic acids, various polyethers (e.g. polyethylene glycols), branched polyvinylethers, and polyalkoxyamines, have been claimed to exhibit anti-foam properties in lubricating oils.

6.3.7 *Foam studies*

Characterisation of foams employs one of two techniques, static or dynamic. In the static method a foam is generated by sparging a gas into a liquid under controlled conditions. The foam decay is then monitored against time. The half-life of the foam is termed the *foam lifetime*. In the dynamic method the foam is generated continuously and the equilibrium volume measured. With care, a linear relationship between foam volume and gas velocity can be determined. The gradient of this linear response is the foam lifetime. Not surprisingly, foam lifetime and viscosity show an identical dependence on temperature.

Traditional methods used for the study of foams employ one of three methods of generating a foam, namely single capillaries, sintered glass spargers and diffuser stones. Single capillaries give uniform bubbles but are relatively slow. The other methods generate a rapid foam with considerable turbulence, which adversely affects the reproducibility of the results. A turbulent-free foam can be created by using a uniform mesh to generate a foam of uniform bubble size.

Direct measurements of single film thickness using optical methods can yield useful information on the critical film thickness.

6.4 Corrosion inhibitors

A corrosion inhibitor is defined (ISO 8044) as 'a chemical substance which decreases the corrosion rate when present in the corrosion system at a suitable concentration without significantly changing the concentration of any other corrosive agent'. This is a useful definition as it enables us to distinguish between the function of antioxidants and overbased detergents on the one hand, and corrosion inhibitors on the other.

Antioxidants and detergents In the course of its normal usage a lubricant is susceptible to oxidation. The process of refining strips the lubricant of its natural antioxidants (chiefly polycyclic aromatics and sulphur and nitrogen heterocyclics). Oxidation results in the formation of undesirable peroxides and acids. Antioxidants (e.g. hindered phenols and zinc dialkyldithiophosphates (ZDDPs)) are added to prevent the formation of these corrosive products. In addition, moisture and acidic combustion by-products invariably enter the crankcase. To protect the engine from corrosive wear by mineral acid formed in this way overbased detergents are incorporated in the lubricant.

Corrosion inhibitors The conditions that influence the onset of corrosion are the entrainment of atmospheric oxygen, moisture from the combustion of fuel, and stop-start running coupled with temperature cycling. In the marine diesel engine, the problem is exacerbated by contamination with fortuitous saline. Corrosion inhibitors are added specifically to cope with this electrochemical process. These additives operate by creating a physical barrier, in the form of a dense hydrophobic, monolayer of chemisorbed surfactant molecules, which prevent access of the water and oxygen to the metal surface.

6.4.1 *The process of corrosion*

An electrochemical couple is formed between the ferrous metal and its oxide layer, invariably present. The anodic partial reaction generates electrons:

$$Fe \longrightarrow Fe^{2+} + 2e^- \qquad (6.1)$$

In neutral or alkaline conditions prevalent in the lubricant environment, the corresponding cathodic partial reaction is:

$$O_2 + 2H_2O + 4e^- \longrightarrow 4OH^- \qquad (6.2)$$

Further oxidation results in the formation of hydrated ferric oxide or Fe(III) hydroxide, i.e. rust. The corrosion potential (E_c) and corrosion current (I_c) for the cathodic and anodic reaction can be represented by an Evans type polarisation diagram (Figure 6.5). Corrosion inhibitors interfere with the anodic or the cathodic partial reaction, or with both, resulting in a reduction in the corrosion current.

6.4.2 *Corrosion inhibitors*

In the selection of an appropriate corrosion inhibitor, an important con-

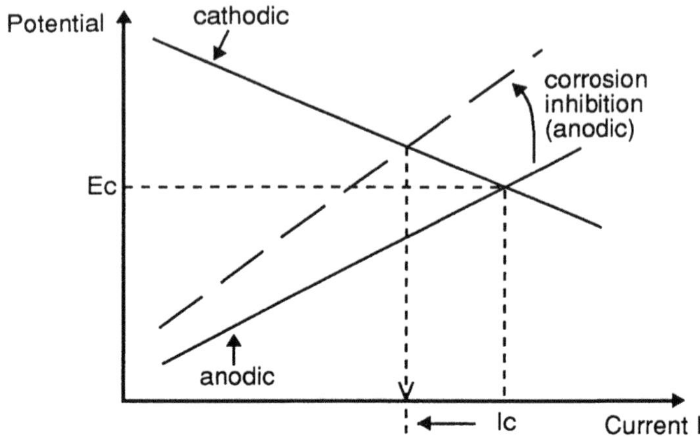

Figure 6.5 Evans type polarisation diagram.

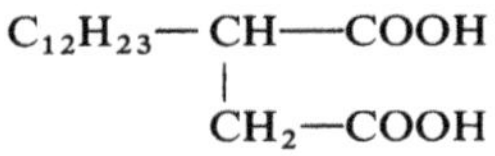

$$
\begin{array}{l}
C_{12}H_{23}-CH-COOH \\
\qquad\quad\;\; | \\
\qquad\quad\; CH_2-COOH
\end{array}
$$

Figure 6.6 Dodecylsuccinic acid.

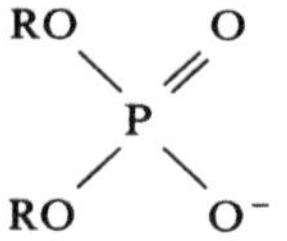

$$
\begin{array}{ccc}
RO & & O \\
& \diagdown \;\; /\!/ & \\
& P & \\
& / \;\; \diagdown & \\
RO & & O^-
\end{array}
$$

Figure 6.7 Phosphate esters (R = alkyl).

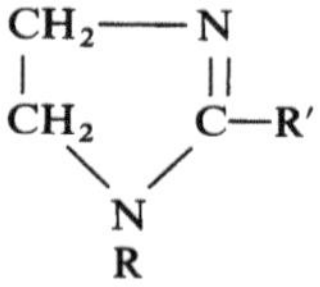

$$
\begin{array}{ccc}
CH_2 & - & N \\
| & & \| \\
CH_2 & & C-R' \\
& \diagdown \;\; / & \\
& N & \\
& R &
\end{array}
$$

Figure 6.8 Alkyl imidazoline.

sideration is the problem of adverse competition with other additives designed to adsorb on the liquid–metal interface. Extreme pressure and/or anti-wear agents compete for the same sites as the corrosion inhibitor. Fatty amines are good corrosion inhibitors in this type of environment. However, their adverse effect on the performance of ZDDP additives often prohibits their use. Half esters or amides of dodecylsuccinic acid (Figure 6.6) and phosphate esters (Figure 6.7) or thiophosphates are frequently employed. A

$$\begin{array}{l} \quad CH_3 \\ \quad | \\ RN{-}CH_2{-}COOH \end{array}$$

Figure 6.9 Sarcosines.

combination of inhibitors is sometimes used, for example fatty carboxylic acids or the dimer/trimer analogues of the unsaturated acids (e.g. oleic) used in conjunction with an amine, (e.g. ethanolamine or alkyl amine) or an amide (e.g. alkyl imidazoline (Figure 6.8)) and sarcosines (Figure 6.9). Shorter chain carboxylic acids and amines are used as volatile corrosion inhibitors.

6.4.3 *Corrosion testing*

The ultimate evaluation of the corrosion protection afforded by a particular lubricant formulation is the Sequence IID engine test. This test simulates short trips in winter conditions. Such trips promote corrosion/rust on the engine components due to formation of condensed water in the engine. The test lasts for 32 hours, after which a rust rating is determined.

A variety of bench tests, which attempt to evaluate the oxidative stability and/or the corrosion protection of a particular formulation, exist. These invariably involve a coupon test in which a coupon of one or more metals (chiefly copper and steel) is subjected to wet conditions for a fixed period at an elevated temperature. The assessment is carried out on the basis of weight loss and appearance of the coupon.

Reference

SAE (1989) Engine oil viscosity classification. *SAE J300*, June 1989.

7 The formulation of automotive lubricants

A.J. MILLS and C.M. LINDSAY

7.1 Introduction

Producing a lubricant can be a simple process. Refineries produce lubricating basestocks of an appropriate viscosity. Additive companies produce chemical components and 'additive packages' which contain a specially developed blend of these components. Chemical companies produce polymers which help to maintain the viscosity of the oil at elevated temperatures. Sometimes two or three of the above are produced by the same company. The information on which basestocks, additives and polymers can be used together can be obtained from the additive companies who attempt to play a co-ordinating role in the lubricating industry, supplying additive packages and recommending the basestocks and polymers required for the formulation. Contacting an additive company can be the first step in the formulation of a lubricant. Following this, the necessary business agreements or contracts can be decided, arrangements made for the additive package, basestocks and polymer to be supplied, blended, and packaged, a quality control system developed, and the result is a lubricant that can be sold and used. This product would be a generic lubricant suitable for some but not all applications. However, this is unlikely to be what the lubricant supplier, the distributor to the consumer, or the consumer actually wants.

7.2 What type of lubricant is wanted?

The first step in developing a new lubricant is to decide what type of lubricant is desired. This sounds obvious but it is easy to forget in the enthusiasm to start putting formulations together in a can. The type of lubricant is defined by its application. A typical application, one of many automotive applications, is an engine oil for a gasoline engined passenger car.

Passenger car engines may run on either gasoline or diesel fuel, may be carburettored or use one of a range of fuel injection systems, may be turbocharged or super-charged, or they may even be two-stroke engines (which are rare outside Eastern Europe). Cars also need transmission lubricants for the gearbox. This may be transverse or longitudinal, may have highly loaded or highly offset gears, may be manual or automatic and will have a range of clutches and synchromesh cones present. After the transmission comes the axle which may be a front wheel drive transaxle or a rear axle, often with offset gears and possibly with a limited slip differential and a four wheel drive facility. Cars also need brake fluid, and often clutch fluid and power steering fluid. Automotive lubricants are also required for the commercial vehicle sector, which uses large direct injection diesel engines and has more highly loaded transmissions and axles, and buses which are designed to operate at low speeds. There are the off-highway and agricultural vehicles which may have complex hydraulic systems, power take off systems and oil-immersed brakes. Motorcycles need special lubricants particularly if they are two-stroke, involving deliberate addition of lubricant to the combustion chamber and increasingly high power outputs.

These examples illustrate the many different performance aspects of a lubricant. An interesting corollary of this is that there is also no such thing as a bad oil, just an inappropriate one. If an excellent transmission oil is put into an engine then severe damage will probably occur very rapidly. The oil is not a 'bad' oil, just disastrously inappropriate. The above list of applications is also an oversimplification in that it is often necessary to have many different lubricant variants for each application. For example, if a high viscosity engine oil that is designed for use only in equatorial climates is used in a cold climate then premature engine failure is the likely result. It is even the case that an oil that is suitable for one engine may be unsuitable for a different engine that serves a closely similar purpose. For example, the top performance heavy duty diesel engine lubricants marketed in Europe are not the best lubricants to use in North American manufactured heavy duty diesel engines. The problems encountered where both such engines are used in one truck fleet will be discussed in more detail later.

7.3 Why there are so many types of lubricant

Originally a lubricant was simply the non-volatile or vacuum distilled base-stock that came out of the oil refineries. There were several viscosities that were used according to the season of the year. The unstabilised lubricant had to be changed frequently and this was normally covered by lubricant changes between seasons: a light oil was used in winter and a heavy oil used in summer. The same oils were suitable for virtually all applications. As automotive technology developed, speeds, loads and temperatures increased and there were problems in different parts of different vehicles at different times. These problems could, of course, be overcome by redesigning the vehicle hardware but that did not help the vehicles that were produced before the problems were identified. The most cost-effective solution was to modify and 'strengthen' the lubricant enabling it to be used for a specific application in both new and existing vehicles. In this way engine oils were developed which had better resistance to oxidation, better wear protection, better engine cleanliness, better resistance to corrosion, etc. Similarly gear oils were developed which had improved wear protection, axle oils were developed for highly loaded sliding contacts in offset gears and so on. This method of lubricant development has been referred to as the 'basketful of broken parts' method because lubricants were only changed when there was a hardware problem in the field. It is easy to criticise this method as being reactive rather than pro-active and as not being in line with today's view of quality; however, with hundreds of vehicle manufacturers each making a small number of vehicles it was not possible to foresee every condition to which a vehicle might be subjected and mistakes were bound to occur. In addition, reliability expectations were much lower than they are today.

The result of this has been the need for different lubricants for different applications and even different lubricants for the vehicles of different manufacturers. This makes the development of new lubricants extremely expensive particularly if the new lubricant is radically different from current established technology. As well as developing the new lubricant it is necessary to prove its performance through a wide range of hardware tests as specified by the hardware manufacturer, by the industry bodies, and by the end-user. Field trial demonstrations may also be required. By the time a new range of engine lubricants has been developed, proven and launched, the overall investment may be approaching £5 million.

It is clear, therefore, that before any formulation work is begun it is necessary to carry out a detailed commercial review of the performance that is or is not required. Several commercial options are likely to be generated. The technical options and feasibilities of producing such lubricants can be fed back into the commercial options and through an iterative process it should be possible to identify a performance profile for the oil that has a chance of being achievable at certain development and formulation costs, and within a certain timescale. This procedure is extremely complex as it continually generates new ideas but the time spent in identifying the performance required of a new lubricant is time well spent. It is not uncommon for this phase to take as long and to use up as much human resource as the formulation programme itself.

7.4 Multifunctional lubricants

To a vehicle operator the ideal lubricant may well be one that can be used with confidence in all major applications in all vehicles. Special lubricants have been developed for some applications, for example universal plant oils for the construction industry, super tractor universal oils (STUOs or STOUs) for agricultural use, and heavy duty diesel engine oils with moderate transmission and clutch performance for off-highway applications. In each case the one oil can be used in almost every area of the vehicle and there is a simplified inventory with virtually no opportunity for misapplication. (Misapplication can be a major concern where on-site maintenance is carried out on complex equipment by personnel who have no qualifications in vehicle maintenance.) However, the price paid for such multifunctionality is the compromise between different aspects of performance because such lubricants are unlikely to meet the latest requirements in all aspects of performance. Passenger car engine requirements are the fastest changing and are the most likely to be compromised. However, if a lubricant claims to be multifunctional or universal then it is likely to be applied in any vehicle. Consumer studies have shown that most motorists believe that all engine oils are suitable for all cars and they do not read their owners' manuals or check to see if the oil is of the correct quality. In the above examples there is a fairly closed environment which limits where the lubricant will actually be used. The plant contractor, farmer or engineer may use the oil in his Jaguar, Mercedes or Range Rover, for which it is not designed, and there is a good chance that the vehicle will be none the worse for it. It would, however, be irresponsible to introduce such an oil to the open market-place as misapplication would be commonplace and field problems at a level of only 5–10% are totally unacceptable. The lubricant industry has a responsibility to protect the vehicle operator from anticipated misapplications and therefore must produce oils with a wide margin of safety in all possible areas of application. The lubricant industry does not only sell lubricants, in effect it sells protection, safety and insurance against equipment failure. This only rarely appears in the lubricant description but suitability for use is implicit in the act of supplying the lubricant. Lubricant technology is complex and responsibility for it cannot be devolved to the end-users as many of them would not have the expertise to handle it.

Although there are constraints on how universal a lubricant can be it is clear that it must have a relevant degree of universality for its area of application. Heavy duty truck engine lubrication provides an interesting example of this. Historical factors have led to engines being designed differently in North America compared to Europe. The two-stroke diesel engines built by Detriot Diesel and used particularly extensively in buses, have proven to be an attractive powerplant with a high power to weight ratio, low maintenance, and low initial cost. These engines prefer low ash oils since high ash oils can lead to exhaust port blocking in two-stroke engines. The North American lubricant industry adopted the strategy of marketing universal oils for both four- and two-stroke engines and these were low in ash compared to their European equivalents. The designers of successive generations of North American four-stroke engines used the lubricants available in the market for their developments and therefore their engines are designed around low ash oils. In Europe there has been no such constraint and the ash level of the oil has been allowed to increase in order to provide certain benefits with the result that European engines are now designed around high ash oils. Trucks and engines are exported around the world in this international business and there are truck fleets that use both types of engine. In such a fleet there is a need to satisfy both types of engine and the oil must therefore be universal. In other fleets, however, one type of engine will dominate and here there may be a benefit in supplying an oil optimised for the dominant type of engine, for example a high ash European oil may allow the operator to extend the engine oil drain interval in a European fleet, or a low ash North American lubricant may extend the engine life of a North American fleet.

The art of developing and formulating a new lubricant lies in identifying and building in all the desired performance benefits for the end-user while making it tolerant to anticipated misapplication.

7.5 Definition of lubricant performance

7.5.1 *Broad performance definition—lubricant classifications*

Once it has been decided to produce a lubricant that is, for example, suitable for passenger car engines, the next stage is to decide what type of oil this should be. In addition, as the lubricant will be sold to workshops which service passenger cars of all types, its performance will have to be communicated to the workshop owner, the engineers and the potential customer.

The performance level of the oil is specified by the car manufacturer and stated in the vehicle handbook. The viscosity of the oil that can be used, according to the climate, and also the recommended 'class' of the oil will be given. This class can be defined by a specification system defined by the car manufacturer or by generic classification systems developed by a group of car manufacturers in co-operation with the lubricant industry and its suppliers. Examples of these classifications are the API (American Petroleum Institute) system from North America and the CCMC (Comité des Constructeurs d'Automobiles du Marché Commun; now ACEA (Association des Constructeur Europeén d'Automobile)) system from Europe.

The API system is the original and best known system. It is the sole definition of oil quality specified by the North American vehicle manufacturers and is often used as a definition of acceptable quality throughout the world. It is therefore a good example to review. The API system is based on a 'tripartite' arrangement, i.e. three organisations working co-operatively. These are the SAE (Society of Automotive Engineers) who broadly represent the vehicle manufacturers, the API (American Petroleum Institute) who broadly represent the oil industry, and the ASTM (American Society for Testing Materials) who represent the scientific community in general. When the SAE demonstrates the need for improved lubricant quality the tripartite actions the ASTM to develop a test that simulates the problem and correlates with the known field performance of specified reference oils. When the ASTM has developed a test and demonstrated its repeatability, reproducibility, and relevance the SAE establishes pass/fail criteria for the acceptability of performance. The API then develops a new performance classification and communicates this to the industry. The lubricant companies can then develop and commercialise lubricants that meet and claim this new performance and the vehicle manufacturers can specify the use of this quality lubricant if the vehicle warranty is to be maintained.

The API system has served the industry for many years and is the most widely recognised definition of oil performance. API performance is stated in terms of 'S' (service) and 'C' (commercial) qualities referring to passenger car gasoline engines and heavy duty diesel engines respectively. Qualities have advanced from SA through to SG and from CA through to CE. Each successive new performance level takes priority over previous performance levels. In this way the most recent API quality oils, i.e. API SG oils, are deemed to be suitable for all gasoline passenger car engines whatever API S quality level is specified in the vehicle handbook. The API has a 'doughnut' symbol that can be licensed for use on literature and packages by applying to the API and paying a small fee. This symbol is as shown in the vehicle handbook.

Checks are made to ensure that oils using the API doughnut symbol are consistent with the performance claimed. These are carried out in North America by a team who arrange for lubricants to be purchased from retail outlets, repackaged and coded, and independently analysed for fundamental physical properties and chemical content. The results are published and reported to the API. Any oil that is inconsistent with its performance claim will be challenged. Several brands of oil have been removed from the API register in this way.

These systems define tests and performance levels that the oil must be capable of achieving if the lubricant marketer is to claim this level of performance. The systems are self-certifying in that there is no need to present the test results to anyone, in fact there is no legal need for any tests to be run providing the marketer believes that the oil will meet the requirements. In practice companies will normally run these tests on the basis of meeting their own development quality criteria or the criteria that they have established for quality standards, such as ISO 9001, under which the company operates. An exception to this may occur if there is a very minor change in the formulation, in which case a much more limited test programme would be run.

A classification system that does not start to develop a new performance level until a major field problem has been proven when using current quality lubricants results in a 'basket of broken parts' being presented by the vehicle manufacturers to the tripartite groups; i.e. the quality of the lubricant recommended in the handbook for a vehicle has been demonstrated to be inadequate in protecting the engine under some driving conditions. This is why improved, higher classification oils should be used in preference to the handbook recommendation as soon as such oils become available. There is a responsibility on the lubricant suppliers to upgrade their main brands of oil to the new performance classification as soon as possible and not to wait until the new oil quality is specified in vehicle handbooks. Unfortunately the development and formulation costs of a higher performance lubricant are always higher than for the previous quality oil and there is a temptation to continue marketing lubricants of a lower performance level and price for the cost-conscious motorist. In many cases this is a false economy to the motorist as the lower quality oil has, by definition, been shown to allow engine damage on the road. Unfortunately the standard of knowledge of the 'do-it-youself' motorist is sometimes not high enough for this to be recognised and it is assumed that if an oil is marketed then it must be of an appropriate quality. The continued marketing of historic quality oils on the basis of a need to lubricate 'classic', 'vintage' and 'veteran' vehicles that were designed to use lubricants which were only minimally additivated and that could be damaged by today's higher detergency oils is poor justification. The number of vehicles still on the road that were designed to use such lubricants is extremely small and this market is better targeted through interest groups.

The API classification system by definition meets the requirements of the American motor industry and due to the international nature of the passenger car business it is also relevant in most areas of the world. There are, however, different driving habits and financial factors around the world and these put different demands on the lubricant. The prime example is the higher speed limit conditions in Europe compared to the US. The absence of any speed restriction on German Autobahns leads to extremely severe driving conditions causing high engine temperatures which in turn affect oil consumption and engine cleanliness. This results in an unacceptable level of engine failures in the field and several European OEMs (Original Equipment Manufacturers) have set up their own approval systems. Prime areas of concern are limits on oil volatility (leading to oil starvation), limits on oil sulphated ash content (which can lead to preignition) to supplement preignition engine tests, and engine tests for high temperature engine cleanliness and piston ring sticking. A further feature of driving outside the US is the attraction of diesel fuelled engines. These operate at higher temperatures than gasoline engines and have to survive in the presence of the highly acidic blow-by gases that are produced by the combustion of the high level of sulphur present in the fuel. The reasons for the attractiveness of diesel cars in Europe compared to the US are the considerably higher prices of gasoline in Europe and the more stable European climate which reduces the problems of fuel line and filter blockages from wax in the fuel. European OEMs have therefore included performance tests in diesel engines as part of their lubricant approval systems.

It is possible to produce separate lubricants for gasoline and diesel engines but there is always a risk of incorrect application of a diesel lubricant to a gasoline engine, and vice versa, which could cause field problems. There are also benefits in having a single lubricant that can be used in both types of engines.

More recently the European OEMs have identified other areas of concern including low temperature wear, the prevention of 'black sludge', and the protection of non-metallic components (in particular oil seals). These have also found their way into OEM approval systems.

The establishment of a system which allows a lubricant to be used in some cars but not in others again leads to misapplication risks and confusion, and therefore there is an attraction in formulating oils that can be used in all cars. For this reason an international system to cover the general requirements of all the European OEMs was developed, resulting in the CCMC classification system. This system responds to the demands of the European car industry and is continually under development along the lines of the latest OEM requirements. Most European OEMs are now satisfied by and specify the CCMC quality oils, however, they often retain their own approval systems for special 'preferred' oils.

In Japan the OEMs try to maintain tight control of the lubricants used. Each OEM has its own brand of oil and the use of these is strongly recommended. These oils are often referred to as 'genuine' oils. It is not normally possible to gain international approval for such lubricants as the tests used to evaluate the oils are often both proprietary and based on specific Japanese driving conditions. Some correlation work has been carried out in exchanging international reference oils and there has been shown to be little correlation between ostensibly similar API and proposed Japan Automobile Standards Organisation (JASO) tests. It is clear that special effort must be made to formulate oils that meet both API and proposed Japenese requirements. The CCMC system has now been operating for several years and underwent a major update in 1989. The JASO classifications are being established as Japanese equivalents to the North American system and will be introduced in the next few years.

There is also now a recognised concern that the API system cannot keep up with the rate of change of engine technology that has been occurring in the US during the 1980s and is likely to continue during the 1990s. New oil classifications need to be established more rapidly. There is also a more modern understanding of the meaning of 'quality' and it is no longer acceptable to wait until field problems arise before any action is taken. A modern system must be pro-active rather than reactive. This need has been stated by the MVMA (Motor Vehicle Manufacturers Association) which represents directly the North American OEMs including some transplant companies such as the American manufacturing parts of Volvo and Honda. The MVMA is holding detailed discussions with the API and the SAE in an attempt to establish a more responsive system. There is a similar concern from the North American heavy duty diesel engine and vehicle manufacturers and a group called the EMA (Engine Manufacturers Association) has been established to represent directly the needs of the heavy duty diesel OEMs. The North American tripartite system may be due for modernisation but its product, the API classification system, remains the most important international description of oil quality.

There are other lubricant approval bodies as well as the above and these mainly have military support. The US Military have the well known specifications MIL-L 46152 and 2104 which refer to administrative and tactical vehicles respectively. There are also national specifications for European and other countries, and also NATO requirements. These specifications are sometimes demanded by purchasers who are not content with the API and CCMC self-certification system and who wish to be assured that the oil's performance has been reviewed by an independent authority.

The performance level of the lubricant is, however, not the first aspect that a consumer considers, usually the oil's viscosity is the prime concern. The viscosity of an oil is defined by the SAE J300 classification which specifies viscosity at low temperatures to define the W (winter) grade and at 100 °C to define the main grade. Oils can be formulated to contain polymers (viscosity improvers) that increase the oil's viscosity more at high temperatures than at low temperatures making the oil 'multigrade'. This gives rise to viscosity grades such as 15W–40, 20W–50, 5W–30. The selection of viscosity grade is primarily determined by climatic conditions but in the US the OEMs prefer the use of 5W–30 oils for all the most recent engines as this justifies the use of these oils for official fuel economy testing.

Based on the above information it is possible to define a lubricant in terms of broadly accepted industry criteria. For example, a 15W–40 oil meets API SG/CD, CCMC G4 and PD2, and is approved by Volkswagen and other OEMs. Such an oil is, however, very much standard. It will meet the basic service fill requirements of most cars under most conditions but, as most motorists assume that any lubricant sold will be satisfactory, this is no more than is expected. If it is intended to produce a lubricant which goes beyond the minimum level of expectations, the lubricant performance needs to be defined more strictly.

7.5.2 *Detailed performance definitions—special applications*

Formulating oils with performance above and beyond the minimum is normally expensive in terms of development cost, formulation cost and development effort. The enlightened motorist, however, increasingly finds the benefits of such oils attractive.

(i) The most common form of special oil is the 'leichtlauf' or light-running oil which was defined in Germany. Such an oil must be suitable for starting the vehicle engine at lower temperatures than a conventional oil and therefore it must be a 5W multigrade or a 10W multigrade oil. This will offer easier starting and more rapid engine lubrication during cold starts. The lubricant must also, of course, offer all the features required for high temperature driving and therefore special, more expensive, basestocks are an essential part of the formulation. This leads to an increase in the cost of the oil. The discerning motorist following the recommendations of the OEM for wide climate protection is increasingly aware of the good sense in paying extra for such oils. The intention of such lightrunning oils is also to reduce fuel consumption by speeding up engine warm-up and reducing the resistance to oil motion, particularly at low temperatures. The cost saving that accrues by reducing fuel consumption by as little as 1 or 2% is a useful bonus for the motorist and can easily make up for the extra cost of the oil. This fuel economy also contributes to a reduction in consumption of natural resources and emissions to the environment. There is, however, no European fuel economy test and therefore it is difficult to communicate the benefit to the motorist.

(ii) The fuel economy aspect of a lubricant has been developed more aggressively in the US where economic measures to reduce dependency on imported crude oil led to the establishment of legislation mandating an overall reduction of fuel consumption for each OEM's cars on an individual company basis as sold in the US. This CAFE (corporate average fuel economy) legislation established targets in terms of miles per gallon that had to be met and a timescale of when each successively severe target came into force. Failure to comply meant that the OEM had to pay a tax according to how much the company exceeded the target. This tax, sometimes referred to as a 'gas-guzzler' tax, was a powerful incentive for the OEMs to develop more fuel efficient vehicles. It was soon identified that the engine lubricant could play a role in achieving the CAFE targets and the OEMs began to design their engines and to demonstrate their fuel economy using lower viscosity and low friction (or friction modified) lubricants. It became essential to have these lubricants available in the marketplace and to have a method of proving and identifying them. A fuel economy test known as the five car test and later its replacement, known as 'sequence VI', were developed to measure fuel economy and relate it through a correlation equation to over-the-road fuel economy benefits under typical US driving conditions. Two levels of performance have now been established, tier I at 1.5% and tier II at 2.7% EFEI (engine fuel efficiency index). The nomenclature is cumbersome as it is difficult to state that fuel economy benefits will be found in all cars under all driving conditions and therefore fuel economy itself cannot be guaranteed to every single motorist. It is, however, legitimate to state that overall in the marketplace the use of such an oil will lead to reduced consumption of fuel and therefore will conserve energy. These energy conserving oils are increasingly being recommended by the North American OEMs and are growing in popularity in the marketplace. Unfortunately these oils cannot be transferred directly to Europe as the high volatility of such mineral oil derived lubricants causes oil consumption problems that can lead to engine damage under the higher speed, and therefore higher engine temperatures, that are typical in Europe. In addition, European engines are designed to operate on lubricants with a higher viscosity at operating temperatures and conditions, as represented by the HTHS (high temperature high shear viscosity at 150 °C and 1 000 000 s^{-1}) viscosity, and this is typically not met by mineral oil based lubricants. The likely result of using such lubricants would again be engine damage.

(iii) Another potential problem is coking of lubricant in the turbocharger leading to premature turbocharger failure. This phenomenon occurs when the heat that builds up in the engine soaks into the turbocharger after a hot engine is turned off. Temperatures during this period are much hotter than when the engine is running. This initially surprising effect causes oils with low thermal stability to form coke resulting in turbocharger failure. The development of a test method and the formulation of an oil to survive under these conditions allows the marketing of oils that can be guaranteed to be suitable for all turbocharged engines. The oil also offers superior thermal protection to drivers of normally aspirated (non-turbocharged) cars who may drive at high speeds.

(iv) The introduction of smaller engined cars in North America as a response to the CAFE legislation made the engine and the lubricant run hotter. The lubricant was also subjected to higher shearing forces due to the higher engine speeds. This combined effect can cause the VII (viscosity index improver) to break down and allow the oil to become thinner. This can reduce the viscosity of the oil to a level below that specified for its viscosity grade or below the viscosity specified or preferred by the OEM. This is a particular concern that occurs in low viscosity lubricants in North America where the use of long polymer chain (high molecular weight) VIIs offers some benefit in fuel economy testing. A lubricant that meets all the standard requirements and also offers the maximum possible protection against high temperature and high speed VII breakdown offers the user superior engine protection.

(v) The occurrence of a field problem may provide an opportunity to develop oils with extra performance. An example of this is the ‘black sludge' problem that occurred in both North America and Europe at a similar time but in rather different forms. Black sludge was not caused by the lubricant but by a combination of engine design and fuel effects. The result was to block oil-ways and oil-filters and to starve the engine of lubricant. The problem was in due course rectified in terms of improved engine and fuel technology. However, sensitive engines remained on the road and improved lubricant technology was required to overcome the problem. By formulating particularly strong lubricants it was possible to offer motorists and workshops a guarantee against the problem. Lubricants that were less strongly formulated could not provide the same guarantee of protection.

(vi) The installation of catalytic converters on cars to reduce the emission of environmentally damaging gases raised the concern that catalyst material could be poisoned not just by the lead present in traditional gasoline (hence the requirement for unleaded gasoline) but also by the phosphorus components present in the engine oil. Several OEMs, notably the Japanese ones, introduced strict limits on the phosphorus level of their lubricants. Legislation was considered by some nations and enacted, for example, in Australia. Testing on small engines confirmed that a problem was possible and the phosphorus level of some lubricants was reduced by reformulating the lubricant to maintain performance at a lower level of phosphorus. These lubricants can now offer the reassurance that the potential problem of catalyst poisoning by the engine lubricant has been minimised. Much work has since been carried out to try to quantify the effect of lubricant-derived phosphorus on catalyst durability and it now appears that phosphorus poisoning is not the main factor involved. Over the years, however, the performance lifetime demanded of the catalyst has been increased and is now expected to be 100 000 miles for cars in North America. This may make the catalyst compatibility of the lubricant a higher priority. Durability concerns over a period of 100 000 miles are difficult to prove or disprove, however a motorist offered the benefit of an improved safety margin may find it attractive. The catalyst compatibility aspect of the lubricant formulation is an example of pro-active lubricant development: the lubricant was developed before a field problem occurred. In Europe, where high speed driving may exacerbate catalyst poisoning, there has been a general reduction of engine lubricant phosphorus content in response to the concern and it will probably never be possible to determine how much of a problem might have been caused by the higher phosphorus content oils.

(vii) The concept of pro-active development of lubricants can also offer benefits to the motorist. If oils are formulated only according to the established classifications and specifications, they are always trying to fight previous battles. Making sure that old problems do not recur is laudable, however, it is to be hoped that the engineering developments of yesterday that gave problems under some unpredicted driving conditions are now better understood and will not be repeated in future engine designs. It is better to formulate lubricants that will perform particularly strongly in aspects that are likely to occur in future engine designs. To some extent these future engine designs are available on the drawing board, in prototypes, on the racetrack and, more immediately, on the road in the more technically advanced vehicles. Even if the exact designs are not known it should be possible to identify the aims and strategies of future designs and the areas where new demands may be put on the lubricant. It is clear that future vehicles must continue to be more fuel efficient, more environmentally friendly, more rewarding to drive, and that engines will become smaller, lighter, more complex and in many ways more difficult to lubricate.

The options for formulating lubricants are clearly enormous and the identification of the detailed performance profile in terms of both technical performance and commercial positioning and profile is a fundamental part of the formulation process. It is not uncommon for the time spent in deriving and devising the product profile to be similar to the time that is taken to carry out the main formulation development programme. Once the formulator has a clear picture of the performance required from the oil and carried out some initial scoping or screening test work, the formulation approach can be established.

7.6 Lubricant formulation—the physical phase

Having defined the performance profile required for the oil it is possible to start selecting the componentry. The first point to appreciate is that it is essential to match up the different components of the formulation. It is not just a principle of selecting the finest ingredients available and combining them. Such a principle has long been abandoned by chefs in preparing meals because it is the overall balance of the final product that is critical. The overall lubricant formulation is a balance of many different aspects of performance which need to be considered as a whole and not as single, separable component selection effects.

7.6.1 *Basestocks*

The first formulation decision to be made is the choice of the basefluid or basefluid mix that will be used. There are three fundamental types of basefluid: mineral, modified mineral (EHVI, extra high viscosity index), and synthetic.

(i) Mineral basefluids are essentially a selected fraction of crude oil with some components removed in order to improve performance. Virtually every molecule present in the basestock is present in the crude oil.
(ii) Modified mineral oils, such as hydrocracked or hydrotreated basestocks, are produced from selected fractions of the refining process that undergo a severe treatment causing some of the molecules to rearrange. The resultant basestock still contains many of the molecules that were present in the original crude oil.
(iii) Synthetic basestocks do not use molecules which were present in the original crude oil. Instead the basestocks are synthesised by the chemical reaction of a very limited number of well defined components. For example, PAOs (polyalphaolefins) are derived from ethylene, and esters are synthesised from acids and alcohols. In this way the synthetic basestock can be tightly controlled and can provide performance that is equally tightly controlled to meet targeted performance. Synthetics can provide special performance such as exceptional low temperature behaviour or high temperature cleanliness. The special performance is obtained because of the nature of the synthetic molecule or the absence of unwanted components (for example, unstable elements) that are present in mineral oils. The synthetic basestock can also have some potential disadvantages due to the absence of naturally occurring components (such as low inherent antioxidancy due to the absence of sulphur and nitrogen containing components which are present in mineral oil derived basestocks). A combination of different synthetic basestocks or additive chemistry will need to be used to compensate for the missing components.

Basestock selection depends on whether the level of performance required can be achieved by the use of mineral oil basestocks. If the target performance is achievable then mineral basestocks will be used on the basis of minimum cost. If, however, the performance cannot be achieved with mineral basestocks then modified or synthetic basestocks will be required. The extra performance is most commonly expressed in terms of the physical properties of the lubricant. However, the use of certain basestocks allows the formulation of lubricants that give further performance benefits such as improved high temperature stability and engine cleanliness.

7.6.2 *Viscosity index improvers*

The choice of the VII, like the choice of the basefluid, will be made on the basis of economics if no other overriding benefit is evident. The first decision is therefore to determine how the VII can contribute to the overall formulation balance.

(i) If there is a need to meet difficult physical requirements then the choice of VII can be critical. Some VIIs (for example OCPs, olefin copolymers) typically cause considerable thickening of the oil at low temperatures and this makes the achievement of 5W and 10W lubricants with low volatility more difficult and more expensive to formulate. Under such circumstances an alternative VII with equivalent high temperature behaviour but reduced low temperature thickening can offer considerable benefits in reducing formulation costs. This makes styrene butadiene, polyisoprene and PMA (poly methacrylate) VIIs very attractive formulation components.

(ii) The use of dispersant VIIs is beneficial if the dispersant function on the VII gives less low temperature thickening than the equivalent dispersancy from the additive system or if there is an economic advantage.

(iii) Another fundamental property of the VII is its behaviour when exposed to the extremely high shear forces that exist between surfaces moving at high relative speeds and with small clearances. This shearing makes the lubricant behave as if it were thinner than expected from its viscosity grade (as defined by its kinematic viscosity at 100 °C) and its laboratory 'cold cranking' performance. There are two different effects, a temporary one, which is reversed when the shear forces are removed, and a permanent one which involves the rupture of chemical bonds in the polymer that do not reform when the shear forces are removed. Temporary shear thinning can be beneficial in that it reduces the resistance to flow at these critical points and may therefore offer reduced friction and hence fuel economy benefits without losing out in terms of film strength. The maintenance of film strength is due to the alignment of polymer chains in the VII with the direction of oil flow. The viscoelastic reaction to this helps to keep the moving surfaces apart. Temporary shear thinning and viscoelasticity are fundamental properties of polymer solutions and can also give benefits in terms of controlling oil consumption by providing an expansive force when the shear forces are removed (for example, at piston reversal where speeds instantaneously become zero). Permanent viscosity loss is more of a problem in that the lubricant viscosity can be reduced to such an extent that it is too thin to protect the engine under all operating conditions. Some OEMs design their engines to run on oils that have a certain level of permanent shear loss and are less concerned by this than others. This is the case in North America where fuel economy is a major issue and the OEMs design around low viscosity oils and do not have to worry about the high speeds that occur in Europe. The shear stability requirements for Europe are therefore more demanding than for North America, as can be seen from the CCMC classification requirements. One principle that is often adopted is to formulate lubricants that will 'stay in grade' after the relevant shear stability test has been carried out. This makes good sense since the engine is designed to run on the recommended viscosity lubricant and there are already other factors, such as condensation of exhaust bypass gases, intake air condensation, and fuel dilution, that can reduce the viscosity of the lubricant in the engine. Most VIIs are available in a choice, if not a range, of shear stabilities and can be selected on this basis. The nature of polymers means that for a given chemical type the higher the molecular weight, the higher the thickening power and the lower the treat cost and the shear stability. It makes sense therefore to select the VII which will give the required shear stability without exceeding it. There is also some concern in having high levels of polymer in a lubricant since the polymer may form deposits under the extreme temperature conditions that occur at the top of the piston and these deposits may inhibit the performance of the piston rings. The tendency to form deposits depends on a combination of the level of the polymer required, its solubility in the basefluid mix, and its thermal and thermo-oxidative stability. All these factors need to be considered in the selection of the VII to be used in the formulation.

Having established well-defined performance targets and some idea of the basestocks and VII combinations that will be appropriate for the lubricant, it is possible to enter the initial 'blend study' phase of the development. The interaction between formulation components is an ill-defined one and therefore a matrix approach followed by iterative optimisation is normally required. The initial matrix is based on prior knowledge which may take the form of a detailed computer database or hundreds of separate similar calculations for conventional formulations. It may equally be entirely experimental if unique components are being utilised. The performance of VIIs in different basestocks can be completely different from its performance in mineral oil derived basestocks. The basestock or VII supplier may advise on the initial matrix based on their prior experience otherwise the formulator will need to devise a matrix based on chemical similarities and gut reaction.

It is essential to obtain fully representative samples of the relevant basefluids and VIIs that are to be used in the study. Every basestock and VII is manufactured within certain limits of performance and a sample that is taken at random from production may or may not be typical of future production. It is important therefore to obtain samples that are representative of current production with mid-point behaviour. Such samples may be available from the supplier, or, if they are not, typical properties should be detailed from production samples and the components carefully analysed to identify any deviation from the norm. Such deviation can be taken into account in defining the physical property targets required from the blend study. Any natural variation must also be taken into account. This is especially important where the physical properties are difficult to achieve as the formulation must be capable of meeting the targets with every delivery of basestock and VII. In such an instance the blend study targets may need to be considerably more severe than the requirements of the classifications and specifications.

All blend studies must be carried out on fully formulated lubricants and therefore an additive system must be included either as a proprietary package or as a 'cocktail' of additive components. Account must be taken of whether the VII has any dispersancy performance and, if it has, a reduced dispersancy additive system is required. Beyond this the exact nature of the additive system is not critical in the initial blend study phase providing it represents modern additive componentry and is broadly appropriate for the defined lubricant purpose. It is also often desirable to include the use of a supplementary pour point depressant to ensure performance in the particular basestock/VII/additive package combination.

When the blend matrices have been produced and evaluated for their physical properties, a decision can be taken as to the formulation closest to the required targets. If any formulation is sufficiently close then the additive chemistry can be developed with it. It is quite likely, however, that no formulation is sufficiently close to the target and it will be necessary to carry out a further iterative blend study. The interactive behaviour of basestocks, VII and additive is not governed by any rigorous equations. Ultimately there will be one or more formulations that can be compared for cost and performance. Once the correct formulations have been identified, the detailed development on the additive system can be started.

As the additive system is developed, it will be necessary to check out the physical properties of the new blends. The information gained in the initial blend development should make this a simple operation and it should only be necessary to check out a few properties for most blends and to carry out full property analysis on only the most serious candidate formulations.

7.6.3 *Additive package*

Hundreds of different chemical components have been found to enhance certain aspects of lubricant performance in some engineering applications. Some chemical components initially appear to be beneficial for all applications but this is normally incorrect. For example, antioxidants that are beneficial at low temperatures can cause heavy deposits when they break down at higher temperatures. Some antioxidant types also can be aggressive towards seal materials, or they can be volatile, incompatible, hazardous to health, etc. Even the most innocent improvement in performance can have serious effects in other areas. In practice the strengths and weaknesses of different antioxidants are broadly understood; some are not used at all while others are used only in limited applications. The technology that is normally responsible for separating the applications of lubricants is anti-wear or extreme pressure (EP) chemistry. The distinction between anti-wear and EP is difficult to define but anti-wear is in general based on the prevention of damage caused by moderate and intermittent loadings whereas EP is the protection from shock loadings and continually applied heavy or sliding loads. Two types of chemistry have evolved around these needs. Anti-wear protection is typically provided by 'ash containing' components such as zinc dialkyl dithiophosphate (ZDDP) and engines are now designed around chemistry of this type. This chemistry has, however, been improved upon in driveline (gear and axle) applications where 'ashless' sulphur and phosphorus chemistry is used. The reference to ash containing or ashless chemistry is historical and there are now claimed to be ashless alternatives to ZDDP. In practice ZDDP-free engine lubricants that have demonstrated good field performance are rare and for some engines the use of zinc-containing lubricants is mandated. The ash containing/ashless division also remains for highly loaded gears in hypoid axles as only GL5 oils based on ashless sulphur–phosphorus chemistry have been demonstrated to give adequate performance. The first question in formulating an oil is therefore, 'will it need to lubricate engines or gears or both?' If the answer is 'both' then the question becomes 'which engines and which gears?' As the demands become more extensive in demanding highly loaded hypoid axle performance combined with engine performance then the formulation becomes more extensive, expensive, and complex.

7.6.4 *Small scale (and analytical) tests*

Lubricant classifications and approval systems specify many performance characteristics that involve small scale and analytical tests. There is no clear dividing line between these tests, however, it is common that tests involving delicate equipment and measurements are located in the chemical or physical test laboratory and the more highly loaded tests are located in a mechanical test laboratory. The most important tests are controlled through the industry by monitoring groups. They organise 'round robin' tests to establish test reproducibility and repeatability and to identify any drift in test results with time or between laboratories, taking action to keep the test under control. These tests include physical measurements of lubricant properties, e.g. viscosities and volatilities, the stability of the lubricant, e.g. its oxidation resistance and shear stability, and the performance of the lubricant, e.g. wear protection.

In terms of lubricant evaluation it is normal to carry out the small scale tests at the start of the development programme.

7.6.5 *Screening tests*

In order to develop the lubricant it is necessary to consider the strengths and weaknesses of the selected additive basestock and VII system relative to the performance that is required. One method for identifying these strengths and weaknesses is to use screening tests. Screening tests are typically small scale laboratory or engine tests that are cheap to run and can be carried out at a high rate either by being of short duration or by being easy to complete in large numbers. It is important that the screening tests mimic as closely as possible the phenomena that occur in the engine so that improvements in screening test performance can be relied upon to represent improvements in performance in the engine. This is clearly a complicated and subjective issue as the closer the screening test is to the engine the more expensive and difficult to operate it becomes. There are two approaches to this issue reflecting two schools of thought.

The first school of thought believes that lubricant performance can only be predicted by using engine tests (or field tests) and that screening tests are generally a poor indication of lubricant performance. In this case, therefore, the defined basestock and VII system are used with a candidate additive system in the most critical engine tests. The additive system is then modified according to the results obtained. This then becomes an evolutionary method of additive system development with the advantage that the formulation is immediately subjected to the rigour of engine tests. The disadvantage is that the availability, duration and cost of the engine tests will restrict the number of options that can be considered in any development programme. This method is therefore best suited to development programmes that are very close to existing areas of knowledge.

The second school of thought uses screening tests to identify potential candidate formulations before any engine tests begin. For example, it is possible to compare the performance of basestocks by using a marginal performance additive system in a series of screening tests. (If the additive system is too strong then the tests will not be able to distinguish the behaviour of the basestock from the additive.) The results are then compared with a database of screening test results on basestocks that have been evaluated previously. The assumption is then made that the basestock will have similar behaviour to a basestock with comparable physical and chemical properties. If such a system is operated there is a need to have a database comparing the performance of different basestocks using the same additive system. In this way the screening tests can be correlated with engine test performance and a database built up which relates basestock properties with screening test performance.

In practice a combination of these approaches, and possibly a third approach, usually applies. The third approach monitors the results of engine tests and estimates what must be done to improve the performance by a specified amount in order to achieve the target performance. This requires a large database on the results of engine tests that have been run on known formulations. The system is particularly useful in taking corrective action when an oil fails an engine test and the additive system may need to be boosted in order to achieve the required performance.

7.6.6 *Engine tests*

Before an engine test programme begins, it is important to have a strategy defined to cope with carrying out tests in an environment where the statistical control is extremely complex. This may sound like a euphemism for tests that have poor reproducibility but in practice engine tests operate under acceptable statistical control. However, they are prone to step changes in severity due to unanticipated changes in test operation, e.g. a change in fuel origin, component production process, surface finish, etc. Such changes are inevitable in the real world of fuels and production engineering and reflect the difficulty of testing the performance of a lubricant in complex engines.

The most informative and unbiased information on engine test behaviour is obtained from the independent engine test houses that run tests on a financial basis. It is in their interest to run engine tests exactly according to the test methods as they have no interest in the actual results of the tests. This information is normally available to customers as a service and it often shows trends in test severity that are quite simply astonishing. The information is provided on a confidential basis and therefore cannot be quoted directly, however, the general features are normally reflected in the general exchange of opinions in the industry and these are free to whoever is prepared to listen to and consider them.

Pass rates for oils on a single test can vary from over 60% to less than 15% without any identifiable change having been made to the test. Minor changes in the rating procedure following an engine test can halve or double the pass rate. It should also be borne in mind that if the test pass rate does fall development programmes are delayed and this creates intense pressure to run only the strongest lubricants. Unfortunately there is not normally any improvement in the pass ratio due to the testing of stronger lubricants, indicating that the change in test severity is test technology related rather than lubricant related. Historically there has been a temptation to carry on running tests until a passing result is finally obtained but this is a very expensive practice that is difficult to defend. More importantly it may stand in the way of running engine tests to try to identify the origin of and solution to the problem. The establishment of an engine test philosophy at the start of the test programme is important in deciding on strategy as the test programme develops.

It is easy to say that a lubricant ought to pass each test first time but since most engine tests run at an average pass rate of below 75%, even for the passing reference oil, the statistical chance of passing a series of 10 or 15 tests first time is very low. (A 75% chance run 10 times has only a 6% chance of successful completion.) The purpose of lubricant development is, however, to produce lubricants that will give performance in vehicles on the road and not just to pass engine tests irrespective of engine test severity at the time. Statements like 'we are not in the business of failing engine tests' and 'if it does not pass the first time then we give it a second shot before reformulating' in effect mean that certain engine tests will not be run until they return to normal severity. This can delay product development but may get away from the situation of running engine tests that are over-severe until finally a passing result is obtained. Instead priority is placed on getting the test running normally again. This is a sensible route towards improved test quality control but it is not an easy concept to explain to a lubricant marketer working towards a launch deadline. A further effect that compounds the problem is that since there is no short-term solution to the problem there is nothing to be gained, and much to be lost, by crying wolf over an apparent test severity problem. (There is always the slight possibility that one particular test-house or additive company test facility has become severe while the rest of the engine test industry has remained on target severity.)

Failures in engine tests must be expected and put into the context of how the individual test stand, and test stands in general, is running. It is tempting to respond to a failing engine test by strengthening the additive system but there is an inherent risk that the strengthening may for some unidentified reason weaken some aspect of performance that had previously proven to be acceptable. There is thus a risk associated with being too willing to respond to a failing engine test by modifying the additive system. This is where experience is valuable in understanding when the test result is a figment of the engine test and when it is attributable to the lubricant itself. This is a double-edged responsibility and refers to engine tests that are running light as well as those that may be running severe.

If OEM approved engine tests run into technical or operational problems they can be reviewed with the OEM and an agreement reached on what is required to demonstrate satisfactory performance and hence gain approval. This may not be a desirable situation to arrive at but it is an honest and constructive solution that can be made to work given goodwill from both sides. Even more difficult is the situation in which a test required for an industry classification starts to run severe. There is no one who can officially review, interpret, adjust or waive such an engine test requirement.

There is also a less publicised situation that occurs when, for whatever reason, the engine test runs light and there is the opportunity for poor oils to gain approval. While this is legitimate, it is necessary for the morality involved to be questioned. Ultimately there is a responsibility for the lubricant supplier to ensure that the lubricant meets the spirit as well as the letter of the engine tests as any weakness in performance is likely to cause engine failures on a statistical, if not catastrophic, scale. This responsibility is one that is particularly important to those lubricant companies which have a large share of the market and in particular those who focus on supplying workshops. If these oils do have any weakness then field failures would soon become evident.

7.6.7 *Other aspects of performance*

The final stage in the formulation process is to assess other aspects of performance which are important to the proper function of the lubricant formulation. In particular the lubricant must be tolerant and robust, i.e. it must not be sensitive to the normal variations that may occur in its manufacture, storage or application.

Lubricants are meta-stable at low temperatures as the polymeric components have a tendency to separate out. This is particularly likely to occur where cooling rates are slow, for example in drums or tanks stored in unheated areas. It is advisable to establish a test that will provide a rapid indication of any concern before the lubricant is commercialised (and preferably at a much earlier stage of the lubricant development programme).

The lubricant must not be applicable only to specific engines, driving conditions or engine tests and the performance of the oil must be maintained throughout its useful life. It is not acceptable to use a component as a 'fix' to achieve satisfactory performance in an engine test if the performance does not translate into field performance. The oil must not be abnormally sensitive to contaminants such as blow-by gases or condensation.

Another area of concern is the compatibility of the lubricant with other lubricants with which it may reasonably be expected to come into contact in the engine. The major source of this contamination is clearly through topping up of the lubricant by other lubricants. Again it is important that the mixture maintains its integrity and performs satisfactorily in the engine under all normal driving conditions.

If there is anything unusual about the formulation, for example a novel combination of basestocks, a novel VII, or a new chemical component, it is essential to check out such performance before, during and after lubricant development. It is totally unacceptable to release an oil onto the market if the robustness of the formulation has not been confirmed.

8 Industrial lubricants

C. KAJDAS

8.1 Introduction

8.1.1 *General aspects of industrial lubricants*

Industrial lubricants comprise a wide variety of products which, depending on their application, differ widely in their chemical and physical properties. With respect to the properties, one can say that industrial lubricants involve all classes of lubricants applied in practice. They include gases (mostly air), various kinds of liquid products (mineral oils, animal and vegetable oils, synthetic oils, water based fluids, etc.), greases (simple soap greases, complex soap greases, greases with pigments, minerals, polymers and other materials) and solid lubricants. The latter comprise (i) inorganic compounds, e.g. molybdenum disulphide, boron nitride, tungsten disulphide and many other chemicals and materials, (ii) solid organic compounds and materials, e.g. phthalocyanine and tetrafluoroethylene, (iii) chemical conversion coatings, and (iv) soft metals.

Furthermore, industrial lubricants make use of many additives which comprise practically all the known additive classes used in other types of lubricants and, additionally, numerous additives that have been developed specifically for industrial lubricants, particularly for water based fluids. Thus, the importance of additives in the formulation of industrial lubricants is difficult to overestimate.

High speed and lightly loaded plain bearings need a low viscosity plain mineral oil. The viscosity of the oil is essential for ensuring hydrodynamic lubrication. Higher loadings and lower speeds require higher viscosity oils. From the point of view of the chemistry of lubricants, these oils are the simplest, being composed of crude oil components. They mostly include isoparaffinic, naphthenic, naphthenic-aromatic and, to some extent, aromatic hydrocarbons. All the ring structures are substituted by alkyl chains. The viscosities of these hydrocarbons depend on their molecular weights. For more details on mineral oils see chapter 1.

Apart from this simple lubrication of plain bearings, the lubrication of tribological elements being rubbed under mixed and/or boundary friction requires lubricants in the form of very complex mixtures of appropriate mineral base oils and a number of possible additives.

The selection of additives to formulate industrial lubricants involves consideration of the requirements of the equipment to be lubricated or the metal/process type of a metalworking operation. Although hundreds of products have been used as industrial lubricants, and some other lubricants (e.g. engine oils) may be applied in lubricating some industrial equipment, equipment and lubricant manufacturers usually recommend lubricants for particular applications.

In terms of quantities, industrial lubricants represent the largest group (over 50%) among the lubricants. Lubricating oils are the most important type of industrial lubricant.

In this chapter it is not possible to discuss all specific details associated with industrial lubricants. Since industrial lubricating oils based on mineral and synthetic base oils are somewhat similar to motor oils (see chapter 7) and are quite broadly presented in some books (Booser, 1983; Klamann, 1984), it seems reasonable to describe them here in general. Metalworking lubricants have also been described in a number of books (Olds, 1973; Kalpakjian and Jain, 1980; Booser, 1983; Schey, 1983; Klamann, 1984; Nachtman and Kalpakjian, 1985).

8.1.2 *Classification of industrial lubricants*

Industrial lubricants include a very large number of specialised products from all the lubricant classes. Therefore, with the exception of some lubricant groups such as hydraulic fluids, it is rather difficult to classify them as clearly as, say, engine oils.

The majority of industrial oils for which viscosity is a significant criterion, for example, fluid bearing lubricants, are classified according to ISO 3448 viscosity grades (Table 8.1). This classification consists of eighteen viscosity grades in the range 1.98 mm^2/s to 1 650 mm^2/s. Each grade is represented by the whole number which is obtained by rounding off the mid-point viscosity at 40 °C, expressed in mm^2/s. This classification does not imply any quality evaluation.

So-called general machinery oils classified according to ISO viscosity grades encompass mineral oil products from light oils for spindle lubrication, to heavy black oils for lubrication of wire ropes. This category includes products that cover the full range of speeds and loads used in industrial machinery. Usually these products contain mineral oil raffinates of suitable thermochemical stability, providing long-lasting lubrication of the machinery, especially bearings. Sometimes these oils are included in a group of machine lubricating oils together with slideway oils, multipurpose oils, and lubricating oils for precision instruments and watches (Klamann, 1984).

Another category consists of metalworking lubricants which, in many cases, are treated separately in conjunction with their tribology (Kalpakjian and Jain, 1980; Schey, 1983). These lubricants are not classified under the

Table 8.1 Viscosity grades for industrial lubricants according to ISO Standard 3448

ISO viscosity grade (ISO VG)	Kinematic viscosity at 40 °C, mm^2/s		
	Minimum viscosity	Maximum viscosity	Mid-point viscosity
2	1.98	2.42	2.20
3	2.88	3.52	3.20
5	4.14	5.06	4.60
7	6.12	7.48	6.80
10	9.00	11.00	10.00
15	13.50	16.50	15.00
22	19.80	24.20	22.00
32	28.80	35.20	32.00
46	41.40	50.60	46.00
68	61.20	74.80	68.00
100	90.00	110.00	100.00
150	135.00	165.00	150.00
220	198.00	242.00	220.00
320	288.00	352.00	320.00
460	414.00	506.00	460.00
680	612.00	748.00	680.00
1000	900.00	1100.00	1000.00
1500	1350.00	1650.00	1500.00

viscosity grade system since properties other than viscosity (e.g. cooling and lubricating properties) are much more important, and a viscosity grade is not required as part of the identifying mark.

A further classification approach might include the following categories: production engineering lubricants (e.g. machine tool and metalworking lubricants), lubricants for engineering components (refrigerator, heat transfer and transformer oils) and plant maintenance lubricants (compressor, hydraulic, turbine and gear lubricants).

To simplify the problem of industrial lubricant classification, the names of the lubricants such as bearing lubricants, compressor lubricants, hydraulic lubricants, gear lubricants, metalworking lubricants, etc. are derived from their application. Such classification is consistent with other lubricant categories, i.e. lubricants for internal combustion engines, aviation lubricants, and marine lubricants.

8.2 Bearing lubricants

8.2.1 *Bearings*

Bearings are the most important machine elements used in all branches of industrial machinery. They permit smooth, low-friction linear or rotary motion between two surfaces. Bearings function by applying a sliding or rolling action. Bearings based on sliding action are called plain bearings, whereas those involving rolling action are referred to as rolling-element bearings or antifriction bearings.

Bearings can be lubricated by gases, liquid lubricants, greases, or solid lubricants. The main function of the lubricant is to keep the surfaces apart so that no interaction can occur, thus reducing friction and wear. Bearings lubricated by gases include aerodynamic and aerostatic bearings (externally pressurised feed). Generally, in externally pressurised bearings the solids are separated by a fluid film supplied under pressure to the interface. The fluid may be a liquid, in which case the mode of lubrication is called hydrostatic.

As the lubrication of plain bearings is more variable compared to that of rolling-element bearings, the former are also referred to by the lubricating principle involved. For example, a specific class of plain bearings is the so-called full fluid-film bearings, which include hydrodynamic (self-acting) and hydrostatic (pressurised feed) bearings. In full fluid-film bearings the load is supported by pressures within the separating fluid film and there is no contact between the solids. In the hydrodynamic lubrication the pressure is developed by the relative motion and the geometry of the system. The friction coefficient, f, in a plain bearing is related to the lubricant dynamic viscosity, η, the bearing load, W, and the sliding velocity, V, by the following equation:

$$f = k\left(\frac{\eta V}{W}\right)$$

This dimensionless term is known as the hydrodynamic factor. Figure 8.1 representing the Stribeck curve in the log scale shows that a compromise must be made with respect to lubricant viscosity, between the friction losses in the

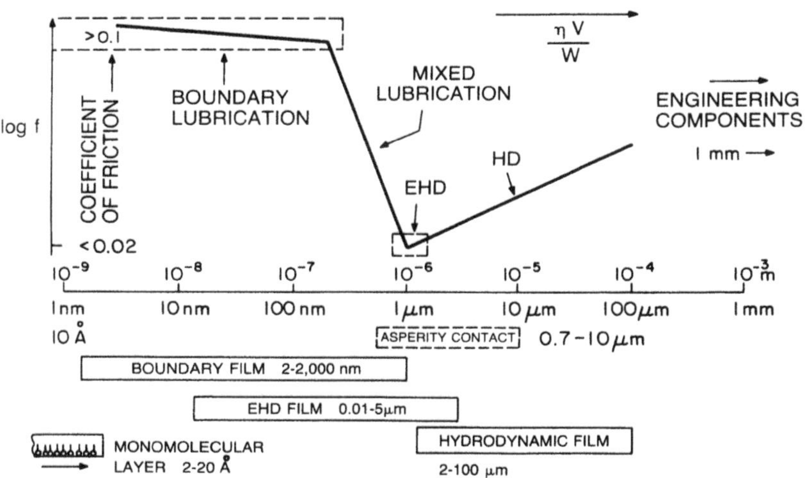

Figure 8.1 Comparison of lubrication film thicknesses with average size of lubricant molecules (redrawn and adapted from Booser, 1983).

region of hydrodynamic (HD) lubrication, and bearing wear when passing through the regime of mixed friction. Both hydrodynamic and hydrostatic bearings operate with infinite service life below some critical value of load and speed.

Another region of bearing lubrication is the elastohydrodynamic (EHD) lubrication. It occurs naturally in rolling-element bearings and other high-contact-stress geometries, such as gears, and is a requirement for their successful operation. Two additional factors must be considered in this lubrication region. One is that when a lubricant is squeezed under the advancing contact zone of a rolling element, it becomes more viscous, according to the formula $\eta_p = \eta_a \exp \alpha (p - p_a)$, where η_p is the viscosity at pressure p, η_a is the viscosity at atmospheric pressure p_a, and α is the pressure viscosity coefficient.

This viscosity increase enables the lubricant to withstand the high contact stresses and to inhibit contacts between the surfaces of mating elements. The other factor to be considered is that the surfaces deform elastically, thereby enlarging the area of support. This leads to more favourable lubrication. The friction coefficient for this lubrication region is the smallest (see Figure 8.1). Thus with EHD lubrication, the material properties of both the lubricant and the solids are of importance.

Rolling-element bearings may be lubricated by liquid and solid lubricants, but are more usually lubricated by greases.

8.2.2 *Gaseous lubricants*

Gaseous lubricants belong to the simplest and lowest viscosity lubricants known. They include air, nitrogen, oxygen and helium, and are applied in aerodynamic and aerostatic bearings.

Since the chemical properties and the aggregate state of most gases remain unchanged over a wide temperature range, gaseous lubricants offer several advantages over liquid lubricants. First, they can be applied at very high and very low temperatures. Their chemical stability eliminates any risk of contamination of the bearing by the lubricant. This is of importance for the machinery used in many branches of industry, primarily in the food, pharmaceutical and electronic industries.

A useful property of gases is that their viscosities increase with temperature, whereas the opposite is true of liquids. This results in the load-carrying capacity increase of gas-lubricated bearings with increasing temperature. However, the relatively low viscosity of gases generally limits the load-carrying capacity of self-acting (aerodynamic) bearings to 15–20 kPa. It is possible to achieve better bearing performance with gaseous lubricants than with liquid lubricants. This is due to the very low viscosity of the gases which results in smaller heat generation by internal friction.

8.2.3 *Liquid lubricants*

As the hydrodynamic behaviour of plain bearings is totally dependent on the viscosity characteristics of the lubricant, typical liquid bearing lubricants are straight mineral oil raffinates of various viscosity grades. Usually, the lubrication of bearings for machine tools requires mineral oils of ISO VG 46 or 68. For fast-running grinding spindles with plain bearings, mineral oils or ISO VG 5 or 7, depending on the bearing clearance and speed, are used. Bearings operating under high loads need lubricants of ISO VG 68 or 100. The service life of the bearing can be increased if the viscosity of the selected liquid lubricant at operating temperature exceeds the calculated optimum viscosity. On the other hand, increased viscosity also means an increase in operating temperature. In practice, therefore, the extent to which lubrication can be improved in this way is often limited. The chemical compositions of these oils differ from typical base oils in that they contain somewhat more aromatic hydrocarbons and heterocyclic compounds. These components serve as natural oxidation inhibitors. An increase in viscosity for oils derived from the same crude oil does not significantly change their chemical composition. Generally, the difference lies in the increasing chain length of the paraffinic hydrocarbons (mostly isoparaffins), and in the aliphatic substituents of naphthenic and aromatic rings, together with a slight increase in the number of naphthenic and aromatic rings. More highly refined mineral oils and oxidation inhibitors are used for applications where higher temperatures or longer service periods require better ageing stabilities.

The use of tribological additives is required for bearings operating with frequent changes in the direction of sliding or rotation, or frequent stops and starts; when metal-to-metal contact is imminent a lubricant's ability depends on the effectiveness of anti-wear and/or extreme pressure additives to prevent catastrophic wear or damage. Three groups of additives are of importance here, namely anti-wear, extreme pressure additives, and friction reducing agents (for more detail see chapters 6 and 12). It is important to note that, in many cases, bearings are lubricated as part of the overall requirement to lubricate a given piece of equipment, e.g. engine oils, compressor oils, gear oils, etc.

It is necessary to bear in mind that lubricating oils of higher viscosity, used to ensure satisfactory lubrication, result in increased energy demands due to high internal friction.

Practically every synthetic oil of adequate viscosity and good viscosity–temperature behaviour can be used as a bearing lubricant. For example, polyglycols are very good bearing lubricants for mills and calenders in the rubber, plastics, textile, and paper industries. However, in most cases, the synthetic oils that have been specifically developed for lubricating particular equipment are also used to lubricate its bearings.

Specific cryogenic fluids such as liquid oxygen, hydrogen and nitrogen form another group of liquid bearing lubricants. Usually bearings operating in these fluids are sufficiently cooled but not sufficiently lubricated by them. This is due to very poor viscosity of the cryogenic fluids.

8.2.4 *Greases*

Greases are almost universally used to lubricate ball and roller-element bearings. They are widely applied in roller-element bearings in machine tools, electric motors, household appliances and many others. To provide adequate consistency and high dropping point, thickening agents in the form of lithium or sodium soaps, and mixed calcium and sodium soaps are used (for more details see chapter 11). Normally, bearings are lubricated with greases of the consistency grades 1 (soft), 2 (creamy) and 3 (almost solid), selected according to speed, bearing sealing, position of the bearing, working temperature, starting torque, etc. These lubricants, which must prevent rusting under wet conditions and under the catalytic effect of metals, are characterised by good oxidation stability. It is important that a bearing assembly be protected against the environment in such a way that no moisture or impurities can penetrate the bearing. Sealing also prevents the grease from leaking out.

Sodium petroleum sulphonates or calcium dinonylnaphthalene sulphonate can be applied as rust preventing additives. The catalytic effect of copper can be retarded by 2-mercaptobenzothiazole. Depending on temperature, various oxidation inhibitors may be used. 2,6-di-tert-butyl-4-methyl phenol is an effective inhibitor at temperatures below 120 °C. Phenyl-alpha-naphthylamine is widely used in greases at low temperatures and above 120 °C. Phenothiazine is effective especially at temperatures above 150 °C. For greases used in the food industry, special chemicals such as dilauryl thiodipropionate or citric acid are used as oxidation inhibitors.

Fluorocarbon polymers are sometimes used as thickeners in grease formulations, making use of their excellent temperature stability, lubricating properties, and chemical inertness.

8.2.5 *Solid lubricants*

8.2.5.1 *General description* Bearings used under vacuum, at very high temperatures, or under very high radiation cannot be lubricated by liquid lubricants or greases. In these and many other cases, solid lubricants, which may be defined as any solid material employed to reduce friction and wear between two moving surfaces, are used.

Generally, the solid material is interposed as a film between sliding and/or rolling surfaces. Simply stated, an adequate solid material is required for the special lubrication requirements of extreme operating conditions (very high and very low temperatures in a wide range, e.g. −200 to 850 °C, corrosive atmospheres, etc.) Such materials normally have a layered crystalline structure which ensures low shear strength and thereby minimises friction. According to Lipp (1976) the shear strength between the crystalline layers is weak, setting up a mechanism of low friction due to slippage of the crystalline layers under low shearing forces. Examples of layer-lattice solids are molybdenum disulphide, graphite, boron nitride, cadmium iodide and borax. Solid lubricants are used primarily in powder form or as bonded solid film lubricants.

Any good solid film lubricant is expected to provide strong adhesion to the bearing substrate material, full surface coverage and good malleability. The solid film lubricant should also be chemically stable and prevent corrosion, taking into account the operational and environmental conditions. Many solid film lubricants have poor wear resistance, since any breaks in the film are not self-healing as is the surface coating formed by a liquid lubricant.

Advanced solid film lubricants perform reliably in many specific applications and much experience relating to a better understanding of their limitations has been gained. The most commonly used solid lubricants encompass the following four materials: molybdenum disulphide, graphite, polytetrafluoroethylene and fluoroethylene propylene (Lipp, 1976). Another group of materials, the self-lubricated materials, are related to solid lubricants and are particularly important for bearings.

8.2.5.2 *Solid lubricant powders in non-bonded films* Solid lubricants used in powder form are primarily tribological additives employed in pastes, greases and suspensions. Molybdenum disulphide and graphite pastes are used for highly loaded journal and rolling bearings as well as for ball joint pivots. Normally, such pastes contain high concentrations (20 to $> 50\%$) of the solid lubricant. As molybdenum disulphide can reduce oxidation resistance of a carrier lubricant, oxidation resistant mineral oils and polyalkylene glycols are used as the liquid components. At higher temperatures, silicone oils are employed. MoS_2 provides increased load-carrying capacity and improved boundary friction properties when added to most greases. On the other hand, MoS_2 can reduce both oxidation resistance and corrosion resistance of grease. This negative effect can be largely overcome by addition of suitable inhibitors. The addition of MoS_2 to a grease provides increased grease life in selected, but not all, applications. For example, Lipp (1976) found that non-MoS_2 containing greases provided longer lives in airplane flap track needle bearings than did greases with 5% MoS_2. Powders of fluorocarbon polymers, such as polytetrafluoroethylene and fluoroethylene, added to greases ensure low friction on wearing surfaces.

Suspensions of the solid lubricant in the form of a colloidal or partially colloidal dispersion in a carrier liquid are used. Mineral oils, polyglycols or silicone oils are used as the carriers. Such suspensions can be used for the impregnation of sintered bearings or added to running-in oils for bearings.

The most important advantage of solid lubricants used as thin film burnished onto wearing surfaces is the provision of low friction. In powder form, MoS_2 is more effective than graphite. The application of graphite is very limited because of galvanic corrosion problems and high friction characteristics under vacuum. Burnished film can also be used as a break-in lubricant for shafts, rollers and constructional elements of ball bearings. An advantage of burnished films is that they generate a minimum amount of debris. Thus, burnished films are used for close tolerance bearings.

8.2.5.3 *Bonded solid films* Bonded solid film lubricants are essentially a combination of solid lubricant powders dispersed in adhesive binders diluted with solvents; this permits their application by spraying, dipping or brushing techniques to form a film. Such films are often described as lubricating paints, in which the pigment is substituted by a solid lubricant called the lubricant pigment. Among the organic binders, thermosetting resins, such as phenolic, epoxy, acrylic, alkyd, silicone, vinyl and polyurethane resins, are used. Polyimides, water glass, and ceramics, (e.g. $AlPO_4$) are used at higher temperatures. Bonded solid film lubricants can operate in extreme environments of temperature, pressure and vacuum and provide low friction. Furthermore, they can prevent adhesion and wear between mating elements. They require little or no maintenance and can simplify the design of bearings.

Bonded solid film lubricants are used preferably for sliding surfaces subjected to high loads and low sliding speeds. Hence they are often used in journal bearings where the required wear life is within the capability of the film and where weight and space considerations are critical.

Some general limitations of bonded solid film lubricants (Lipp, 1976) may be summarised as follows (i) solid film lubricants will not last indefinitely, (ii) they may promote corrosion and are not corrosion preventive coatings, (iii) they have no capacity to remove frictional heat, (iv) in most cases relubrication is not possible, (v) application techniques and surface preparation must be meticulously controlled or performance is jeopardised, and (vi) inspection techniques are not always satisfactory.

Molybdenum disulphide bonded films alone or in combination with Sb_2O_3, ZnO or TaS_2 are suited for low sliding speeds at high loads. Although bonded solid films are widely applied in many areas, their limited wear life, and other limitations which affect their performance, restrict their range of applications. To improve their performance, Bartz *et al.* (1986) performed investigations on bonded films containing molybdenum disulphide, graphite and antimony-thioantimonate [$Sb(SbS_4)$] as single components or as two- or three-component systems. It was found that optimum formulations of MoS_2/graphite, $MoS_2/Sb(SbS_4)$ or MoS_2/graphite/$Sb(SbS_4)$ resulted in a longer wear life than that of each component alone. $Sb(SbS_4)$, which exhibits no lubricating properties whatsoever, improves the tribological behaviour of MoS_2 and graphite.

8.3 Compressor lubricants

8.3.1 *General description*

In industry, compressors are widely used for compressing many types of gases in order to store and/or transport them, either as compressed gas or as liquids. There are three main types of compressors, namely reciprocating, rotary and turbo-type air compressors. Reciprocating compressors employ a piston and cylinder with valves. Rotary compressors can be further divided into two types: sliding vane and screw. The former operate by trapping gas in a succession of cells, the latter by compressing gas in intermeshing screws. Reciprocating and rotary compressors bring lubricant and gas into intimate contact over a large area of exposed surfaces under high pressure and high temperature. Such conditions promote chemical reactions. Consequently, when compressing air, lubricants with excellent oxidative stability must be used. Turbo-type air compressors operate continuously by the velocity of gas. In these compressors the lubricant does not come into contact with compressed air, therefore the requirements are less severe.

There are three main groups of compressor lubricants (i) lubricants for gas compressors, (ii) lubricants for refrigerators, and (iii) lubricants for vacuum pumps. Depending on the application, various classes of lubricating oils, including mineral oils of various levels of refining, semi-synthetic, and/or synthetic oils are used. Some of the oils for gas compressors may contain a wide variety of additives.

8.3.2 *Lubricants for gas compressors*

There is a wide range of compressor oils with viscosities ranging from about $4\,mm^2/s$ at 100 °C to over $20\,mm^2/s$ at 100 °C. Generally they cover the ISO VG 22, 46, 100, 150 and 460. The criteria for the selection of the lubricant depend primarily on the gas to be compressed, and on the design of the compressor which determines the final compression temperature.

The purpose of the oil in the pressurised section of the compressor is to reduce friction, prevent wear, improve sealing of the pressurised spaces, and provide cooling. In terms of function, some similarity between air compressor oils and engine oils may be easily found in certain types of compressor oil. This similarity is clearly recognised since heavy duty oils of appropriate viscosities are used to lubricate mobile compressor units on vehicles or in workshops. It should be remembered, however, that the lubrication of the pressurised section is difficult, particularly in the presence of oxidising and aggressive gases, due to the significant safety hazard involved. Air is the most frequently compressed gas. Lubricants for reciprocating air compressors must provide low carbon-forming tendencies, as the formation of carbonaceous deposits in the air discharge system is the prime cause of fires and explosions. Usually the fire and the explosion risk can be controlled by the selection of good quality petroleum-based lubricants. Fire-resistant halogenated fluids are used as compressor lubricants for oxygen compressors.

The higher the final compression temperature of the air at the outlet from the compression section, the higher are the requirements relating to oxidative stability of the lubricant and its resistance to sludge formation. The oxidation process under pressure may be catalysed by finely dispersed iron oxide which also lowers the ignition temperature of the lubricant.

Gas compressor lubricants based on mineral oils contain (i) oxidation inhibitors to prevent varnish and sludge formation and retard corrosion of alloy bearings, (ii) rust inhibitors, (iii) polar compounds to improve lubrication in the presence of water, and (iv) anti-wear and anti-foam agents.

Gas compressor lubricants of the highest requirements are made from specially selected, highly refined base oils with a low coking tendency, and appropriate additive package including a corrosion inhibitor. Thus, the chemistry of gas compressor lubricants exhibits a considerable similarity to that of engine oils.

Synthetic oils are widely used as gas compressor lubricants. Polyalkylene glycols are used for the compression of hydrocarbon gases, e.g. natural gas and refinery gases. The application of this synthetic lubricant prevents a viscosity drop in the pressurised section and washing away of the lubricant which may cause dry running. Polyalkylene glycols are adequate lubricants for compression temperatures up to 200 °C without forming any kind of deposits in service. Other synthetic gas compressor lubricants include (i) diesters and polyesters, (ii) polyalphaolefins, and (iii) dimethyl silicones and fluorinated silicones. These synthetics, along with polyalkylene glycols, are most widely used since they meet the requirements for most compressor applications.

For long term hydrolytic stability and exceptional water separation, polyalphaolefins are the preferred rotary compressor lubricants. According to Miller (1989) energy cost savings through conversion to the polyalphaolefin lubricant yielded a nine month payback for the extra cost of the lubricant. Additional benefits included longer lubricant change intervals, extended oil filter life, extended overhaul intervals and less downtime. Polyalphaolefin lubricants are used in conjunction with additives specifically formulated to increase chemical and thermal stability, and to reduce corrosion. These lubricants have found increasing acceptance for use with ammonia gas.

8.3.3 *Lubricants for refrigerators*

Compared with lubricants for gas compressors, lubricants for refrigerators are of lower viscosity with ISO VG ranging from 15 to 100. The characteristic features of these lubricants are good low-temperature properties, and behaviour towards the refrigerant. Lubricants for refrigerators are widely employed in refrigerating industrial goods in cold storage plants and ships. Furthermore, they are employed in sealed refrigerators and air conditioners with lifetime refrigerant charges. Ammonia is used as a refrigerant in large plants. Another group of refrigerants are the fluoro- and chlorofluorohydrocarbons which, unlike ammonia, are oil miscible.

Refrigerating compressors operate with the exclusion of air, therefore oxidation of the lubricant is low. Consequently lubricants for refrigerators are generally free from additives. This is very important since the presence of additives would create the possibility of additive degradation products reacting with the refrigerant.

Generally, refrigerator lubricants are highly refined petroleum oils. To provide them with low pour temperatures, naphthenic base crude oils are used. Oils from paraffinic crude oils have to be deeply de-waxed otherwise hydrocarbon solids may precipitate in the cold halogenated hydrocarbons. This may occur when the lubricant is diluted with the refrigerant at a temperature which exceeds the cloud point of the lubricant. This phenomenon may cause plugging of control devices and lines of the refrigerating plant, and affect the heat exchange.

Although some methods employ immiscible systems, the lubricant is often required to have good miscibility with the refrigerant. Thus, miscibility is an important property of a refrigerator lubricant, and is controlled by both the chemistry and viscosity of the lubricant. Paraffinic oils are less miscible in the cold than are naphthenic or aromatic oils (Klamann, 1984). However, the chemical stability of highly aromatic oils is sufficient only in exceptional cases. Consequently, high viscosity index isoparaffinic hydrocarbons are becoming popular as lubricants for refrigerators. Immiscibility increases with increasing viscosity, but this effect is less pronounced than that due to the chemistry of the lubricants.

Semisynthetic oils, being mixtures of mineral oils and synthetic alkylbenzenes, or pure synthetic alkylaromatics can also be used as good refrigerator lubricants. These oils have good thermal stabilities, low pour point and favourable miscibility properties with refrigerants. Semisynthetic and synthetic refrigerator lubricants are especially suitable for refrigerator systems operating at very low evaporator temperatures where Refrigerant 22 is used.

8.3.4 *Vacuum pump lubricants*

Both the vacuum section and the moving parts of vacuum pumps may be lubricated by the same lubricants as those used for gas compressors. However, this applies only to industrial equipment producing low or medium vacuum. To produce high vacuum, lubricants must provide better properties, particularly in relation to vapour pressure and sealing of the vacuum section, therefore more viscous lubricants are required. The best petroleum-based vacuum pump lubricants are narrow oil fractions of sufficiently high viscosity.

Such oils possess high flash points, and do not contain low-boiling components which may affect the final vacuum achieved, and also produce oil mist, leading to the formation of condensates in adjacent areas.

Usually special lubricants with low vapour pressures at the working temperature are employed for the lubrication of high and ultra-high vacuum pumps. These are produced from naphthenic or paraffinic mineral oils by molecular distillation, and have to provide good oxidative and thermal stability. Synthetic esters are another group of lubricants for vacuum pumps, the most widely used being di-*n*-butyl and di-*n*-octyl phthalates.

8.4 Hydraulic lubricants (fluids)

8.4.1 *General description*

Hydraulic systems (i.e. hydraulic power transmission equipment) serve a wide range of purposes where multiplication of force is required, or where correct and reliable control gear must be provided. Developments in automation have significantly extended the use of hydraulic equipment. Typical hydraulic equipment consists of a circulating system, comprising a pump, series of control valves, and hydraulic motors. With respect to lubrication, the pump is the most critical component in the hydraulic system. Usually sliding vane pumps, piston pumps and gear pumps are used. Although lubricants for hydraulic systems must also reduce friction and prevent wear of the mating elements, especially pump components, in practice they are usually called hydraulic fluids.

Hydraulic fluids represent one of the most important groups of industrial lubricants, being widely used in industrial hydraulic systems, particularly machine tools, steering gears, etc. Hydraulic fluids are also applied in land, sea and airborne transport, as well as in brake systems. The existence of greatly differentiated hydraulic systems, their operation in various environmental atmospheres and sometimes at extreme temperatures, requires a multitude of products of consistently varied properties. As in the case of gear or engine oils, this situation has called for the classification of hydraulic lubricants. The ISO 6743/4 classification (Table 8.2) classifies hydraulic fluids according to type. It clearly reflects the chemistry of these lubricants, ranging from straight mineral oils, additive treated products that include all the most important additives, emulsions and water-based fluids, to synthetic oils. The classification also takes into account some important properties of these lubricants. Apart from the properties that are typical for all fluid lubricants, such as viscosity, viscosity index, oxidation resistance, tribological and anti-corrosion properties, the important features of hydraulic fluids are compressibility (to transmit power successfully a hydraulic fluid must be virtually incompressible), compatibility with seals, air-release and anti-foam properties

Table 8.2 Classification of hydraulic fluids for hydrostatic systems according to ISO Standard 6743/4

Fluid symbol	General fluid characteristic
HH	Non-inhibited refined mineral oil
HL	Rust and oxidation inhibited refined mineral oil
HM	HL type with improved anti-wear properties
HR	HL type with improved viscosity index
HV	HM type with improved viscosity index
HG	HM type with antistick-slip properties
HS	Synthetic fluids with no specific fire-resistant properties
HFAE	Fire-resistant oil-in-water emulsions containing maximum 20 mass % of combustible materials
HFAS	Fire-resistant solutions of chemical in water containing minimum 80 mass% water
HFB	Fire-resistant water-in-oil emulsions
HFC	Fire-resistant water polymer solutions containing minimum 35 mass % water
HFDR	Fire-resistant synthetic fluids based on phosphoric acid esters
HFDS	Fire-resistant synthetic fluids based on chlorinated hydrocarbons
HFDT	Fire-resistant synthetic fluids consisting of mixtures of HFDR and HFDS
HFDU	Fire-resistant synthetic fluids of other types

(to avoid gas entrapment and increased compressibility), filterability and, as with engine and gear oils, shear stability in the case of non-Newtonian fluids. Filterability is associated with an extremely important requirement of hydraulic fluids, namely their purity. High quality hydraulic fluid must be absolutely free from impurities, especially those causing wear. Some hydraulic fluids (e.g. HR, HV and HG), possess high viscosity indices ranging up to 150, and are required when changes in temperature occur within the working range of hydraulic systems.

The selection of the hydraulic lubricants and the specifically required properties depends on the hydraulic system operating conditions. In systems with a high leakage rate, non-toxic hydraulic lubricants with adequate biodegradability must be employed in order to avoid polluting the environment. Hydraulic lubricants designed for the food and pharmaceutical industries require products that comply with the foodstuff regulations.

In terms of their chemistry, hydraulic lubricants can be divided into four classes (i) mineral oil based products, (ii) synthetic lubricants, (iii) emulsions, and (iv) water-based fluids.

8.4.2 *Mineral oil based hydraulic lubricants*

These hydraulic lubricants are the most important and widely used, consisting of HH, HL, HM, HR, HV and HG products. Since viscosity is one of the most important properties of these lubricants, they are produced in a wide range of viscosities (ISO VG 10 to 100). For a given application, this property should be kept as low as possible to provide a rapid response of the hydraulic system, that is, the oil should be sufficiently fluid for efficient transmission of power. On the other hand, a minimum viscosity level is required in order to eliminate or minimise leakage losses and to ensure lubrication of all the tribological elements of the hydraulic system.

Mineral oil based hydraulic lubricants are mostly raffinate fractions of paraffinic crude oils that provide adequate viscosity temperature characteristics. Straight mineral oils are products of the HH group. The hydraulic lubricants with improved performance characteristics incorporate oxidation and rust inhibitors (HL group), and in addition, anti-wear agents (HM group). Lubricants of the HM type containing viscosity index improvers form the HV group of the mineral oil based hydraulic lubricants. Products of the HR group contain oxidation and rust inhibitors plus viscosity index improvers. The HG group of lubricants is formulated with both the above, and antistick-slip additives, i.e. friction modifying additives. Special oils may also contain anti-foam and demulsifying additives.

When particularly low pour points are required, the lubricants are formulated using raffinates obtained from naphthenic crude oils. Such oils possess poorer viscosity–temperature characteristics. However, at higher pressures the naphthenic oils show a significantly greater increase in viscosity than the paraffinic oils. Consequently, these oils can compensate or even exceed the viscosity losses due to heating.

8.4.3 *Synthetic hydraulic lubricants*

Several classes of synthetic oils, such as polychlorinated biphenyls, phosphoric acid esters, polyglycols and silicones, are suitable as hydraulic lubricants. They mostly belong to the fire-resistant lubricants which are of importance for coal mines, steel mills and foundries, especially when the hydraulic systems operate close to hot areas, e.g. furnaces. Polychlorinated biphenyls are the most fire-resistant; however, due to their high toxicity, they are no longer used. Silicones have very poor lubricity and are very expensive, whereas polyglycols with high flash points do not fully meet more stringent fire-resistant requirements. At present, therefore, synthetic hydraulic lubricants are almost exclusively based on trialkyl or triaryl phosphates, or on mixtures of these. They are available with a wide range of viscosities and adequate low temperature properties required for hydraulic lubricants.

The phosphoric acid triesters possess excellent anti-wear properties (see chapter 12) and sufficient oxidative stability. Their disadvantage arises from their ability to hydrolyse, which depends significantly on the structure of the ester; short chain alkyl esters and aryl esters hydrolyse more rapidly than the long chain esters.

8.4.4 *Emulsions and water-based fluids*

Emulsions fall into HFAE and HFB groups of hydraulic fluids, with the HFAE oil-in-water emulsions being more important. These have a high water content, exhibit very low flammability, and are widely used due to their low cost. However, in the case of oil-in-water lubricants, the wear behaviour is poorer than that of pure oils and, in addition, their protection against rust/corrosion and bacterial attack may provide problems. A certain improvement in these properties can be achieved using rust/corrosion inhibitors and bactericides and, if necessary, anti-wear additives. Suitable emulsifiers ensure good stability of the emulsions.

Water-in-oil emulsions (HFB group) provide better lubricating properties and corrosion protection, and generally require less bactericides and rust/corrosion inhibitors.

Water-based fluids containing polymer thickened aqueous solutions (HFC group) are superior to the oil-in-water emulsions in respect to wear behaviour. These fluids are mixtures of glycol, water-soluble polyethers and polyglycols with rust/corrosion and oxidation inhibitors. In order to be accepted as a fire-resistant fluid, the water content must be at least 35%. Special polyethers with molecular weights of between 20 000 and 40 000 can be used as shear-stable thickeners for water-based hydraulic fluids. Pentaerythritol, trimethylolpropane, or glycerol initiated ethylene oxide-propylene oxide polyethers are examples of such thickeners. The chemistry of synthetic thickeners for water-based hydraulic fluids, their thickening effect and shear stability, as well as their technical properties are described in detail by Rasp (1989).

8.5 Industrial gear lubricants

8.5.1 *General description*

Compared with automotive hypoid gears, gears employed for industrial equipment usually operate under conditions of moderate loading. Consequently, they place lower demands on the load-carrying capacity of the lubricants. Their typical function is to increase or decrease the speeds of shafts or to change the direction of drive. Spur gears connect parallel shifts. Worm, bevel, or crosshead helical gears are used to change direction of drive. All of these gears perform their function either with or without alteration in speed. Employing worm gears, large transmission ratios can be achieved in one stage and in a small space. The main functions of a gear lubricant are wear prevention and friction reduction by providing a lubricating film between the gear mating elements. In the case of enclosed gears, the lubricant also has to carry away the heat developed during friction. The selection of industrial gear lubricants is described by the American Gear Manufacturers Association in AGMA 250.2 standards which classify industrial gear oils into 11 types.

8.5.2 *Lubricants*

In industrial gears where the conditions of tooth engagement are not excessively severe, and hence the danger of scuffing is small, mineral base oils may be used. The choice of oil viscosity depends on the transmitted power and pinion speed. Generally the lubricant viscosity decreases as the speed increases, and increases as the power increases. This relates mostly to low-loaded spur gears. In cases where conditions of the tooth engagement are particularly severe or where gears are subjected to shock loads, the use of tribological additives is essential.

For highly loaded spur, helical, worm and bevel gears, lubricants containing tribological additives such as sulphurised and phosphorus-containing additives are used. These provide excellent anti-wear and extreme pressure protection over a wide range of conditions.

High viscosity residual lubricants with good adhesion are used for open gears. Industrial gear lubricants for more severe conditions are based on polypropylene glycol, their characteristics being high load-carrying capacity, very high viscosity index and low pour point. They also possess low frictional characteristics which provide more power transmission. Special greases and semi-fluid gear lubricants are also employed. For example, a lithium soap/synthetic oil based lubricant provides outstanding low steel/bronze frictional characteristics in wide temperature ranges, and long life properties.

8.6 Turbine lubricants

8.6.1 *General description*

There are three main categories of turbines (i) gas turbines, (ii) steam turbines, and (iii) water turbines. Gas turbines are mostly used in aircraft (see chapter 9), though some aircraft-derived gas turbines are used for propulsion of naval vessels or for industrial purposes, e.g. stand-by generation of electricity. Industrial gas turbines are usually robust. The same is true of steam turbines. However, through the years steam turbines have become increasingly compact in design, and are run under more vigorous conditions.

8.6.2 *Industrial turbine lubricants*

In gas turbines no moving parts are involved in the combustion process, therefore the turbines make little demand on the lubricant and present few lubrication problems. Gas turbines for industrial applications have evolved from steam turbine practice, thus they are similar in design and lubrication requirements, except that operating temperatures are much higher. Accordingly, steam turbine lubricants can often be used in industrial gas turbine lubrication. For particularly high temperatures, synthetics are required.

In steam turbine lubrication systems, a pump transports the lubricant from the storage tank through a filter and oil cooler to all lubrication points. Consequently, the lubricant must not only provide reliable lubrication, but must also serve as a coolant, hydraulic fluid, gear lubricant for geared turbines, and prevent rusting of the turbine components. Since the oil is in intimate contact with steam, condensation water, metals, and air, it requires high oxidative stability and satisfactory separation from water. In order to prevent malfunction of the turbine hydraulic system (governors and other control gears) a low tendency to air entrainment is essential.

These requirements have been met by special turbine lubricants, usually formulated from highly refined paraffinic base oils, with high viscosity–temperature characteristics, and adequate oxidation and corrosion inhibitors. Such oils have excellent oxidation stability, very high demulsibility and resistance to foaming, and are also able to rapidly release entrained air. The ISO VG 32 and 46 are usually used for turbines without transmission. ISO VG 68 is applied only for integral gear lubrication due to its higher viscosity and increased lubrication properties. However, lower viscosity oils, e.g. ISO VG 46, with improved wear protection require less energy for the circulation and ensure a better heat removal from the bearings.

High quality steam turbine lubricants usually satisfy the demanding requirements of modern, high-output steam turbines, including their geared units. Such lubricants can also be used for industrial gas turbines and water turbines.

8.7 Metalworking lubricants

8.7.1 *General description of metalworking processes*

One of the purposes of metalworking processes is that of creating a new shape. Usually the processes bring into contact two solids: the tool and the workpiece. The contact involves either the plastic flow of metals (metal forming processes) or the creation of a new shape by controlled removal of excess material (metal cutting processes). The creation of new shapes by metalworking processes involves high friction, high temperatures and tool wear. Consequently, metalworking lubricants influence both the effectiveness of these processes, and the overall efficiency of the manufacturing operation.

Plastic deformation processes allow shape change by deforming the workpiece without fracture due to its ductility. Schey (1977) distinguishes steady-

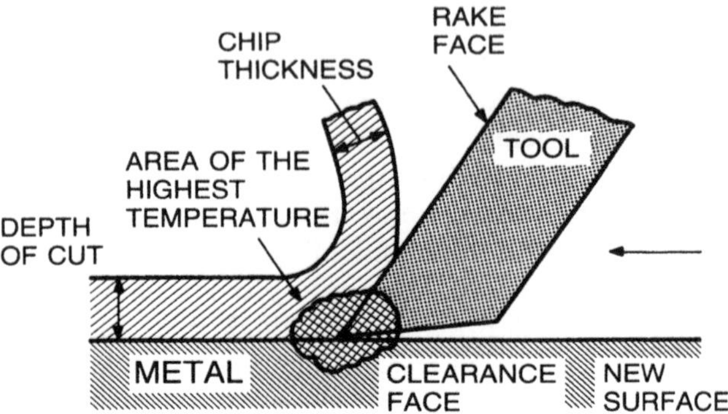

Figure 8.2 Basic cutting action and chip formation (redrawn and adapted from Lindsey and Russel, 1984).

state and non-steady-state processes. The distinction is highly significant in terms of lubrication. In steady-state processes (e.g. rolling), the surface of the workpiece arriving at the deformation zone should be lubricated with a fresh lubricant film. In non-steady-state processes, the pre-applied film must suffice in the course of deformation, and re-application is not usually feasible unless the process is interrupted. Some processes have a transitionary character; for example, deformation in non-steady-state at the beginning and end of extrusion, but steady-state characteristics during the greater part of long billet extrusion.

When the shape is obtained by removing material in the form of chips, the operation is referred to as metal cutting. Figure 8.2 shows that this involves two important processes (i) the formation of a chip from the workpiece by the tool, and (ii) movement of the chip across the face of the tool. Thus, it is extremely important to provide a lubricant that reduces friction and removes heat as rapidly as possible. Figure 8.2 also shows that the area of the highest temperature generated during cutting is at the tool tip. The heat involved is related to the frictional heat and the heat generated during the deformation of the metal.

8.7.2 *Lubricant types for metal forming processes*

Metal forming processes encompass rolling, extrusion, drawing (e.g. drawing tubes, tube bending, deep drawing, wire forming), forging, and sheet-metal forming. Normally the operations are performed under both ambient and high temperatures, and referred to as either cold-working or hot-working processes. The hot-working processes generally include bulk-deformation processes, known as primary metalworking processing. Secondary metalworking processes involve cold operations, such as rolling, drawing and extrusion.

The lubricant used in a given metal forming process should primarily reduce friction which, generally, affects tool life, metal flow, energy consumption, heat evolution, and surface finish. Usually the lower the friction coefficient, the higher the reduction in forces and power required. For example, good lubrication in forging reduces die wear and forging load, prevents die seizure and helps release the forged part from the dies. Furthermore, it improves material flow in the dies and acts as a thermal barrier between the workpiece and the die surfaces. However, in rolling, the coefficient of friction between the work rolls and the rolling stock should be neither too high nor too low. Clearly, low friction reduces the energy consumption, the heat evolved and roll wear, but too low friction causes slipping which may damage the surface of the rolling stock and has a negative effect on the forming process.

Liquid lubricants, suspensions, pastes, greases, solid lubricants and coatings are the lubricants of choice for metal forming processes. Since the metal formed may vary widely in composition, from almost pure metals (e.g. aluminium, beryllium, copper, magnesium, nickel, titanium) to very complex alloys (e.g. stainless steels) a wide variety of liquid, semi-solid and solid lubricants, as well as their combinations, can be used for special forming operations. The lubricants most commonly used in these processes include mineral oils, mineral oils plus fatty oils (compounded oils), emulsions, synthetic oils, graphite suspensions, esters, fatty acids, fatty compounds, lanolin, tallow, paraffin waxes, polymers, conversion coatings (involving the reaction of the surface of the workpiece material with selected anions), copper plating for refractory metals, graphite, molybdenum disulphide, salts, glass, soaps, aqueous solutions, bentonite, lime, mica, talc, and others. Additives used for metal forming lubricants primarily include extreme pressure agents (e.g. sulphurised fats and oils, sulphochlorinated fats and oils) and chemicals used to formulate emulsions. Fatty acids or fatty compounds, as well as graphite or molybdenum disulphide can also be used as additives.

Lubricants for metal forming play a critical role in many manufacturing processes and contribute to the success or failure of the process, productivity, worker safety, energy consumption, and environmental impact. The complex problems of lubricating metal forming processes may be illustrated by the aluminium rolling process, which includes both hot and cold rolling. Generally aluminium or aluminium alloy sheets are made by hot rolling of large ingots to a thickness of about 3 to 17 mm, which are then reduced further by cold rolling.

Oil-in-water emulsions are mostly employed as lubricants and coolants for hot rolling. Such emulsions contain approximately 2 to 5% of an oil concentrate. The concentrate may consist of 70 to 80% naphthenic or paraffinic oil, 15 to 20% of anionic and/or nonionic emulsifiers and 5 to 10% ingredients (Klamann, 1984). The ingredients may comprise many compounds of the following additive groups: load-carrying (tribological) additives, oxidation inhibitors, wetting agents, bactericides, and defoamers. They may include oleic acid and/or other fatty derivatives, phosphoric acid esters (e.g. tricresyl phosphate), ethanolamines, and other compounds.

During hot rolling, the most important function of the lubricant is cooling; however, it should also reduce energy consumption, provide good surface quality of the rolling stock, and achieve maximum thickness reduction of the rolled material. Thus, friction is also of importance.

During cold rolling, hot rolled strips are reduced at ambient temperature to a sheet or foil. Since aluminium tends to transfer to the rolls during cold rolling (the same applies to the hot rolling process), the primary function of the lubricants for cold rolling is to reduce this tendency. Highly refined, narrow distillate fractions of paraffinic crude oils are mostly used as base oils for these lubricants. Annealing temperature in the aluminium industry is approximately 320 °C, thus, it is convenient to use mineral oil fractions with boiling range not exceeding this temperature. Consequently, these base oils fall into kerosene cuts. To provide the cold rolling lubricants for aluminium with suitable frictional characteristics and load-carrying capacity, a package of special additives must be added to the low-viscosity mineral base oils. The additives used comprise mainly fatty alcohols, fatty acids, and fatty esters. Fatty alcohols ensure better performance of the lubricant since they do not affect the annealing properties of aluminium.

Tribochemical reactions of these additives with aluminium are very interesting, especially from the point of view of organometallic products resulting from alcohols. The aluminium rolling process produces abrasive particles and organometallic compounds, mostly called soaps. Chambat *et al.* (1987) found that the soaps result from a reaction between fatty additives and fresh unoxidised metal surfaces produced during plastic deformation. Fatty acid soaps are well known and broadly discussed in the literature. However, few papers deal with direct reactions of fatty compounds with aluminium. Montgomery (1965) clearly showed that esters, alcohols and ethers can react with an uncontaminated fresh aluminium surface. It has been suggested that, during lubrication, alkoxide formation from alcohols and soap formation from esters can occur. Chambat *et al.* (1987) described the products resulting from tribochemical reactions between aluminium and foil-rolling kerosenes containing fatty acids and fatty alcohols as additives. The organometallic products were isolated and analysed and it was suggested that during plastic deformation fatty acids react directly with aluminium to yield polymeric soaps. Fatty alcohols can lead to a similar reaction but on a smaller scale. In this case, a part of the alcohol was assumed to be changed into acid which can react to give an ester and a blend of hybrid soaps of the general formula

$$(RCOO)_x\text{—Al}\begin{matrix} \diagup (OR)_y \\ \diagdown (OH)_z \end{matrix}$$

where $x = 1, 2$ or 3 and $x + y + z = 3$.

On the other hand, the negative-ion-radical lubrication mechanism of alcohols, proposed by Kajdas (1987), provides a clear explanation of the formation of aluminium alkoxides (i.e. the $Al(OR)_y$ part of the hybrid soap) from alcohols. This mechanism is based on the low-energy electron (2–4 eV) emission process from the aluminium surface. The action of the emitted electrons on alcohol molecules produces negative and radical-negative ions, including $R–CH_2–CH_2–O^-$ ions which provide alkoxides by reacting with the positively charged sites on the deformed aluminium surface. Furthermore, this model elucidates the finding described by Mori *et al.* (1982), showing the formation of aluminium methoxide and aluminium butoxide on the aluminium surface prepared by cutting under high vacuum. These experiments were conducted in a special vacuum chamber. A small lathe was constructed in the chamber and an aluminium disc was mounted on a drive shaft which was rotated using a magnetic assembly. Reactants were introduced into the chamber through the variable leak valve. The components in the chamber were analysed by means of a mass spectrometer. It was found that the rate of chemisorption was proportional to the cutting speed and that chemisorption took place not only during, but also after cutting. Hydrogen concentration increased during cutting.

The reaction mechanism can be explained in terms of the Kajdas negative-ion-radical model as follows: the cutting process is associated with the electron emission (similar to the emission caused by plastic deformation) which is responsible for the rate of chemisorption. Thus, the exoemission process is the main factor governing the reactivity of methanol and butanol towards aluminium. It is known that the exoelectron emission rapidly reaches a maximum during rubbing and then, after rubbing, decays with time. This explains why chemisorption takes place after cutting. Hydrogen radical recombination is responsible for the hydrogen formation during cutting.

The above examples demonstrate the complexity of metalworking lubrication. They also provide evidence that the mechanical activity at metalworking surfaces tends to promote chemical reactions and produce surface chemistry that may be entirely different from that observed in static studies, i.e. those activated predominantly by temperature. In other words, the metalworking process itself initiates and accelerates chemical reactions which would otherwise take place only at much higher temperatures or not at all. According to Rowe and Murphy (1974), such enhanced surface activity might be represented as follows:

$$\text{Enhanced reactivity} = \text{Exoelectron} + \text{Elevated temperature} + \text{High pressure}$$

8.7.3 *General lubricant types for metal cutting processes*

There are three major types of metal cutting lubricants: oils, emulsions, and water-based products. Oils are derived from petroleum, vegetable and animal sources. Emulsions contain an oil, such as mineral or compounded oil, in the form of suspended droplets dispersed with the aid of emulsifiers. Water-based products, also known as water-soluble fluids, chemical fluids or synthetic fluids, do not contain oil, but only water-soluble chemicals.

Although many alterations have been made in the formulation of metal cutting lubricants since their introduction, the basic functions have remained unchanged. They comprise (i) cooling of the workpiece and chips to improve tool life and permit higher cutting speeds, as well as to reduce distortion and maintain dimensional accuracy, (ii) lubrication to reduce friction, improve surface finish of the metal cut, protect the newly machined surfaces from rust/ corrosion and build-up of undesirable deposits, and to reduce both tool wear and the tendency for vibration and chatter, (iii) safety in use for the operators, and (iv) removal of the chips produced from the cutting zone by flushing.

The hardness of the metal is an important property. As hardness increases, so does the tool wear, and thus lubricating properties of cutting fluids become significant. Almost as important as the type of metal is the severity of the machining operation in cutting fluid selection (Figure 8.3). Turning is the least severe operation, and heat is easily removed by the cutting fluid and the unconfined flow of the chip. Internal broaching is the most severe operation. Each cutting tool is in contact with the workpiece during the entire cut, and with each succeeding cut, the tool is buried deeper in the workpiece. This makes it difficult for the cutting fluid to reach the tool tip during machining. Consequently, cutting fluids used in broaching must have very good anti-weld and lubricating properties to protect the cutting tool and ensure proper finish and accuracy. Figure 8.3 shows that increasing operational severity is usually accompanied by decreasing cutting speed. More severe operations demand more active cutting fluids. This means that cutting fluids with additives, particularly of the extreme pressure type, must be used under severe

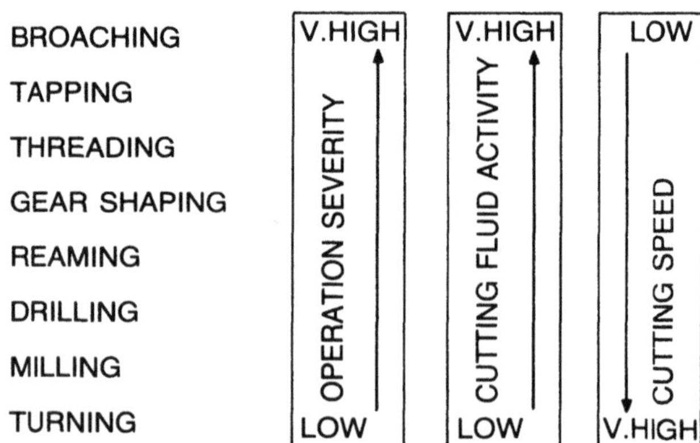

Figure 8.3 Relative severity of some metal cutting processes (redrawn and adapted from Weindel, 1981).

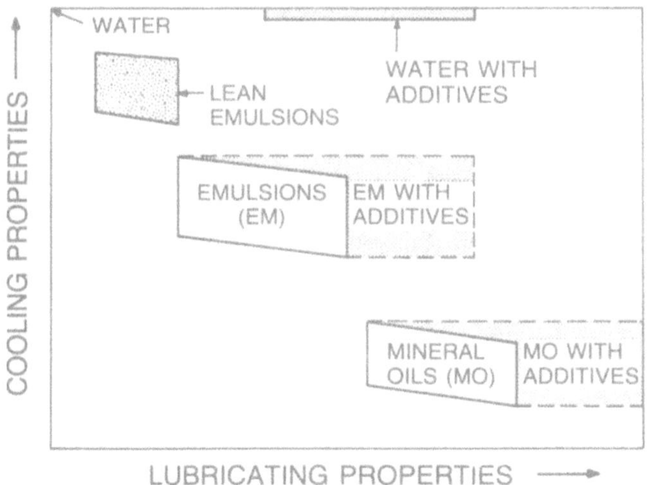

Figure 8.4 Lubricating and cooling properties of metalworking liquids.

operational conditions. According to Hunz (1984) today's cutting fluids contain a wide variety of speciality chemicals designed to supply a definite degree of lubricity, surface activity, stability and anti-weld properties.

The influence of operational severity on the cutting fluid requirements may be summarised as follows:

(i) High cutting speeds normally generate higher temperatures. In these cases, low cutting fluid activity is required, but cooling of the cutting area becomes crucial and thus metal cutting lubricants that possess great cooling power are of importance. Water is one of the best coolants known, but is a very poor lubricant. Consequently, water-based products containing wetting agents, extreme pressure additives, corrosion inhibitors and other chemicals are the frequent choice of metal cutting fluids for turning, milling, drilling and other processes of low operational severity. Figure 8.4 shows that some metal cutting fluids having excellent cooling properties may also provide good lubricating properties.

(ii) In lower speed processes that involve high tool wear and friction, such as gear cutting, tapping, or broaching, lubrication is critical and cooling is of lesser importance.

In such cases, associated with very high operational severity, oil-based lubricants with additives, mainly of the extreme pressure type, are most frequently used.

The chemistry of additives used in metalworking lubricants has recently been described in detail by Kajdas (1989).

8.8 Specialities

8.8.1 *Process oils*

Process oils are not typical lubricants and are mostly used as processing aids in manufacture. They are generally additive-free mixtures of crude oil hydrocarbons and include products such as (i) medicinal white oils, (ii) technical white oils, (iii) bright process oils, and (iv) dark process oils. Medicinal white oils are composed exclusively of isoparaffins and alkylnaphthenes. Technical white oils are less refined products than medicinal white oils and are composed of saturated hydrocarbons, though they may also contain a slight amount of aromatic compounds. Bright process oils include both yellow raffinates and brown distillates. Dark process oils are extracts from solvent refining of mineral base oils.

Process oils are widely used in various industrial processes, including rubber, plastics, pharmaceutical, food, cosmetics, printing ink, textile and other industries.

8.8.2 *Textile oils*

These oils are used in the fibre and textile industries either to lubricate the textile machinery, or as components of process oils used for the working of natural fibres, the production and processing of synthetic fibres, or the finishing of intermediate or final products. Textile oils are often made from technical white oils and oxidation inhibitors, plus agents ensuring removal of the oils by washing, even after a long period of use. High quality products also contain additives that assist in removing oil stains from the fabrics.

8.8.3 *Slideway oils*

Slow moving slides and tables in machine tools are subject to a jerky motion due to alternating slipping and sticking of the sliding surfaces. Consequently, specially developed lubricants are required to prevent and/or reduce the stick-slip phenomenon. Slideway oils usually contain polar surface-active compounds, mostly fatty acid derivatives, together with anti-wear additives and oxidation inhibitors. The polar surface-active agents form oriented boundary layers which prevent adhesive friction.

8.8.4 *Cylinder oils*

These products are highly viscous mineral oils produced from raffinates of high boiling vacuum residues, or from brightstocks and fatty oils. The latter improve water displacement characteristics of the cylinder oil and provide better adhesion of the lubricating film to the cylinder walls. Cylinder oils, also known as compounded cylinder oils, are used for lubricating cylinders, valves, and other moving parts of steam engines.

8.8.5 *Other lubricants and related products*

Coupling fluids are low viscosity lubricants with high oxidative stability. Low viscosity gear oils and engine oils may also be used for simple transmissions.

Lubricating oils for precision instruments and large clocks must have excellent oxidation stability, good anti-wear and anticorrosion properties, and retain their viscosity over a long period of time. Sometimes small amounts of gel-forming soaps, e.g. aluminium soap, are added in order to reduce the creeping effect.

Wire rope lubricants include straight mineral oils, bituminous materials, adhesive compounds, and water displacement agents. A wide variety of product types is used depending upon the severity of service conditions. Where wire ropes are used in exposed locations, the lubricant should be resistant to weathering.

Efforts have also been made to cover several applications with one oil. Thus, some multipurpose lubricants have been proposed, e.g. combinations of hydraulic lubricant, metalworking fluid and gear lubricant. Usually, multipurpose lubricants require relatively high levels of additive treatment.

Finally, there are oils whose function is not related to lubrication, such as heat transfer oils, quenching oils and insulating oils. The latter encompass a wide variety of products, including transformer oils, cable oils, condenser oils and switch gear oils.

References

Bartz, E.J., Holinski, R. and Xu, J. (1986) Wear life and frictional behaviour of bonded solid lubricants. *Lubr. Eng.* **42** 762–769.

Booser, R.E. (ed.) (1983) *CRC Handbook of Lubrication — Theory and Practice of Tribology*. Two Volumes. CRC Press Inc., Boca Raton, Florida.

Chambat, F., Mashermes, M. and Hendricks, H. (1987) Organometallic compounds produced during aluminium cold rolling. *Lubr. Eng.* **43** 522–527.

Hunz, R.P. (1984) Water based metalworking lubricants. *Lubr. Eng.* **40** 549–553.

Kajdas, C. (1987) About an anionic-radical concept of the lubrication mechanism of alcohols. *Wear* **116** 167–180.

Kajdas, C. (1989) Additives for metalworking lubricants. A review. *Sci. Lubr.* **1** 385–409.

Kalpakjian, S. and Jain, S.C. (eds.) (1980) *Metalworking Lubrication*. American Society of Mechanical Engineers, New York.

Klamann, D. (1984) *Lubricants and Related Products*. Verlag Chemie, Weinheim, Deerfield Beach (Florida) and Basel.

Lindsey, A.R. and Russel, J.C.D. (1984) In *Lubrication in Practice*. Robertson, W.S. (ed.) Marcel Dekker, New York and Basel.

Lipp, L.C. (1976) Solid lubricants—their advantages and limitations. *Lubr. Eng.* **32** 574–584.

Miller, J.W. (1989) Compressor lubricants. *Syn. Lubr.* **6** 102–122.

Montgomery, R.S. (1965) The effect of alcohols and ethers on the wear behaviour of aluminium. *Wear* **8** 466–473.

Mori, S., Suginoya, M. and Tamai, I. (1982) Chemisorption of organic compounds on a clean aluminium surface prepared by cutting under high vacuum. *ASLE Trans.* **25** 261–266.

Nachtman, E.S. and Kalpakjian, S (1985) *Lubricants and Lubrication in Metalworking Operations.* Marcel Dekker Inc., New York and Basel.

Olds, W.J. (1973) *Lubricants, Cutting Fluids and Coolants.* Cahners Books, Boston.

Rasp, R. (1989) Water based hydraulic fluids containing synthetic components. *Syn. Lubr.* **6** 233–251.

Rowe, C.N. and Murphy, W.R. (1974) In: *Proc. Tribology Workshop.* Ling, F.F. (ed.) National Science Foundation, Washington D.C.

Schey, J.A. (1977) *Introduction to Manufacturing Processes.* McGraw-Hill, New York.

Schey, J.A. (1983) *Tribology in Metalworking: Friction, Lubrication and Wear.* American Society of Metals, Metals Park, Ohio.

Weindel, H.F. (1981) Elements of selecting and using metal-cutting fluids. *Tooling and Production* **43** 66–71.

9 Aviation lubricants

A.R. LANSDOWN

9.1 Introduction

There are three factors which dominate all aspects of aircraft design. The first is the need for the highest possible reliability. With a few minor exceptions, aviation is the only activity in which it is impossible to stop to investigate or rectify a failed mechanism on the spot.

The second factor is the need to minimise the weight and volume of all components. This results in high specific loading in all mechanisms, and therefore in high specific power dissipation, so that operating temperatures are high.

The third factor is the extreme range of environmental conditions encountered, from temperatures of $-60\,°C$ on the ground, or even $-80\,°C$ in the stratosphere, to over $200\,°C$ skin temperatures in supersonic aircraft, and with pressures from 1 bar down to less than 100 millibars (Air Ministry, 1960).

As a result of these factors the lubrication requirements of aircraft are so critical that only in a few cases can lubricants developed for non-aircraft use be used in aircraft. This has not always been the case; the mineral oil or castor oil lubricants used in the earliest aircraft were all standard automotive or marine products.

World War I led to the recognition of the need for special lubricants in aircraft engines (Air Board, 1918). Before then, aircraft rarely climbed higher than a few thousand feet and the mechanical reliability of engines was so poor that lubricant reliability was not a limiting factor. By 1918, however, aircraft were flying regularly as high as eighteen thousand feet, and flights often lasted as long as five hours.

The high flights brought problems of low temperatures, with a resulting need for a high level of refining and good viscosity–temperature characteristics. Long flights necessitated clean oils with low deposit formation. The castor oil in rotary engines gave no problems, for reasons which will be explained later, but the long-range bombers and flying boats did not use rotary engines and their needs led to a steady improvement in the quality of engine oils.

The divergence between ordinary automotive engine oils and aircraft engine oils widened during the 1930s when there was a steady increase in the use of additives in automotive oils. These were considered undesirable for aircraft use and aircraft engine oils remained largely additive-free.

With the introduction of gas turbines for aircraft propulsion, it was no longer possible to base aircraft engine lubrication on mineral oil lubricants and a completely new class of lubricant had to be devised for aircraft use. For the first time lubricants developed for aircraft use spread downwards, accompanying the spread of aircraft gas turbines into industrial and marine use. By this time, however, lubricant technology had advanced to the stage where a new class of lubricant could be developed on a sound scientific basis, even if some uncertainties and misunderstandings still persisted.

The recognition of the critical nature of aircraft lubrication brought a need for detailed specifications to control the quality and performance of the lubricants and for over 40 years practically all lubricants used in aircraft have had to meet stringent specification requirements.

The great expansion of international air transport after 1945 and the formation of major military alliances led to great pressure for the standardisation of aircraft fuel and lubricant specifications. The logistic problems of supplying large numbers of different lubricants, and the potential hazards of using the wrong lubricant, have also led to great efforts to reduce the number of lubricants permitted to be used in aircraft. The result is that most aircraft lubrication requirements are now met by a relatively small number of closely-controlled high-quality products.

9.2 Lubrication of rotary engines

In the aviation context, the term 'rotary engine' refers to the class of reciprocating piston-engines in which an assembly of radially-mounted cylinders rotates around a stationary crankshaft.

Rotary engines were a major factor in aircraft propulsion for only about ten years. During that short period, however, they made a vital contribution to military aviation in World War I. The first rotary engine to fly in an aircraft was a 7-cylinder Gnome used by Louis Paulhan in a Voisin in June 1909 (Gunston, 1986). By 1917 they were used in thousands in many of the best British and French scout (fighter) aircraft but by 1920 production had virtually ceased, although some remained in service until about 1930.

Because of the difficulty of providing a controlled fuel or lubricant feed to the rotating cylinder assembly, the lubricant was supplied in the fuel feed. The high centrifugal forces caused rapid loss of lubricant from the piston/cylinder interface so the technique of dissolving a mineral oil in the fuel, as in modern small two-stroke engines, would have left an inadequate oil film on the cylinder walls. It was therefore standard practice to use a vegetable oil insoluble in the fuel, and the usual choice was castor oil. To overcome the problem of rapid lubricant loss, the concentration of castor oil in the fuel was often as high as 30%. Castor oil and other vegetable oils have very good boundary lubrication characteristics, and the standard of lubrication in all the engine components was generally good. The disadvantage of these lubricants is their poor thermal and oxidative stability.

The oil supply in rotary engines was a total-loss system and oil consumption was very high. Even in the most economical engines, built by the Le Rhone company, oil consumption was about 4 l/h in a 90 hp engine. As a result of this high consumption, the rapid degradation of the oil was not a problem but the engine surfaces rapidly became lacquered with a varnish-like film of polymerised degraded lubricant.

Small quantities of aviation-grade castor oil lubricant are still manufactured (e.g. Castrol R40) for use in the small number of antique rotary-engined aeroplanes still able to fly.

9.3 Lubrication of conventional aircraft piston engines

Apart from the rotary engines described above, piston engines can all be classified as radial or in-line. In-line engines may have either one bank of cylinders, horizontal or vertical, or they may have two or more banks in various arrangements. Radials may have one, two or four rings of cylinders each containing from three to nine cylinders mounted radially about an axis parallel to the direction of flight.

The drive to minimise weight led to increases in the power : weight ratio of piston engines of 1:3 or 1:4 hp/lb in 1918 to over 1:1 for the Rolls-Royce Merlin and Bristol Centaurus by 1945. However, after the development of successful gas turbines, the use of large piston engines decreased rapidly and by 1970 few of more than 400 hp were being manufactured.

It is probably universal in automotive engines for the oil to be contained in a sump which also encloses the crankshaft. Such a system is not always suitable for aircraft engines because turbulence or manoeuvring can lead to rapid changes in the magnitude and direction of acceleration. For radial engines the use of a sump is impossible. Many piston engines are therefore lubricated on the 'dry sump' principle in which oil returning to the crankcase or other collection points is removed by a scavenge pump and transferred to a separate oil tank usually via a cooler. However, small horizontally-opposed in-line engines commonly operate with a wet sump. The oil is then fed to the various lubrication points by a pressure pump which has a lower capacity than the scavenge pump. This ensures that oil is efficiently scavenged and cannot accumulate in the engine.

The earliest piston engines used either mineral or vegetable oil lubricants, and in many cases it is now difficult to find out which was preferred. By 1919 the situation had generally stabilised; castor oil was always used for rotary engines, but mineral oil was widely used in other engines. Some engines could use either (Cirrus, 1929).

The choice between mineral oil and castor oil in those early years was not as obvious as it now appears. Castor oil had excellent boundary lubrication characteristics and was therefore more forgiving of poor design features which led to inadequate hydrodynamic film formation. Less avoidable problems, such as excessive dynamic loading and the temporary problem of re-starting a hot engine, also benefited from the better boundary lubrication of castor oil. Mineral oils, on the other hand, were inferior to those of even 20 years later. Their boundary lubrication characteristics were inferior to castor oil while their viscosity–temperature characteristics and oxidation resistance were also poor by modern standards. The total replacement of castor oil by mineral oil was therefore not approached until the 1930s, either for aircraft or automotive use. Even then castor oil based lubricants were probably the most widely used in racing.

The growing dominance of mineral oil was due to improved refining, greater availability and lower cost. Improvements in refining led to the availability of large quantities of clean mineral oils with good viscosity–temperature characteristics and far better oxidation resistance. For the high cylinder temperatures of aircraft engines these two factors were all-important.

Compared with automotive engines, aircraft piston engines do not suffer from water contamination or low temperature sludge formation because of their higher operating temperatures, and they are less prone to corrosion because aircraft fuels are low in sulphur and oil consumption is relatively high. Their main problems are oxidation and foaming which are associated with dry sump operation.

It is important in aircraft engines to avoid solid deposits because the relatively lengthy supply pipes are sensitive to blocking by deposits and the motion of the aircraft tends to prevent solids from settling out in the tank. Early lubricant additives tended to produce solid decomposition products and were therefore considered unacceptable for aircraft use. Oxidation resistance was ensured by the use of highly-refined solvent-extracted base stocks, and antioxidants were only used for a few particularly demanding engines.

In recent years the development of ashless additives has reduced the risk of solid deposit formation, and ashless dispersants, antioxidants, and anti-foam agents are now permitted in some engines. In spite of this, most aircraft piston engines can still operate satisfactorily on straight mineral oils and some engine manufacturers will not approve any use of dispersant oils in their engines.

The viscosity characteristics of the oil are important. High viscosity is needed at high operating temperatures because of the high specific power and consequential high bearing loads. Good viscosity–temperature characteristics are obtained by the use of highly-refined paraffinic basestocks, and ashless dispersants can give some viscosity index improvement. In spite of

Table 9.1 Oil dilution instructions for Beechcraft 18

Starting temperature expected (°C)	Oil dilution period (min)
4 to −12	$1\frac{1}{2}$
−12 to −29	3
−29 to −46	5

Oil temperature 40 °C or below before dilution
Idle at 1000 to 1200 rpm
Add 1 min dilution for every 5 °C below −46 °C
Maintain oil temperature below 50 °C during dilution

this, oil viscosity at low temperatures is usually too high to allow the engine to respond satisfactorily when increased power is required. It may even be too high for the engine to be started at all. This leads to several constraints on engine operation.

(i) All piston engines must be run at low power after starting until a specified oil temperature is reached above which the power can be increased.
(ii) In an extended glide descent, the power must be increased at regular intervals, commonly every thousand feet of descent, to warm up and circulate the oil. Failure to do so can result in the engine failing to respond when power is again required at the end of the glide and many forced landings and even crashes have resulted.
(iii) It may be necessary to exercise the propellor pitch control at regular intervals when an aircraft with a constant-speed or variable-pitch propellor is operating in a cold environment at a steady speed and power setting. If this is not done, the small volume of oil present in the pitch-change mechanism may become too viscous to flow when pitch change is again required.
(iv) Where air temperatures are very cold and there may be doubt about whether an engine can be restarted after a shut-down, many different techniques are used to warm either the oil or the engine. One widely-used procedure is to dilute some or all of the oil in the engine with petrol before shutting down. Table 9.1 shows a typical oil dilution schedule for the Wasp engines of a Beechcraft 18. Use of oil dilution makes it necessary to warm up the engine carefully on re-starting in order to evaporate off the petrol before its boiling point is reached.

Table 9.2 shows typical properties of the common grades of straight mineral oil used in aircraft piston engines.

9.4 Lubrication of aircraft turbine engines

Aircraft gas turbine engines, colloquially known as jet engines, are much

Table 9.2 Typical properties of piston engine oils

Property	Viscosity grade				
	65	80	100	120	20W/50
Viscosity at 100 °C (cSt)	11	15	19	23	20
Viscosity at 40 °C (cSt)	95	130	200	270	140
Viscosity index	110	105	100	100	150
Pour point (°C)	−25	−23	−21	−20	−30
Flash point (°C)	230	240	250	260	230
Ash (wt%)	0.001	0.002	0.002	0.002	0.001
Total acid number (mg/g)	0.02	0.02	0.02	0.02	0.12
Sulphur (wt%)	0.3	0.3	0.4	0.4	0.1
Density at 15 °C	0.882	0.889	0.891	0.894	0.878

simpler than piston engines in design and construction, and in their lubrication requirements. There are no moving parts in the combustion chambers so the lubricant is not exposed directly to combustion temperatures. The main moving parts are the compressor and the turbine, which are in steady rotation, so the problems of reciprocating loads are avoided.

All gas turbine engines have three basic components; an air compressor to supply air to the combustion chambers, the combustion chambers themselves, and a turbine which drives the compressor and is itself driven by the combustion gases. This arrangement is most clearly seen in an early turbo-jet such as the Rolls-Royce Nene. Modern gas turbine engines can be conveniently divided into three classes with a fourth under development.

The earliest successful aircraft gas turbine engines were the true jet engines, more specifically called turbo-jets. In these engines the whole of the propulsive force is provided by the jet thrust and the turbine is designed to extract only enough power to drive the compressor and some auxiliary components.

In the second successful class, the turbo-props, or prop-jets, the turbine is designed to abstract a high proportion of the power from the combustion gases in order to drive a propellor which provides most of the propulsive force. Turbo-props are much more economical than turbo-jets at aircraft speeds below about 450 mph where propellors themselves retain their efficiency. The gas turbines used in helicopters are similar in some respects where the bulk of the power is abstracted by the turbine to drive the rotors.

The third class of gas turbine includes the bypass and fan engines, jointly classified as turbo-fans. In these engines the turbine abstracts more power than is required simply to supply compressed air to the combustion chambers. The surplus is used to drive a low pressure compressor, supplying additional airflow which bypasses the combustion chambers and combines with the combustion gases downstream of the turbine. This ensures that the jet efflux approaches the optimum conditions of high pressure and low speed. These engines, like the turbo-jets, are most effective at high airspeeds, but are more economical than the turbo-jets.

Nevertheless, the fuel economy of turbo-fans is not considered adequate in a world of diminishing fossil fuel availability and excessive carbon dioxide production. A fourth class of gas turbine, the propfan, is therefore being developed, which has similarities to both turbo-props and turbo-fans. This turbine will provide significant propulsive force from both the jet efflux and a novel form of multi-bladed propellor. It is intended to give much higher fuel economy than the turbo-fans at aircraft speeds which are too high for conventional propellor designs.

The lubrication requirements of turbo-jet and turbo-fan engines are undemanding. The bearings are all rolling contact bearings in steady rotation at high speeds so it is easy to maintain full elastohydrodynamic film lubrication. There is a wide variety of other lubricated components such as couplings, gears, actuators and bearings associated with ancillary components, but these are small, with relatively low loads and powers, and can be kept clear of the high temperature zones of the engine.

For turbo-props, helicopter engines and propfans the most demanding lubrication requirements are those of the reduction gears. These carry high power, sometimes at high torque, and may be subject to fluctuating loading.

The critical requirement for all gas turbine lubricants is their ability to cope with a wide range of temperatures. The hottest lubricated components are the turbine bearings and some interesting design features are used to cool the turbine hub. In spite of this, the oil in the turbine bearings may be subjected to temperatures as high as 280 °C (David *et al.*, 1956). The residence time of the oil during normal operation is short, but after engine shut-down the bearing temperature will often rise even higher because of heat soak from the blades after the cooling air-flows have ceased. A small quantity of retained oil will therefore be exposed to very high temperatures until the bearings cool. At the other end of the temperature range, the oil must flow easily enough to permit engine starting at specified temperatures down to −40 or −54 °C. If an engine must be restarted in flight, the ambient temperature may be even lower, exceptionally down to −80 °C.

The earliest aircraft gas turbines were lubricated with highly refined mineral oils, and some mineral oils were still in use as late as 1958. The thermal stability and low temperature viscosity of even the best mineral oils were never completely adequate for the earliest days of gas turbine propulsion.

The potential of synthetic aliphatic esters as lubricating oils was recognised as early as 1936, and they were particularly investigated for use as instrument oils (Zisman, 1957). They were probably first studied for use in gas turbines in Germany, but by 1950 they had been accepted worldwide in the gas turbine industry. Their advantages included not only high thermal stability and good viscosity–temperature characteristics, but also low volatility, low foaming tendency, lack of corrosiveness and good boundary lubrication.

The main conflict in the early years was over the choice of viscosity and this conflict still exists forty years later. From 1943 the most important gas turbines for several years were based on British designs. These designs included several turbo-props, such as the Rolls-Royce Dart and the Armstrong Siddeley Python and Mamba, and the reduction gearboxes for these engines required higher viscosity than the bearings. British use therefore focused on ester oils having viscosities of not less than 7.5 cSt at the usual reference temperature of 99 °C (210 °F).

In the US, development of turbo-props lagged by several years, and the initial lubrication requirement was only for turbo-jets. These could be lubricated satisfactorily by oils of lower viscosity, and the US Air Force requirement for winter starting at very low temperatures led to the choice of oils with a minimum viscosity of 3.0 cSt at 99 °C.

It has been said (Klamann, 1984) that over 3500 different esters were tested for lubricant use between 1937 and 1944. For aircraft gas turbines the thermal stability requirements restricted the choice to long chain aliphatic esters. The US 3.0 cSt high temperature limit could be met by several different esters of monohydroxy alcohols with dibasic carboxylic acids, but the low temperature viscosity requirements were better achieved by blending. The earliest commercial ester lubricants were therefore blends, usually of alkyl adipates, azelates and sebacates, especially di-(2-ethylhexyl)sebacate.

The British 7.5 cSt requirement could have been met by the use of longer-chain acids and alcohols but acceptable low temperature flow properties would not have been achieved. The solution was to use a slightly more viscous ester blend than used in the 3.0 cSt fluids, together with a polymeric viscosity index improver, generally a polyglycol. From the theoretical aspect the use of a viscosity index improver was satisfactory, since the highest load-carrying capacity was needed at or near the pitch line in the reduction gears where the shear rate was too low to cause significant viscosity loss. Shear rates in the bearings were high, and probably caused significant reduction in viscosity, but in these locations even the base oil viscosity was probably adequate and the shear thinning may even have helped cold starting.

The existence of two fundamentally different classes of gas turbine lubricant is unsatisfactory for standardisation purposes and is a potential hazard, since the two classes are not considered as acceptable alternatives even for emergency use. An attempt was made therefore in the mid-1960s to introduce an intermediate type of oil with a 99 °C minimum viscosity of 5.5 cSt which could replace the 3.0 cSt and 7.5 cSt oils, at least for future engines. This coincided with a new US Navy requirement for a similar oil for new turbo-prop aircraft. The new class of oil was introduced but has not helped to achieve standardisation on one grade. In fact engine manufacturers appeared to have seized with delight on the availability of three alternative viscosity grades for their later engines and all these grades are now in widespread use.

The introduction of supersonic aircraft into service during the 1960s led to higher oil temperatures and a need for improved oxidative stability. The weak

Table 9.3 Typical properties of turbine engine oils

Property	Oil type		
	3.0 cSt	5.0 cSt	7.5 cSt
Viscosity at 100 °C (cSt)	3.0	5.2	7.5
Viscosity at 40 °C (cSt)	14	29	34
Pour point (°C)	−65	−60	−60
Flash point (°C)	225	255	235
Total acid number (mg/g)	0.15	0.2	0.152
Spontaneous ignition temperature (°C)	400	420	390

point for thermal degradation on the diesters is the hydrogen on the beta carbon atom of the starting alcohol. Stability was therefore improved by fully substituting that carbon resulting in a new class of complex esters based on polyhydroxy alcohols such as trimethylolpropane, neopentylglycol and pentaerythritol, with straight-chain monocarboxylic acids up to *n*-nonanoic acid.

Apart from the use of viscosity index improvers, additives used in gas turbine lubricants include antioxidants, anti-wear additives, metal deactivators, corrosion inhibitors and anti-foaming agents.

Like aviation piston engine oils, the oils used in aircraft gas turbines are tightly controlled by specifications. There are basically four classes of specification, all primarily military. Although the major engine manufacturers may have their own specifications, these are generally in line with the military specifications.

The four classes of specification are:

(i) 3 cSt oils, originating in the US Air Force MIL-L-7808 specification and consisting of diesters/hindered esters with a small amount of additives but no viscosity index improver.
(ii) 7.5 cSt oils, originating in the British D.Eng.R.D. 2487 specification, and consisting primarily of diesters with viscosity index improver and small amounts of other additives.
(iii) 5 cSt oils meeting British specification D.Eng.R.D. 2497.
(iv) 5 cSt oils meeting the US Navy MIL-L-23699 specification.

The two 5 cSt oil specifications differ only in detail but are not completely interchangeable. They are based on hindered esters or polyol esters and have higher thermal and oxidative stability than the earlier 3 cSt and 7.5 cSt oils. They are therefore known as Type 2 oils and the earlier oils as Type 1 oils. Some properties of gas turbine engine oils are listed in Table 9.3.

9.5 Aircraft hydraulic fluids

Larger and more complex aircraft usually use hydraulics for most mechanical systems, including flying controls, undercarriage retraction, flap operation, variation of wing geometry, and others. Many of these systems are critical to the safety of flight and the quality of the hydraulic fluid is vital.

The earliest aircraft hydraulic fluids were either castor oil or mineral oil. Currently both types remain, but castor oil is now limited to a very few aircraft and is likely to disappear within a few years.

The most widely-used hydraulic fluid for over forty years has been a mineral oil meeting the British Def. Stan. 91–48/1 (previously DTD 585) and US MIL-H-5606 specifications. It is also known by the NATO designation H-515 and the British Joint Service Designation OM-15. It has a low viscosity of only 13 cSt at 40 °C, and contains up to 20% of a low molecular weight polymeric viscosity index improver, as well as antioxidant, anti-wear and other minor additives. Its most unusual property is its very high level of cleanliness which is necessary for reliability in critical systems such as flying controls and automatic pilot servos. The achievement and maintenance of this high level of cleanliness is difficult, and therefore expensive, so where it is not essential an otherwise identical but not 'super-clean' fluid is used.

Since 1955, several synthetic non-hydrocarbon hydraulic fluids have been used in aircraft. Phosphate ester fluids, especially Skydrol 500A, are widely used in large civil aircraft for their excellent fire resistance. Silicate esters have been used in a few aircraft, either alone as in Concorde or mixed with a carboxylic ester oil as in the Convair B-58. A chlorinated phenyl silicone was used successfully, if briefly, in the BAC TSR-2.

In general, apart from the phosphate esters, there is very little use of non-hydrocarbon fluids, although there is considerable effort in the US to find an acceptable synthetic fluid for use in supersonic aircraft.

9.6 Helicopter gearboxes

Helicopter transmissions involve a variety of gears, many of them high-speed or highly-loaded. They are the most critical gears in aviation because they carry the whole of the power for thrust and lift, and are almost always unduplicated. Their proper lubrication is therefore vitally important; there have been serious helicopter crashes in which lubrication played a significant part.

The most important part of the transmission is the main gearbox, which contains the most heavily-loaded reduction gears transmitting power to the main rotor blades. It is always oil-lubricated but a variety of different oils are used in different helicopters, from 7.5 cSt gas turbine engine oils to automotive mineral oils of similar viscosity.

In turbine-engined helicopters the first reduction stages operate at very high speed and are usually lubricated with the engine oil. The tail rotor gearbox is critical to directional control and stability but carries only about

5% of the total power. It may be lubricated with the same oil as the main gearbox but, because the bearings and gears are relatively small with low linear sliding speeds, they may be grease-lubricated.

9.7 Undercarriage lubrication

The undercarriage systems of most aircraft have three important but distinct lubrication requirements, the wheel bearings, the shock absorbers and the brakes.

Brake systems are in many ways simple conventional hydraulic systems and the fluids used for other hydraulic systems in the aircraft, as described in section 9.5, will also usually be used for the brake systems. In light aircraft, where the brake system is the only hydraulic system, a silicone or mineral oil brake fluid may be used. Because of the pressure to reduce the weight and volume of all components in an aircraft, as well as the high landing speeds, aircraft brakes have to absorb a lot of power in a short time within a small mass. They therefore become very hot and there is a potential risk of ignition if the brake fluid contacts a hot surface. This has been one of the main pressures leading to the use of phosphate ester hydraulic fluids in aircraft.

The heat generated in landing and taxiing large aircraft also makes the wheel bearings very hot. They are always grease-lubricated, but for large fast aircraft the greases must be capable of withstanding very high temperatures. Typical wheel-bearing greases for such aircraft are therefore highly-refined or synthetic hydrocarbon base oils with non-soap thickeners.

The shock absorbers are known as oleo legs and, as the name implies, they are oil-filled dampers usually containing some gas volume to cushion the initial touch-down impact loads. In light aircraft, and previously in most aircraft, the oleo legs use a mineral oil fluid with a free airspace. At higher temperatures and high compression ratios this combination can cause compression-ignition or 'dieselling' and some high-speed aircraft with high undercarriage loads have been found to be particularly prone to this problem. The problem can be avoided either by the use of a silicone-based fluid instead of mineral oil or by using nitrogen instead of air.

9.8 Airframe lubrication

All airframes include a variety of linkages, hinges, pivots, latches, bearings and other components requiring lubrication. Such components are usually unenclosed or too isolated for convenient incorporation in a circulating oil system. They are usually lubricated by grease although oils, anti-seizes and solid lubricants may be used where appropriate.

It is easy, but absolutely erroneous, to consider these types of component as less critical than the engines of an aircraft. An aircraft should very often be able to survive the failure of an engine, but it is much less likely to survive the failure of a flying control or a single flap. Even the failure of an undercarriage lock has caused fatalities.

Just as with engine oils and hydraulic fluids, it is important to minimise the number of different lubricants in use, both for logistic reasons and to avoid misuse. Tight control is therefore exercised over the selection and quality control of airframe lubricants. The variety of these lubricants is too great for detailed coverage, but a few examples follow.

Flying control hinges may operate on plain or rolling bearings, but they are usually grease-lubricated. The bearings are exposed to the full range of environmental temperatures and may be heated by supersonic skin friction or by rapid operation. The greases must therefore be able to withstand a range of temperature from −60 °C to +150 °C. For many years these requirements were met by diester base oils thickened with lithium soap, but more recently there has been a move towards synthetic hydrocarbon oil with a calcium or aluminium complex thickener. It is important with greases to ensure that all products approved to the same specification are fully compatible when mixed.

For highly-loaded pivots, such as variable-incidence tailplanes and the hinges of flying controls, flaps and slats, solid lubrication may be used. One high performance type consists of PTFE with a woven glass fibre reinforcement bonded to a metal backing. This type of bearing is effective at temperatures from −70 °C to over 200 °C and at contact pressures up to 100 MPa (20,000 psi).

Sparking plug threads in piston engines are lubricated with an anti-seize compound consisting of equal parts of petrolatum and graphite, and this material can also be used for other highly-loaded but slow-moving fittings at normal or high temperatures.

9.9 Safety aspects of aircraft lubrication

The potential safety aspects of aircraft lubrication are, of course, obvious. All the components and systems in aircraft which are critical for safe operation involve lubrication. If, therefore, the insistence on correct formulation, specification, selection and application of aircraft lubricants is justified, then it must follow that failure of lubricant or lubrication system is likely to be hazardous.

In practice there is no evidence that lubricant or lubrication failures are a major cause of aircraft accidents. The proportion of aircraft accidents due mainly to mechanical problems is in fact quite small (Civil Aviation Authority, 1987). This is probably due in large part to the fact that a combination of pilot skill and well-designed emergency procedures enable most mechanical failures to be handled without further damage to the aircraft or injury to the passengers.

This is particularly true in the case of engine failures. Most engine failures are caused by fuel starvation or fatigue but a small number are due to lubrication failure. However, it is comparatively rare for an accident to result. In multi-engined aircraft the failure of one engine should not prevent continuation of the flight, but even in the event of complete power failure the controllability of the aircraft would normally not be compromised.

During the early 1980s a three-engined wide-body airliner which had taken off from Miami suffered oil loss from all three engines because the magnetic plugs had been replaced without the oil seals. As a result all three engines failed many miles offshore but the aircraft was successfully glided back to land and a safe landing was made after one engine was restarted in the last few minutes.

A survey (Lansdown and Roylance, 1989) of over 700 aircraft accidents in the United Kingdom between 1984 and 1988 showed only nine which were directly related to bearing failures. One of these was caused initially by galling and one by excessive wear, either of which might have been due to a lubrication fault. One was due to failure of a phenolic bearing. At least four of the remaining six were caused by an inadequate supply of lubricant. In one case this was due to a poor flow path for the grease in an undercarriage support beam pivot, but in the other three cases it appears to have been because of failure to regrease the mechanisms often enough.

Overall it seems clear that the standard of lubricants for aircraft is generally satisfactory and that in the very few cases where lubricant failure results in an accident the cause is likely to be a failure to supply lubricant and not any inadequate quality of the lubricant itself.

9.10 Space lubrication

The greatest constraint on lubrication of spacecraft is imposed by high vacuum. The volatility of oils, greases, and even linear polymers is such that they will outgas (i.e. evaporate) quickly enough to impede their effectiveness in space. In spite of this both high-vacuum oils (mineral and synthetic) and greases have been used successfully in many space applications where efficient shielding has been carried out. Even non-contacting labyrinth seals can be effective in space vacuum provided that the labyrinth clearances are less than the mean free path of the lubricant molecules. Many problems can be avoided by designing spacecraft mechanisms to avoid the exposure of lubricated components to full space vacuum, but where this is not possible solid lubricants are often used very successfully.

Graphite depends for its low friction on adsorbed films of water or other volatile compounds and is therefore not usable in space vacuum. PTFE has been used in some applications, but it outgases relatively quickly.

The best of the conventional solid lubricants for space use is molybdenum disulphide, for which low friction is an inherent property of the crystal structure and is at its best when free of adsorbed substances. In air, the wear life of molybdenum disulphide is limited mainly by oxidation (Salomon *et al.*, 1964). In the vacuum of space oxidation cannot occur and the wear life is very much extended even at temperatures as high as 930 °C (Brainard, 1969).

One interesting lubrication technique for space mechanisms is the use of lead films to lubricate rolling contact bearings. Lead is almost completely ineffective as a lubricant in air because of surface oxidation but in space vacuum where oxidation cannot occur, its low adhesion to steel leads to low friction and good wear life in rolling contact.

References

Air Board (1918) *Manual of Clerget Aero-Motors and their Installation in Various Aircraft.* Air Board Issue No 243.

Air Ministry, Meteorological Office (1960) *Handbook of Aviation Meteorology.* HMSO, London.

Brainard, W.A. (1969) The thermal stability and friction of the disulfides, diselenides and ditellurides of molybdenum and tungsten in vacuum (10^{-9} to 10^{-6} torr). *NASA TN D-5141.*

Cirrus Aero Engines Ltd. (1929) *Notes on the Care and Maintenance of the 85/95 h.p. 'Cirrus' (Mk.III) Aero Engine.* Cirrus Aero Engines Ltd., London.

Civil Aviation Authority (1987) *General Aviation Accident Review 1987 (CAP 543).* See also other CAA and ICAO surveys.

David, V.W., Hughes, J.R. and Reece, D. (1956) Some lubrication problems of aviation gas turbines. *J. Inst. Pet.* **42**(395) 330.

Gunston, B. (1986) *World Encyclopaedia of Aero Engines.* Guild Publishing, London.

Klamann, D. (1984) *Lubricants and Related Products.* Verlag Chemie, Weinheim.

Lansdown, A.R. (1987) Lubricants. *Proc. I. Mech.E. Intl. Conf. on Friction and Wear*, London, 1987, Paper No.C244/87.

Lansdown, A.R. and Roylance, B.J. (1989) The contribution of bearing failures to aircraft accidents. *Proc. I. Mech.E. Seminar on Aerospace Bearing Technology*, Solihull, 17 May 1989.

Salomon, G., De Gee, A.W.J. and Zaat, J.H. (1964) Mechano-chemical factors in MoS-film lubrication. *Wear* **7** 87.

Zisman, W.A. (1957) *A Decade of Basic and Applied Science in the Navy.* US Government Printing Office, Washington.

10 Marine lubricants

B.H. CARTER

10.1 Introduction

The lubrication of marine diesel engines presents its own particular problems. These arise from the sheer size of the engines, their high efficiencies and the fuel they burn. Cylinder bore diameters can be in excess of 1 m and their high efficiency is achieved with firing pressures of 150 bar resulting in liner temperatures of over 200 °C. The fuels burned include poor quality residual fuels with viscosities of 100 cSt at 80 °C, sulphur contents of 4% or more and carbon residues of up to 22% wt. In addition, the crankcase oil charge is simply topped up and only drained if excessively contaminated.

Until recently, any article concerning the lubrication of marine engines would have included a section on steam turbines but, due to the rapid escalation of fuel prices in the late 1970s, their use has been virtually discontinued. The improved design and efficiency of modern marine diesel engines means that steam turbines are no longer used even for cruise vessels or large (250 000 tonnes) crude oil carriers.

Marine lubrication includes the use of ancillary grades such as hydraulic oils, compressor oils, gear oils, grease, etc. Their application is covered in chapters 8 and 11 but a sound knowledge of these and other grades is required by anyone connected with marine lubrication.

The supplier of marine lubricants is faced with considerations not always applicable to other lubricating oils. Any given marine lubricant has to be available at the same quality, at relatively short notice and at literally hundreds of ports throughout the world. In addition, the ship operator frequently requires advice on lubrication and related problems, together with a used oil analysis service.

10.2 Marine diesel engines

Diesel engines can be classified by speed as shown in Table 10.1. The majority of marine propulsion engines are slow speed, two stroke engines and the remainder are mainly medium speed engines. Principal characteristics of the two engine types are outlined in the following sections.

Table 10.1 Classification of diesel engines by speed.

Type	Speed (rpm)	Bore (mm)	Output (bhp/cyl)
Slow speed	60–200	300–1050	750–5500
Medium speed	300–1000	300–700	100–2000
Medium/high speed	600–1500	200–400	100–300
High speed	600–2000	100–200	20–200

10.2.1 *Slow speed engines*

Most slow speed diesel engines run at speeds in the range 60–120 rpm and operate on high viscosity residual fuel oil. Propeller efficiency is inversely proportional to engine speed and their low speed enables these engines to be coupled directly to the propeller shaft without the use of a gearbox. Omission of the gear box aids mechanical efficiency. Slow speed diesel engines are frequently referred to as crosshead engines because of their construction. For each cylinder, the piston rod and connecting rod are linked to a reciprocating block, the crosshead, which usually slides up and down in guides. The crosshead separates the firing cylinder from the crankcase, and a stuffing box completes the seal. The cylinders and crankcase are lubricated separately by cylinder and system oils respectively.

The cylinder oil is fed to the cylinder walls through a number of injection points called quills. Each cylinder can be fitted with between 4 and 16 quills, depending upon the cylinder stroke and bore. The quills are arranged circumferentially at either one or two levels and the oil is delivered to them by pumps which may be hydraulically or engine driven.

The system oil is used for forced lubrication of the various bearings and crosshead guides present in the crankcase. In some engines it is also used for cooling the piston undercrowns.

10.2.2 *Medium speed engines*

Medium speed engines may be defined as trunk piston engines having a rated speed of between 300 and 1000 rpm. In contrast to the slow speed engines, the connecting rod is attached directly to the piston by a gudgeon pin. Their design is similar to the normal diesel engine but is on a considerably larger scale. Most are designed to operate on blended or heavy fuel oil.

When used as propulson units, these engines are coupled to the propeller shaft through a gear box because of their speed. Auxiliary engines, e.g. engines used to provide electrical power when the ship is in port, are normally medium speed engines.

In trunk piston engines, a single lubricant is used for crankcase and cylinder lubrication. All major moving parts of the engine, i.e. main and big end bearings, camshaft and valve gear, are lubricated by a pumped circulation

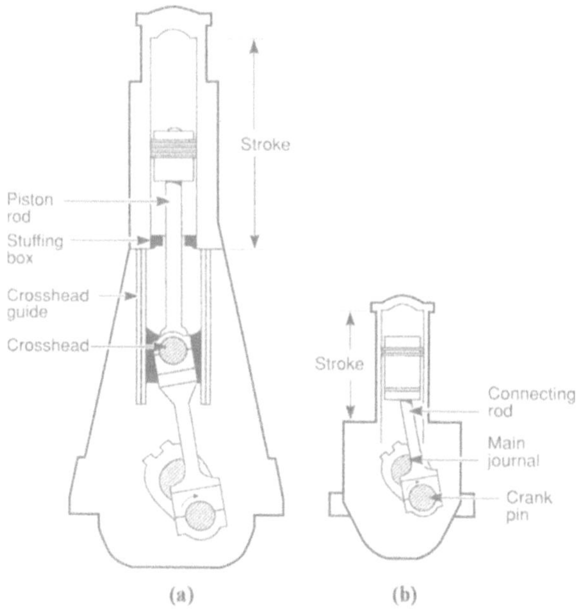

Figure 10.1 Simplified comparison of: (a) crosshead and (b) trunk piston engines.

system. The cylinder liners are lubricated partially by splash lubrication and partially by oil from the circulation system which finds its way to the cylinder wall through holes in the piston skirt via the connecting rod and gudgeon pin. The latest trend is for medium speed engines to have direct cylinder lubrication—as in slow speed engines—with the excess lubricant returning directly to the crankcase.

Figure 10.1 is a simplified diagram illustrating the design features of the crosshead and trunk piston engines. The diagram also indicates the relative sizes of the two types of engine.

Typical characteristics of a selection of crosshead and trunk piston engines are given in Tables 10.2 and 10.3 respectively. The different engine types find

Table 10.2 Characteristics of large bore crosshead engines.

Manufacturer Engine type	MAN B&W L90 MC	Sulzer RTA 84M	Mitsubishi 8UEC 75LS
Bore (mm)	900	840	750
Stroke (mm)	2916	2900	2800
Stroke:bore ratio	3.24	3.45	3.73
Engine (rpm)	74	78	84
Bmep (bar)	16.2	16.6	17
P max. (bar)	125–130	130	135
Output (bhp/cyl)	5310	4760	4000

Table 10.3 Characteristics of medium speed trunk piston engines.

Manufacturer Engine type	MAN B&W L 58/64	MAK M601	Pielstick PC4-2	Sulzer ZA 40S	SWD TM 620	Wartsila Vasa 46
Bore (mm)	580	580	570	400	620	460
Stroke (mm)	640	600	620	560	660	580
Engine (rpm)	428	425	400	510	425	514
Bmep (bar)	22.3	19.9	21.5	24.1	20.4	25
Output (bhp/cyl)	1650	1500	1650	885	1925	1230

Table 10.4 Typical applications for engine types.

Slow speed	Medium speed	Medium/high speed	High speed
Main engines for: General cargo Container vessels Bulk carriers Tankers	Main engines for: Cruise liners Ro-ro ferries Container vessels Auxiliary power for all ships Electric power generation	Tugs Inland waterways Trawlers Locomotives Small ferries Power generation	Small boats Pumps Auxiliary generators Compressors

typical applications as shown in Table 10.4. An impression of the size of a modern crosshead engine can be gained from Figure 10.2.

10.3 Fuel oil

Slow speed crosshead engines operate on heavy fuel oil, with marine trunk piston engines using a range of fuels from marine diesel to heavy fuel oil (General Council of British Shipping, 1983). Although heavy fuel oil contains a complex and variable mixture of refinery distillates and by-products, e.g. vacuum residue, visbroken residue, catalytically cracked bottoms, propane de-asphalted residue, lubricating extracts, slop wax, etc., no standard for marine fuel oil quality existed until the introduction of BS MA100 in 1982. Fuel was purchased simply on the basis of viscosity and density. Since then, International Standard ISO 8217 (1987) and CIMAC (Conseil International des Machines à Combustion) recommendations (1986) have been published. The current British Standard, BS MA100:1989 is identical with ISO 8217:1987 and an extract from Table 2 of this standard 'Requirements for marine residual fuels' is given in Table 10.5. These publications do not include any measure of fuel ignition quality because no official method exists. Various procedures have been proposed of which the most successful is probably the calculated carbon aromaticity index (CCAI) (Zeelenberg *et al.*, 1983), based on viscosity and density measurements. Many combustion studies on heavy

Figure 10.2 MAN B&W type 6L 90MC in the builder's shop: cylinder bore 900 mm; stroke 2916 mm; output 23,460 kW (31,170 BHP) at 78 rpm (courtesy of MAN B&W Diesel A/S Copenhagen, Denmark).

fuel oil have been reported (Ruzicka *et al.*, 1984; David and Denham, 1984; Barnes *et al.*, 1987; Sjöberg, 1987; Negus *et al.*, 1987).

The quality of marine fuel has a direct impact on engine operation, and fuel quality has deteriorated in recent years with the increased use of secondary refining processes (Holbrook and Fabriek, 1989). This deterioration is a direct consequence of the drive to maximise the yield of premium products from crude oil which has produced heavier residues with poorer combustion properties and more impurities. Engine design has been continuously improved to combat the poorer combustion properties (Chapuy, 1986; Schmidt-Sorensen and Sunn Pedersen, 1989; Aeberli *et al.*, 1989). Improved centrifugal separators and procedures have been introduced to reduce impurities such as water, ash and catalyst fines (Bengtsson, 1986; Sprague, 1986; Bantour and Chapuy, 1983). Improved lubricant design can also help engine performance, e.g. increased detergency/dispersancy to solubilise deposits resulting from poor combustion, increased basicity to neutralise sulphur acids and added antioxidants to improve oxidation stability.

Table 10.5 Requirements for marine residual fuels.

Characteristics	Test method	Limit	Designation ISO-F- RMA 10	RMB 10	RMC 10	RMD 15	RME 25	RMF 25	RMG 35	RMH 35	RMK 35	RML 35	RMH 45	RMK 45	RML 45	RMH 55	RML 55
Density at 15 °C (kg/m)	ISO 3675	max.	975.0	991.0		991.0	991.0		991.0		—		991.0	—		991.0	—
Kinematic viscosity at 100 °C (cSt)	ISO 3104	max.	10.0			15.0	25.0		35.0				45.0			55.0	
Flash point (°C)	ISO 2719	min.	60			60	60		60				60			60	
Pour point (upper) (°C)																	
winter quality	ISO 3016	max.	0	24		30	30		30				30			30	
summer quality		max.	6	24		30	30		30				30			30	
Carbon residue, Conradson (%, m/m)	ISO 6615	max.	10		14	14	15	20	18	22		—	22		—	22	—
Ash (%, m/m)	ISO 6245	max.	0.10			0.10	0.10	0.15	0.15	0.20			0.20			0.20	
Water (%, V/V)	ISO 3733	max.	0.50			0.80	1.0		1.0				1.0			1.0	
Sulfur (%, m/m)	ISO 8754	max.	3.5			4.0	5.0		5.0				5.0			5.0	
Vanadium (mg/kg)	—[a]	max.	150		300	350	200	500	300	600			600			600	

[a] An international Standard for the determination of vanadium is being prepared. Pending its completion vanadium shall be determined by the method DIN 51 790, Part 2.

10.4 Base oils

Most modern marine lubricants are prepared from good quality paraffinic base oils although traditionally naphthenic base stocks were preferred. Paraffinic base oils have better oxidation resistance, a higher viscosity index and lower volatility but give harder carbon deposits. However, modern additive technology can modify the hard deposits allowing paraffinic base oils to be used and thereby gaining the advantage of their other superior properties.

Marine lubricants are supplied as monograde oils ranging from SAE 20 to SAE 60, with the SAE 30 and 40 grades predominating for trunk piston engine oils and SAE 50 as the main cylinder oil grade (SAE J-300, 1987). System oils are invariably SAE 30. The complete range of viscosities can be blended from three base stocks which are generally 150 SN, 500 SN and Brightstock. All base oils are screened and approved by the marine lubricant supplier to ensure that the quality of their products is consistent worldwide.

Synthetic lubricants are starting to find limited specialised applications and trunk piston engine oils of 12 and 30 total base number (TBN) are available. They are currently based on polyalphaolefins with the inclusion of esters to improve additive solubility and seal compatibility.

10.5 Additives

The main types of additives used for formulating marine lubricants are:

Alkaline detergents
Dispersants
Antioxidants
Corrosion inhibitors
Anti-wear and extreme pressure additives
Pour point depressants

Many additives are multifunctional and their properties and functions, with particular reference to marine applications, are discussed below.

10.5.1 *Alkaline detergents*

Overbased calcium detergents, with total base numbers ranging from 250–400, form the backbone of the majority of marine lubricants. Although their principal function is to supply alkalinity to neutralise sulphur acids resulting from the high sulphur fuels, they also contribute some detergency. Materials used include calcium sulphonates (natural and synthetic), phenates, salicylates, carboxylates and naphthenates. Except for calcium sulphonates, where all the alkalinity is in the form of micellar calcium carbonate, the alkalinity is built into the detergent molecule itself. Most formulations are based on a balanced blend of calcium sulphonates and phenates.

Low base detergents, such as calcium sulphonates, with a soap content of approximately 40% and TBN of 5–25, are frequently included to maintain engine cleanliness and provide additional protection against rust and corrosion.

10.5.2 *Dispersants*

Ashless dispersants, like detergents, are included to improve engine cleanliness. Most are of the polyisobutylene–succinimide or polyisobutylene–succinate ester type and three basic structures can be identified; mono-succinimides, bis-succinimides and succinate esters. Bis-succinimides are normally used for marine applications because although they are less effective in peptising low temperature sludge found in gasoline engines than mono-succinimides, they give better diesel performance by reducing lacquer formation.

10.5.3 *Oxidation inhibitors*

Both chain terminating oxidation inhibitors, e.g. hindered phenols and amines, and peroxide destroying inhibitors, e.g. dithiophosphate and dithiocarbamates, can be included in marine formulations. Mixtures of phenols and amines are often used for synergy but they must have good high temperature performance. The sulphur-containing oxidation inhibitors also have extremely useful anti-wear properties. Oxidation inhibitors can be used advantageously in some base oils refined from low sulphur crudes and in synthetic base stocks.

10.5.4 *Corrosion inhibitors*

Not unexpectedly, marine crankcase oils are sometimes contaminated with water which is normally removed by the lubricating oil centrifuge. To assist further in protecting against rusting, inhibitors such as alkyl sulphonates, phosphonates, amines and alkyl succinic acids/esters are added. They work by forming a hydrophobic film on the metal but must be selected with due regard to the other additives present.

Non-ferrous metals, although resistant to attack by oxygen and water can be corroded by acids arising from the products of combustion and oxidation of the lubricating oil. Corrosion can be combated in two ways; neutralisation with low base alkaline earth detergents and the formation of a protective barrier. Oxidation inhibitors can also assist by preventing the formation of acids through oxidation of the lubricant.

10.5.5 *Load carrying*

Load carrying requirements are generally less severe for marine than automotive applications, e.g. there are no specialised cam and tappet test requirements. Nevertheless, good load carrying properties are required particularly for crankcase oils. The requirements are achieved through the use of compounds such as zinc dithiophosphates, dithiocarbamates, sulphurised fatty esters, disulphides, sulphurised alkenes, etc.

10.5.6 *Pour point depressants*

The removal of wax in the refining of a base oil is moderately expensive and for this reason paraffinic base oils are normally only produced with pour points of no lower than −12 °C. Pour points of this level are not acceptable for many marine applications, e.g. oil used in deck machinery, and pour point depressants are therefore added, usually in the range 0.1–0.5%. Materials used include polyalkylmethacrylates, alkyl naphthalenes and alkylated wax.

10.6 Properties and formulation of marine lubricants

Typical properties of the three types of marine diesel engine lubricants are summarised in Table 10.6. The three types of oils have quite different performance requirements and these are summarised in Table 10.7.

There is no simple system for classifying marine engine lubricants comparable to the well known API system of CC, CD and CE for automotive diesel lubricants because they are used in such a range of designs, ratings and service applications on engines burning a wide range of fuels. Consequently, the lubricants are developed through a series of laboratory, rig and engine tests culminating in shipboard trials. The results are shared with the engine builder when seeking formal approval.

Tests used in developing the various lubricants are outlined in the next three sections. Many of these tests are used for more than one type of oil.

Table 10.6 Typical properties of marine diesel engine lubricants

	System oil	Cylinder oil	Trunk piston engine oil	Test method
Viscosity grade	SAE 30	SAE 50	SAE 40	—
Viscosity (cSt) @ 40°C	103	218	138	ASTM D445
Viscosity (cSt) @ 100°C	11.5	19.0	14.0	ASTM D445
Flash point, closed cup (°C)	225	210	220	ASTM D93
TBN (mg KOH/g)	5	70–100	20–40	ASTM D2896
Pour point (°C)	−18	−12	−18	ASTM D97

Table 10.7 Performance requirements of marine diesel engine lubricants.

Slow speed crosshead		Medium speed trunk piston
System oil	Cylinder oil	Crankcase oil
SAE 30	SAE 50	SAE 40
Good oxidation stability	Neutralise sulphur acids	Control piston deposits
High thermal stability	Prevent scuffing	Prevent ring sticking
Keep crankcase clean	Remove deposits	Neutralise sulphur acids
Release water and insolubles	Provide film strength	Retain alkalinity
Aid load carrying	Improve anti-wear	Protect bearings from corrosion
Low emulsibility	Compatible with system oil	Assist load carrying
Prevent rust and corrosion		Provide good filterability and demulsification

10.7 System oils

Traditionally, system oils were simply rust- and oxidation-inhibited mineral oils but the use of high sulphur residual fuels has required the introduction of alkaline system oils to provide adequate corrosion protection. A typical system oil now has a TBN of 5 and sufficient detergency to keep crankcases and piston cooling spaces clean. It is also used to lubricate turbo chargers, stern tubes, deck machinery, geared transmissions, etc.

10.7.1 *Demulsibility*

Although the water washing of system oils to remove acids is no longer necessary and rarely practised, it is usual to check that the oil will continue to release water in a series of repeated ASTM D1401 tests. The test is also used to check that the system oil will still release water when contaminated with cylinder oil containing high levels of detergent.

Water shedding is finally checked using a small lubricating oil centrifuge to determine the number of passes required to reduce the water content from, say, 5 to 0.1%. The DG Ships test is used to check emulsion stability and additive stability in the presence of water.

10.7.2 *Rust and corrosion protection*

Protection against ferrous corrosion by sea water can be assessed by IP 135B in which a mild steel pin is suspended in a mixture of oil and sea water for 24 hours at 60 °C. Other tests used include a static water drop test and a hydrobromic acid test.

Bearing metal corrosion is evaluated by the Mirrlees corrosion test in which test coupons are suspended in the candidate oil for 100 hours at 140 °C.

The procedure evaluates the tendency of a lubricant to cause grain boundary attack and incipient corrosion on small specimens of:

(a) white metal
(b) copper–lead
(c) phosphor–bronze
(d) aluminium

10.7.3 *Oxidation and thermal stability*

There are a considerable number of oxidation and thermal stability tests, mirroring the complexity of the processes themselves. Oxidation tests used include the Indianna stirred oxidation test (Japanese Industrial Standard, 1959), rotary bomb oxidation by IP 229 and other standard procedures such as IP 48, IP 280 and IP 306.

Despite their use only in diesel engines, marine oils are often checked for oxidation tendencies in the Petter W-1 gasoline engine, which measures viscosity thickening resulting from oxidation/polymerisation and loss of weight of copper–lead bearings through acidic corrosion.

High temperature coking resistance is checked by the panel coker test where an aluminium plate, held at temperatures ranging from 275 to 350 °C, is splashed with lubricant and the deposited coke is weighed and rated. This procedure has many variations including methods to measure the oxidising effect of fuel oil contamination on lubricant stability.

10.7.4 *Load carrying*

System oils to be used in crosshead engines equipped with power take-off, in addition to the general lubricating oil requirements, must have adequate load carrying properties as specified by the engine builder. The requirements, defined by the FZG test, can be met by the inclusion of low levels of ZDDP because of the low TBN of alkaline system oils. The low TBN is achieved by relatively small amounts of overbased alkaline detergents minimising the competition for metallic surface sites and so allowing the anti-wear agents to function effectively.

Other tests used for screening purposes include four-ball, Timken, pin-on-disc and Cameron Plint.

10.8 Cylinder oils

Marine diesel cylinder lubricants (MDCL) are total loss lubricants. In their brief operational life their main functions are to provide a strong oil film between the cylinder liner and the piston rings, hold partially burned fuel residues in suspension so promoting engine cleanliness, and neutralise acids formed by the combustion of sulphur compounds in the fuel. An MDCL is therefore formulated to combat mechanical, abrasive and corrosive wear.

10.8.1 *Colloidal stability*

The high total base number of MDCL requires the use of large amounts of overbased detergents. For example, a 70 TBN cylinder oil contains approximately 25% wt overbased additives. In order to achieve specific formulation benefits, overbased additives such as calcium sulphonates and phenates are often mixed together. Unfortunately, the mixing of these additives in such high concentrations can cause interactions leading to colloidal instability and deposits, mainly of calcium carbonate. This must be avoided because if deposits occurred in practice they could cause blockages of feed lines, filters and lubricating quills. Marine quality additives with good colloid stability are selected through extended static tests and accelerated centrifuge tests.

10.8.2 *Acid neutralisation*

Despite the high total base number of MDCL, if all the sulphur compounds in the heavy fuel oil were converted to sulphur acids, only a small proportion of the resulting acid could be neutralised because of the low volume of lubricant employed. Typical feed rates for fuel and lubricant are 120 and 1 g/bhp/h respectively. At these feed rates and with a fuel containing 3% sulphur, a 70 TBN lubricant could neutralise less than 1% of the sulphur acids formed if all the sulphur were converted into acids at the quoted feed rates. Fortunately, most of the sulphur is discharged with the exhaust gases as sulphur oxides.

Any acids that do form in the cylinder must be neutralised rapidly. Different overbased additives have different acid neutralisation rates and reaction rates can be increased by small additions of suitable additives. Several laboratory tests have been developed to compare neutralisation rates. Most are dependent on measuring the carbon dioxide evolved when sulphuric acid is added to candidate oils in laboratory glassware, both in the presence and absence of iron.

10.8.3 *Spreadability*

A single cylinder of a super long stroke crosshead engine can have a surface area of up to 7 m^2. Cylinder lubrication is achieved by timed injection through approximately eight lubricating points spaced equidistantly around and near the top of the liner. The injection points are connected by moustached grooves to assist the oil to spread. Approximately 1 g of oil is injected every revolution (or second).

Spreadability is determined in the laboratory by dropping small, known

Table 10.8 Characteristics of Bolnes 3DNL.

Type	2 stroke
Cylinders	3
Aspiration	Turbocharged
Speed (rpm)	500
P max. (bar)	120
Bmep (bar)	11.2
Bore (mm)	190
Stroke (mm)	350
Test duration (h)	75–200
Fuel	Heavy fuel oil

weights of oil onto a flat metal plate held at an elevated temperature and measuring the area covered by the oil. The plates are normally machined from cylinder liners and the temperature is chosen to correspond to temperatures found at the top of the liner in service, e.g. 200 °C. Base oils alone have good spreadability but this is reduced dramatically by the overbased additives present in MDCL. Spreadability of MDCL can be improved by a limited number of additives.

10.8.4 *Engine tests*

There are no standard test engines for assessing marine diesel cylinder oil performance. Lubricating companies have their own engines and procedures although the engine most commonly used is probably the Bolnes 3DNL (see Table 10.8).

The fuel is selected carefully to have high sulphur, Conradson carbon residue and asphaltene conιent but no catalytic fines. There is no set procedure even for the Bolnes engine and at the end of the test the engine is assessed by in-house procedures for liner wear, ring wear and cleanliness. A matrix of tests is run to accommodate cylinder-to-cylinder and test-to-test variations.

Thin layer activation has been used with the large cylinder oil test engines to give rapid screening of candidate oils. Oils can be assessed over periods of 10–12 h and a series of lubricants tested without dismantling or even stopping the engine. Cobalt 56 is a preferred gamma ray source and either ring or liner wear can be measured.

10.8.5 *Field tests*

The final proof of a cylinder oil's performance is established by field testing using land-based and ship engines. Engines are assessed for liner wear, ring wear and cleanliness using a variety of instruments and procedures. Comprehensive engine inspections require the engine to be opened and the pistons lifted. This is not always convenient, particularly for the ship operator, and

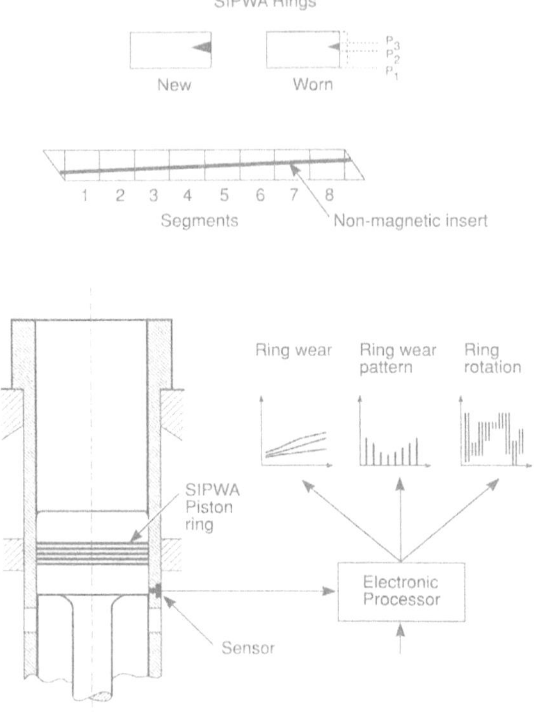

Figure 10.3 Sulzer integrated piston ring wear detecting arrangement (redrawn and adapted from Sulzer Diesel Ltd., original drawings).

some lubricating oil companies use the Sulzer integrated piston ring wear detecting arrangement (SIPWA) to measure ring wear (Hellingham and Barrow, 1981).

SIPWA was originally designed to enable the ship operator to (i) correlate ring wear with fuel treatment and take remedial action if necessary, and (ii) safely pursue an economic fuel purchasing policy. The technique is ideally suited to be a research tool because it monitors ring wear without interfering with the operation of the engine. The patented piston ring is built up from two parts, the normal cast iron ring itself and a circumferential, non-magnetic (bronze) wedge shaped element on the working surface (Figure 10.3). This insert is arranged in a spiral from one butt end to the other and as the ring wears, the width decreases. A wear dependent signal is produced every revolution as the ring passes the detector sited in the bottom of the liner. From measurements P_2–P_3 and P_1–P_2, the ring wear and ring position can be determined. By lubricating a multicylinder engine with two or more cylinder oils simultaneously, the performance of the oils can be compared directly under the same operating conditions.

10.9 Trunk piston engine oils

A manufacturer's range of trunk piston engine oils will include products with total base numbers from approximately 15 up to 40. The appropriate oil is then selected to match the sulphur content of the fuel being used. In service, the initial TBN will fall fairly rapidly but will then stabilise. The stabilised TBN is normally rather more than half the original value of the new oil, ensuring good corrosion protection.

10.9.1 *Filterability*

Considerable care is required to achieve the appropriate formulation balance of detergency and dispersancy. Insoluble material, derived principally from combustion products, must be kept in fine suspension to avoid damage to bearings, etc. but at the same time the oil must release insolubles to the filter and centrifuges (Loddenkemper, 1989). Oils are screened using laboratory filters of known pore diameter and distribution. Before final field trials the lubricants are checked in small marine lubricant centrifuges. Water has a strong influence on dispersancy and can cause insolubles to be deposited prematurely. The effect of water contamination has therefore to be checked.

10.9.2 *Heavy fuel engine tests*

Trunk piston engine oils, even those with relatively low total base numbers, easily surpass the normal diesel engine test performance requirement based on engines such as Caterpillar, MWM-B and Petter AV-B when they are run on distillate fuel. More severe tests have therefore been devised using the same engines operating on heavy fuel oil. There are no standard industry tests and companies have perfected their own test procedures. The heavy fuel engine tests are designed to evaluate the performance of the lubricants with respect to piston cleanliness, ring and liner wear, TBN depletion and insolubles.

10.10 Used oil analysis

Regular testing of lubricants in service helps to safeguard the machinery. For maximum benefit, testing must be conducted rapidly and the tests are chosen to give the maximum information with a minimum number of tests. Supplementary tests can be carried out if the need arises.

Interpretation of results and, where necessary, a recommended course of action is part of the lubricant suppliers responsibilities (Van der Horst, 1987). The reasons for carrying out individual tests and the information they can give are outlined in the following sections.

10.10.1 *Density*

Used for selecting the correct gravity disc for the lubricating oil centrifuge.

10.10.2 *Viscosity*

Changes in viscosity in service arise from either fuel dilution or suspended matter. Assuming no other adverse symptoms, changes in viscosity of ±20–25% can be tolerated. Contamination by marine diesel or distillate fuel reduces viscosity; contamination by heavy fuel oil increases viscosity. In both cases the flash point of the lubricant is likely to be reduced. If increased viscosity is due to carbonaceous insolubles, both viscosity and insolubles can often be reduced by correct centrifuging.

10.10.3 *Flash point*

Both marine diesel and heavy fuel oil have low flash points compared with lubricating oil, e.g. 60–100 °C compared with over 200 °C. The relatively low flash point of heavy fuel oil is caused by the cutter stock, e.g. kerosine or gas oil used to give an acceptable viscosity from the extremely viscous vacuum residues, visbroken residues, etc. Contamination by diesel or heavy fuel oil therefore normally reduces the flash point of the lubricant, although with heavy fuel oil the small amount of volatile matter can be driven off from the lubricating system.

10.10.4 *Insolubles*

The majority of insolubles are derived from the combustion products of the fuel and lubricant. Insolubles can also contain spent additives from the lubricant and general debris such as rust, wear metals and contaminants from water ingress and air intakes.

Heavy duty detergent oils usually include ashless dispersants within their formulation to keep combustion contaminants in fine suspension in order to prevent damage to machinery. Warning limits for the insolubles content of trunk piston engine oils vary with the oil but in some cases levels of 5% can be tolerated. When high levels of insolubles are encountered, the fuel and fuel combustion systems should be examined.

10.10.5 *Total base number*

In service, the total base number, or reserve alkalinity, steadily declines until a plateau is reached. At this point the higher total base number of the new oil, used for topping up, balances the acidic products of combustion. The plateau is usually rather more than 50% of the original TBN and must meet the minimum level set by the engine manufacturer. If TBN declines by more than say 60%, countermeasures must be taken depending on the circumstances. In one-off situations, part of the charge can be replaced with a higher TBN oil, otherwise the continuous use of a higher TBN oil is recommended.

10.10.6 *Water content*

The water content of system and heavy duty crankcase oils can generally be maintained below 0.2 and 0.5%, respectively. If water is detected, the first priority is to establish and correct the source of contamination. This is particularly true for sea water contamination because severe corrosion can occur very rapidly.

10.10.7 *Wear metals*

Inductively coupled plasma atomic emission spectroscopy has made the determination of wear metals very easy and used oils can be scanned for the presence of 20 elements in less than 1 minute. Wear trends can be obtained by comparing wear metals from a series of samples. However, it should be noted that the accuracy of the determination can be limited by the particle sizes present.

References

Aeberli, K., Mikulicic, N. and Schaad, E. (1989) The development of a reliable and efficient two-stroke engine range. *CIMAC Paper D78*, Tianjin, June 1989.

Bantour, M.T. and Chapuy, J.F. (1983) Contribution to the treatment of low grade fuel oils and evaluation of the influences on engine wear. *CIMAC Paper D6.1*.

Barnes, G.K., Liddy, J.P. and Marshall, E.L. (1987) The ignition quality of residual fuels. *CIMAC Paper D75*, Warsaw, June 1987.

Bengtsson, G. (1986) Upgrading fuel treatment systems on-board ship. *Motorship Conference*, March 1986.

BS MA 100 (1982) *British Standards Institution Specification for Petroleum Fuels for Marine Engines and Boilers BS MA 100: 1982.*

Chapuy, J.F. (1986) New PC engines and their adaption with regard to the deteriorating quality of residual fuel oils. *Motorship Conference*, March 1986.

CIMAC (1986) *Recommendations Regarding Requirements for Heavy Fuels for Diesel Engines.* Volume 8.

David, P. and Denham, M.J. (1984) The measurement and prediction of the ignition quality of residual fuel oils. *Trans. Inst. Mar. Eng. (TM)* **96** paper 66.

General Council of British Shipping (1983) *The Storage and Handling of Marine Fuel Oils Onboard Ship.*

Hellingham, G.J. and Barrow, S. (1981) Shipboard investigations with selected fuels of tomorrow. *CIMAC Paper D63.*

Holbrook, P. and Fabriek, W.P. (1989) Residual fuels—money versus quality. *Motorship Conference*, March 1989.

ISO 8217 (1987) *Petroleum Products—Fuels (Class F)— Specifications of Marine Fuels.* First edition, 1987-04-15.

Japanese Industrial Standard (1959) Method of testing the oxidation stability of lubricating oil for internal combustion engines. *K2514*.

Loddenkemper, F.J. (1989) State-of-the-art separation of large diesel engine lube oils. *Motorship Conference*, March 1989.

Negus, C.R., Dale, B.W., Stenhouse, I.A. and McNiven, A.J. (1987) An investigation of the confined combustion properties of residual fuels used in marine and industrial engines. *CIMAC Paper D78*, Warsaw, June 1987.

Ruzicka, D.J., Robben, F. and Sawyer, R.F. (1984) Combustion of residual fuels in a CFR diesel engine. *The Motorship*, July 28.

SAE J-300 (1987) *Engine Oil Viscosity Classification*. June 1987.

Schmidt-Sorensen, J. and Sunn Pedersen, P. (1989) The M C Engine: design for reliability and low maintenance costs. *CIMAC Paper D31*, Tianjin, June 1989.

Sjöberg, H. (1987) Combustion studies and endurance tests on low ignition quality fuel oils. *Trans. Inst. Mar. Eng. (TM)* **99** paper 24.

Sprague, S.W. (1986) Shipboard fuel handling and operating practice. *Motorship Conference*, March 1986.

Van der Horst, G.W. (1987) Used oil inspection, a contribution to large diesel engine reliability. *CIMAC Paper D-64*, Warsaw, June 1987.

Zeelenberg, A.P., Fijn van Draat, H.J. and Barker, H.L. (1983) The ignition performance of fuel oils in marine diesel engines. *CIMAC Paper D13-2*.

11 Lubricating grease

G. GOW

11.1 Introduction

Lubricating grease has been regarded by many prominent members of the oil industry, including the National Lubricating Grease Institute (NLGI), as more an art than a science. One of the primary objects of a new generation of lubrication engineers is to radically change this concept and start speaking the same language as other scientists. Grease lubrication is a complex mixture of chemistry, physics, tribology, rheology and health and environmental sciences. Without an extensive interdisciplinary co-operation and understanding, any ambitious research and development project will eventually meet immovable obstacles.

Lubricating grease is a vital part of a great number of machine components. However, this is often forgotten (or ignored) by product designers. Lubrication specialists are called into design projects at the very last minute and this can result in serious problems which could perhaps have been avoided if the lubricant had been considered as an integral part of the total construction from the start.

Economically speaking, lubricating grease is regarded as a commodity and, as such, it is often a low price product. This means that the potential manufacturing profit is limited by the cost of the resources (both capital and manpower) required to develop and produce the grease. Basic research into the fundamentals of grease lubrication is somewhat scarce in the academic community, presumably due to lack of funding. Most research takes place under the auspices of the oil companies and their laboratory personnel are, by tradition, predominantly chemists. Whilst grease manufacturing is indisputably based on chemistry, the study of grease in practical situations, in which grease acts as a mechanical barrier between two moving surfaces serving to keep them apart, leads invariably into the realm of physics. Looking for chemical answers to physical problems can only lead to questionable conclusions and compromises.

Ten years ago, in trying to visualise the grease industry of the 1990s, Sorli (1980) expressed concern that 'as our industry rationalises and attempts to find the appropriate balance between the specialist grease manufacturers and the large integrated oil companies, the future research and development required for our long term success (could) simply fall between the cracks and we may become increasingly reliant on outmoded processes and recipes.' Some of his concerns were justified in that many grease manufacturers still use formulations and techniques originating from the 1940s. However, some of the most interesting developments in grease technology have occurred in the last ten years.

Lubricating grease is not a new product; evidence from archaeological findings in the Middle East suggests that the art of making grease was known at least 4 000 years ago. However, the current generation of lubricating greases have no more than 40 to 50 years of history behind them. Earle's patents on lithium greases were issued as late as 1942–43 but today well over 60% of lubricating greases produced in Europe are based on lithium technology. However, there is still a long way to go in lubricating grease technology (Jenks, 1985). A comprehensive study of all aspects of grease technology with the corresponding literature references is beyond the scope of this short presentation. There are numerous textbooks available on this subject (Boner, 1954; Boner, 1976; Klemgard, 1937; Klamann, 1984; Dorinson and Ludema, 1985; Lansdown, 1982; Vinogradov, 1989; Erlich, 1984) and, since the basics of lubrication will be dealt with elsewhere in this book, this chapter will focus on illustrating the difference between a grease and an oil and the advantages a grease can offer as opposed to a fluid lubricant.

According to its classical definition (ASTM, 1961), lubricating grease is 'a solid to semifluid product of a thickening agent in a liquid lubricant. Other ingredients imparting special properties may be included'. This definition is somewhat diffuse but it does establish one very important fact regarding composition and properties: grease is not a thick (viscous) oil, it is a *thickened* oil, a multi-phase system consisting of at least two well defined components, a thickener (gelling agent) and a fluid lubricant. A more rheologically based definition is offered by Sinitsin (1974), 'a lubricant which under certain loads and within its range of temperature application, exhibits the properties of a solid body, undergoes plastic strain and starts to flow like a fluid should the load reach the critical point, and regains solid-body properties after the removal of the stress'. This establishes another very important point: grease is both solid *and* liquid, depending on the physical conditions of temperature, stress, etc., and features an additional property compared to a pure liquid lubricant, a *yield value* (σ_0), the threshold level of shear stress or strain. However, modern lubricating greases are so varied in both characteristics and contents that the only absolutely accurate definition is, with apologies to Cheng (1989), 'lubricating grease is what grease manufacturers make'.

11.2 Structure and properties

Lubricating grease is perhaps best described as a viscoelastic plastic solid and, as such, is extremely complicated in physical and chemical characteristics. The separate components are not in chemical solution but co-exist like water in a sponge. This popular illustration, in which the thickener system is the sponge and the fluid lubricant is the water, is of course not a strictly valid scientific description but is useful in conceptualising certain problems arising from the use of this type of material. SEM-photographs of the fibre structure of soaps commonly used in grease manufacturing reveal that the concept of the sponge is not totally invalid (see Figure 11.1) and the problem becomes how to quantatively characterise the different 'sponges'.

Grease is not a thick oil but a thickened oil. The grease matrix is held together by internal binding forces giving the grease a solid character by resisting positional change. This rigidity is commonly referred to as *consistency.* When the external stresses exceed the threshold level of shear (stress or strain)—the yield value—the solid goes through a transition state of plastic strain and turns into a flowing liquid. Consistency can be seen as the most important property of a lubricating grease, the vital difference between grease and oil. Under the force of gravity, a grease is subjected to shear stresses below the yield value and will therefore remain in place as a solid body. At higher levels of shear, however, the grease will flow. It is therefore of the utmost importance to be able to determine the exact level of yield. This is, of course, a time dependent factor and the time scale must be chosen carefully depending on the information required. At present, consistency is measured by means of the penetrometer (ASTM D 217) and greases classified according to the NLGI system (Erlich, 1984). This classification is universal and the great majority of greases are used and chosen solely on the basis of their NLGI grade. However, this method is scientifically unacceptable since the material is exposed to varying levels of shear rate and the resulting figures of penetration should therefore never be compared to each other (Gow, 1988). Penetration measurements can be interesting for production control to ensure that a given product has the same consistency as a previous batch but to use the figures to predict performance under dynamic conditions involves dangerous assumptions.

The distinction between a liquid and a solid can sometimes be very diffuse. Liquids which exhibit both elastic and viscous tendencies under varying conditions are termed *viscoelastic.* There is equipment available which is suitable for testing viscoelastic properties but it is expensive and requires qualified personnel. Depending on the different applications, the grease trade has used certain standard methods for the determination of rheological properties. *Apparent Viscosity* (ASTM D 1092), *Measurement of Flow Properties at High Temperatures* (ASTM D 3232), *Determination of Flow Pressure* (DIN 51805) and, of course, *Cone Penetration* (ASTM D 217) can be given as examples. Other non-standardised tests such as sliding plate systems and the 'Lincoln ventmeter' have also been adopted. In recent years, the approach to rheology has become more modern and the use of, for instance, a constant

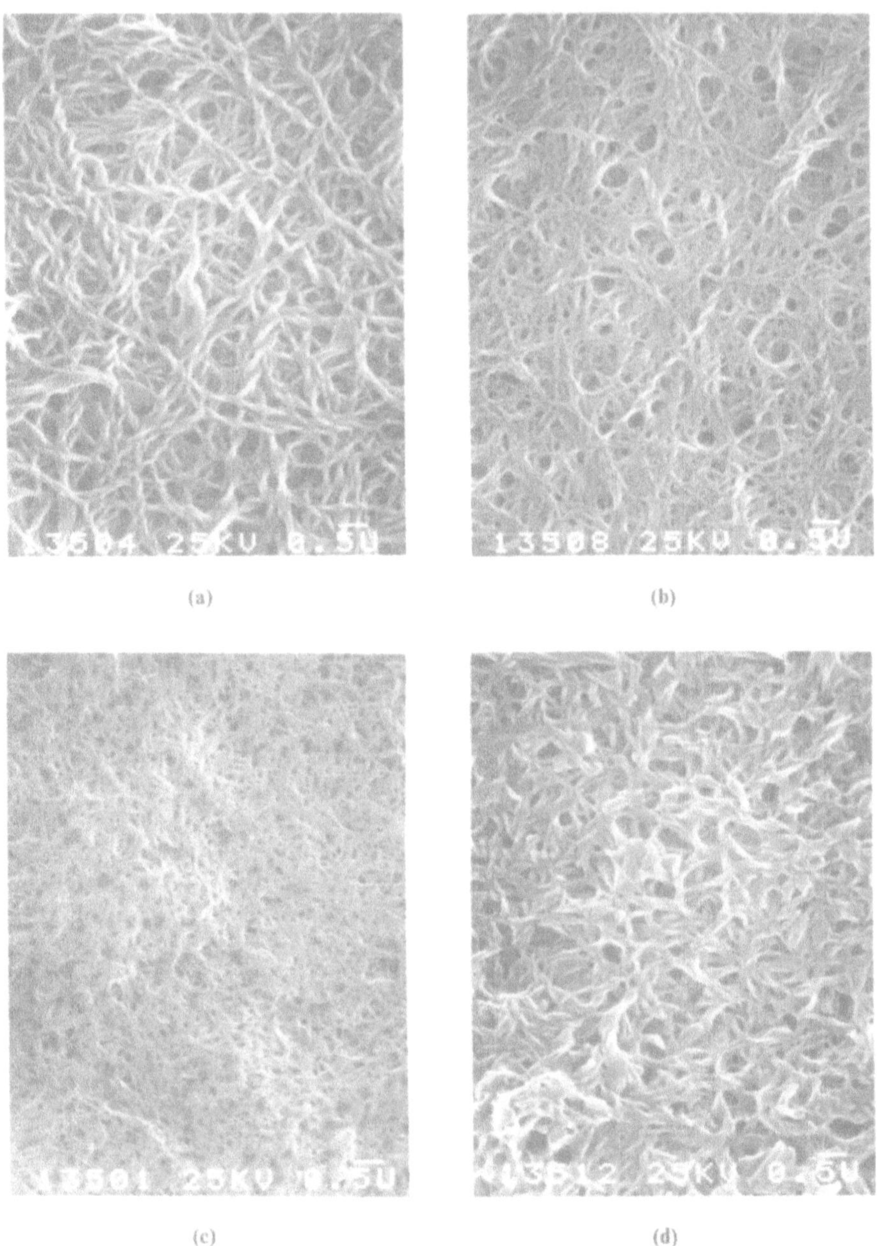

Figure 11.1 SEM photographs of different soap structures in lithium greases: (a) lithium-12-hydroxy stearate in mineral oil (normal 'coarse' structure); (b) lithium-12-hydroxy stearate in mineral oil (finer structure); (c) lithium-12-hydroxy stearate in synthetic ester (very fine structure); (d) modified lithium-12-hydroxy stearate in mineral oil (lithium complex grease) (courtesy of SKF ERC, Holland).

pressure viscometer (Cantin, 1981; Winterburn, 1988), a rotary viscometer with a well defined gap (Gow, 1988; Powell, 1982) an oscillatory viscometer and a controlled stress viscometer (Gow, 1988, 1990; Hamnelid, 1990) are positive developments. An attempt to find one single method of measuring the rheological properties of grease is, of course, as doomed to failure as trying to find one method of determining all the chemical properties. Units of viscosity pertaining to non-Newtonian fluids are meaningless without the corresponding values of shear rate and it is therefore of the utmost importance to define the shear rates in absolute terms. In different pump situations, for instance in centralised lubrication systems, greases are subjected to shear rates perhaps as low as 10^{-1} s^{-1}. On the other hand, in the EHD regimes of heavily loaded bearings and gears, the shear rates may be as high as 10^6–10^8 s^{-1}. Intermediate values (1 to 10^4) are probably less significant. The rheological test systems of the past are unfortunately in the wrong (irrelevant) shear rate range to be able to predict functional performance.

For the exact measurement of yield value (stress or strain), a modern controlled stress rheometer is a useful instrument. For lower shear rates (below 1 s^{-1}), a rheometer capable of both rotary and oscillatory viscometry is necessary. The oscillation mode also permits the measurement of elasticy and the phase angle (δ) in combination with the complex shear modulus (G^*) provides vital information as to the dynamic properties of any given product. Previous investigations have often involved the extrapolation of higher shear rate situations into the lower area and this can lead to very misleading conclusions. Figure 11.2 is an example of rheological measurements on lubricating grease and Figure 11.3 shows the temperature dependence of the yield value. This temperature dependence can be considered as an equivalent to the viscosity index (VI) of fluid oils. For the very high shear rate range, a slit viscometer would appear to be an interesting possibility but few are commercially available. Other types of instruments have been tested by technical universities and institutes. One example is the 'bouncing ball' at Luleå University in Sweden (Jacobsson, 1984; Höglund, 1989) which gives information on the maximum shear strength of greases in relation to the base oil and provides a simple way of comparing different thickener systems. Figure 11.4 is an example of this (Åström and Höglund, 1990).

11.3 Chemistry

The chemistry of lubricating grease is also exceptionally complex but has already been investigated in a detailed way. A concise summary is provided in the NLGI Lubricating Grease Guide (Erlich, 1984) along with a basic description of the most common raw materials used.

Grease contains at least two components, the base fluid and the thickener system. A typical multipurpose grease can contain about 85% base fluid, 10%

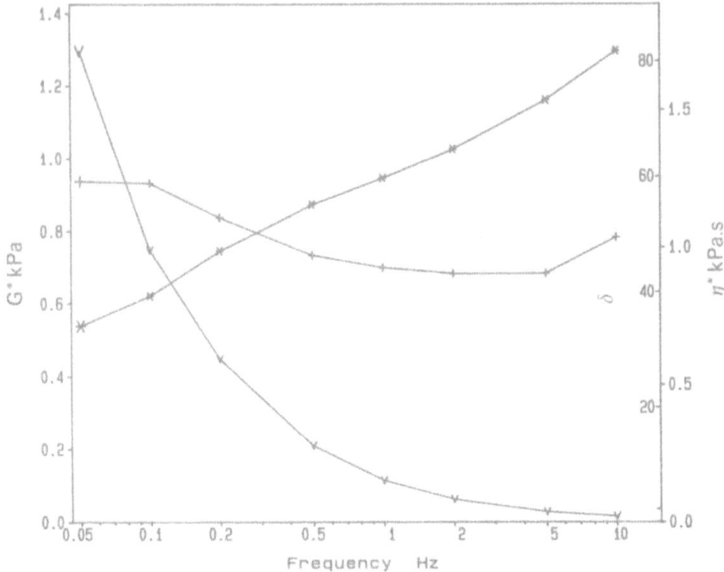

Figure 11.2 The frequency dependence of certain rheological properties of lithium grease at +25°C (×—×) complex shear modulus, G^*; (+—+) phase angle, δ; (v—v) complex viscosity, η^*.

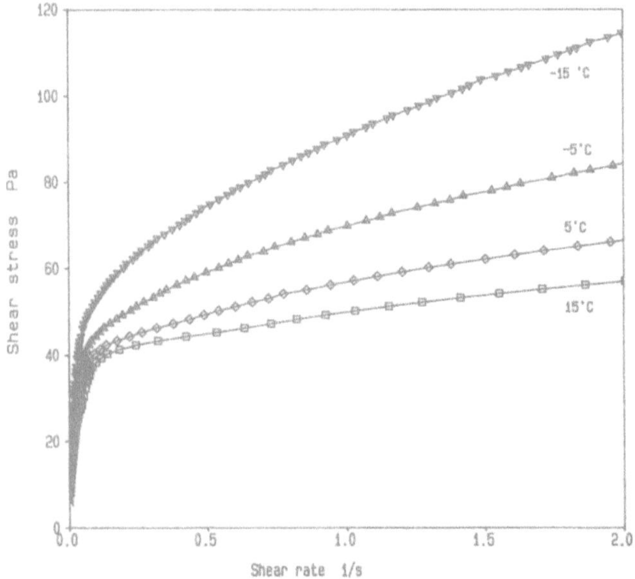

Figure 11.3 The temperature dependence of the yield value of a lithium complex (NLGI 00) grease specially formulated for use in automotive centralised lubrication systems.

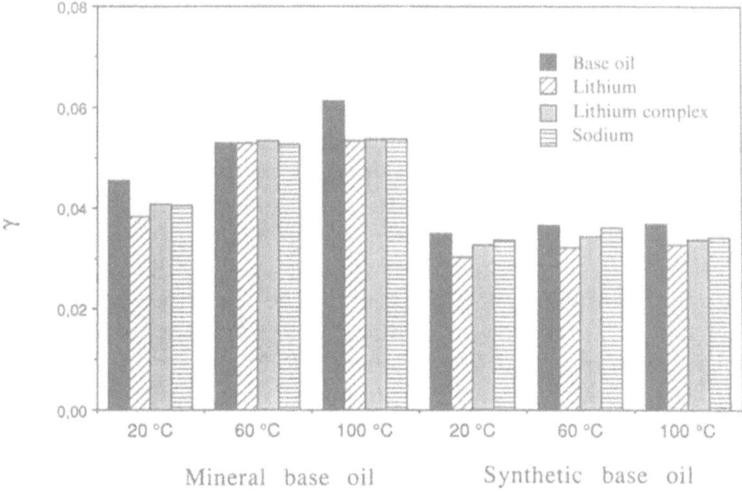

Figure 11.4 The limiting shear stress–pressure coefficients of six different greases and their two base oils (original art work from Åström and Höglund, 1990).

thickener and 5% 'other ingredients imparting special properties' (ASTM, 1961).

The *base fluids* can be divided into two main groups: mineral oils and synthetic fluids (although vegetable oils are experiencing something of a renaissance because of environmental considerations).

Mineral oil is an undefinable mixture of hundreds and thousands of different hydrocarbons depending on the specific crude involved and the method and degree of refining. A balance in solubility parameters between the oil and the thickener system is a vital factor in grease formulation since the strength of the multi-phase system depends on the binding forces within the grease matrix. Aromatic and/or naphthenic oils were predominant in the past because of their high degree of polarity. However, these oils can contain considerable amounts of polycyclic aromatic hydrocarbons and these are possible carcinogens. Today, paraffinic oils are preferred, although severely refined naphthenics have been shown to meet the health and environmental demands of the European Community (Stephens *et al.*, 1985). The oil component must meet many different requirements such as viscosity/VI, evaporation loss, oxidation stability, response to inhibitors, etc., but perhaps the most important property of an oil for grease manufacturing is its solubility capability. This can clearly be seen in Figure 11.1 where photographs (a) to (c) represent the same basic soap chemistry prepared in fluids of varying solubility parameters. There are many ways of classifying solubility with respect to various substances such as additives, polymers, etc. Typical examples are the VGC (viscosity gravity constant), the aniline point, carbon type analysis and the various ways of quantifying mutual solubility on the basis of thermodynamic forces. These types of classification have been developed into multidimensional theoretical models where vectors include polarity, dispersing forces, hydrogen bonding, etc. For optimal solubility between two materials, the solubility parameters should be as similar as possible.

Synthetic fluids are, in contrast to mineral oils, well defined chemical substances with given physical and chemical characteristics. The only real problem at present is the price because grease is a commodity! Synthetic hydrocarbons such as polyalphaolefins and alkylated aromatics, esters, polyglycols, silicone oils, perfluoropolyethers and many other types of fluids are used when the prevailing conditions exclude the use of mineral oil. In addition to a wider temperature range, synthetic fluids provide other advantages such as chemical resistance and plastic compatibility, electrical properties, the lubrication of non-metallic surfaces, environmental acceptability, biological degradability, and so on. The possible advantages and disadvantages are not given here because they are not exclusive to grease.

Different types of *thickener systems* are used to provide a suitable consistency to the finished product. Lansdown (1982) refers to the thickener system as the 'second most important component of a grease'. However, this is assumed to refer only to the quantity since the thickener system is, of course, the vital difference between grease and oil. Greases are often classified according to the type and amount of thickener. In the great majority of cases, different metal soaps are used to provide a network of fibres which act as the 'sponge'. Other systems such as inorganics and polyureas do exist but soaps are by far the most dominant materials. Calcium saponified vegetable oils were quite adequate for the lubrication of cartwheels and other simpler machine elements, and a more sophisticated lubricant was not required until the advent of the steam engine and the industrial revolution. The higher temperatures and speeds which were introduced made the use of conventional calcium soaps difficult. These kinds of soaps have melting points around 100 °C and, in many cases, the evaporation of the water inherent in the soap structure limits their use to a maximum of 50 °C. Sodium soaps with higher melting points (160 °C) were the first generation of high temperature greases but their water solubility also limited their use in many applications. Aluminium stearate provided a tacky substance suitable for chassis lubrication but the complete lack of mechanical stability excluded its use in rolling elements. The lubrication engineer of the 1930s had problems in choosing the optimal grease for a particular application. The development of *lithium soaps* in the 1940s and 1950s revolutionised the grease trade. It suddenly became possible to provide a 'universal' grease which could withstand higher temperatures (melting point 180 °C), was water resistant and, in combination with a particular fatty acid (12-hydroxystearic acid), had an excellent mechanical stability. Lithium 12-hydroxystearate is the foundation stone of the current generation of lubricating greases. The fatty acid is a castor oil derivative and, being a natural (vegetable) raw material, is subject to fluctuations in climate and even the political situation in the countries of origin. (The quality and economics of European grease is heavily dependent on the natural and economic climate in Brazil.) The grease industry would welcome an alternative synthetic material but so far natural products have been easier to obtain than man-made ones.

Mixed base greases (e.g. lithium–calcium, aluminium–barium) are becoming popular but many of these so-called 'new developments' are in fact old-fashioned compromise mixtures. The most important development of the last decade has been the improvement and optimisation of so-called *complex* greases in which different metal soaps and salts are co-crystallised into the same fibre structure. Aluminium, calcium and barium are the classical complexes but lithium and mixed base complexes also have some very interesting properties. The advantage of a complex grease is often the higher melting point but other physical and chemical characteristics are also affected by the complexing agent, e.g. elasticity at low temperatures, water resistance, EP-properties and fire resistance.

Another type of thickener is based on inorganic substances which 'gel' the oil into a solid state. Different types of clays (bentonite, hectorite), silica-gel, PTFE and graphite are examples of this. These materials do not form a fibre structure but consist of electrochemically charged particles (platelets) which hold the matrix together. The fact that these products do not melt is both an advantage and a disadvantage. If properly used, and if relubrication is provided at appropriate intervals, they can perform well at elevated temperatures. However, the base oils in such greases are as prone to evaporation, oxidation and thermal degradation as other oils, and claims that they can withstand extremely high temperatures because the grease does not melt are often found to be false. The lack of fibre structure is an advantage for pumpability in that the product has a low level of elasticity but is a disadvantage in that oil separation under pressure can become a major problem.

Polyureas and polyurea complexes are also used as thickeners, especially in Japan, and these often perform excellently at high temperatures. However, the toxicity of the raw materials limits the possibility of manufacturing this type of product in certain parts of the world.

The properties of a lubricating grease can be modified by the incorporation of *additives* into the basic structure. In the ideal (hydrodynamic) situation, where the base grease keeps the metal surfaces apart, the need for additives may be minimal, but a grease is often much more than a lubricant. It is also expected to perform as a seal, a corrosion inhibitor, a shock absorber or even a noise suppressant and most modern greases contain an additive package of some kind. There are many synergistic effects between the different types of materials available but solubility is once again a decisive factor. To be able to perform in fluid oils, the majority of additive systems have an optimised solubility in mineral oils. This is a neccessity for oils in order to prevent sedimentation and/or phase separation. The same does not hold true for greases. In fact optimised solubility may be a disadvantage. To begin with, grease is a solid body so even completely insoluble substances can be dispersed in the matrix (e.g. MoS_2, graphite and zinc oxide). Very few liquid additives perform well in greases and one explanation offered is that many of them are 'locked' into the oil phase. Most soap structures are highly polar in character and will compete with the additive for the metal surfaces. In many cases, the soap wins and covers the surface with a thin layer of fatty material. The additives dissolved in the oil cannot reach the metal surface and therefore have no effect. Insoluble additives may well be better since they will be dispersed in both phases and at least have a theoretical chance of reaching the metal surface. The best place for the additives to be is on the soap structure itself which is why calcium complex greases perform so well in EP tests (as calcium acetate, a basic component of the soap structure, is also an EP agent).

The production of modern lubricating greases on a large scale involves a complicated saponification process requiring a high level of investment in both capital resources and manpower. The number of commercial grease plants is steadily decreasing because of two major factors: a rationalisation within the grease market and the current cold economic facts of the petroleum business. There are very few hypermodern grease plants, most of the current facilities originating from the 1950s or even earlier. The saponification process is carried out in two main types of production units: batch production in kettles or 'contactors', and continuous production in small reaction chambers. The chemistry is, however, the same in both cases. Most thickener systems are produced by saponifying fatty acids with metal hydroxides in the presence of a portion of the base oil and often some water. The temperature is raised to the reaction temperature, kept there for a suitable period of time and then, in most cases, raised to an even higher level to dehydrate and melt the soap ($> 200\,°C$). The bottleneck of any grease manufacturing is the cooling process since lubricating greases have inferior heat transfer properties. On cooling, the soap crystallises into its characteristic fibre structure and in order to optimise this stage the cooling is often promoted under continual stirring. No matter how smoothly the cooling operation proceeds, the resulting mass is a lumpy mixture and must be homogenised. This can be done in a multitude of ways, e.g. by pressure valves, tooth-colloid mills and high pressure homogenisers. The finishing process, in which the additives are introduced, the consistency adjusted and the whole mixture homogenised, is supplemented with a deaeration unit, filters and a whole array of packing systems. Different greases require different manufacturing methods and each grease plant has its own particular technology.

Modern lubricating greases contain a variety of chemical substances ranging from complicated mixtures of natural hydrocarbons in the base oils through well defined soap structures, polymer solutions and complex organic molecules in the additives, to very simple chemicals such as carbon black and metal powders. If formulations are regarded as proprietary and confidential then the manufacturing processes must be classified as top secret. The 'art' of making grease has probably more to do with the production method than with the formulation!

11.4 Applications

'Grease is better than oil'. This can be considered a rash statement but there is more than an element of truth in the claim. Grease cannot be used as a coolant since it has inferior heat transfer properties and is therefore extremely unsuitable as a motor oil or a metal working fluid. In addition, grease is highly elastic so its use as a brake fluid or in power transmission is more than questionable. As a pure lubricant, however, grease does have many advantages over fluid oil. A grease can be said to have four major 'abilities': lubricating ability, sealing ability, corrosion inhibition ability and carrying (matrix) ability, all pertaining to a material's consistency and multi-phase structure.

Because of its solid character at low shear, lubricating grease stays in place and can act as a reservoir of lubricant. This means that grease lubricated components do not require as short relubrication intervals as fluid oils, and that grease provides a certain degree of lubrication even if maintenance has been overlooked or forgotten over a considerable period of time. Grease has superior high temperature and load carrying properties because of the higher dynamic viscosity and elasticity which the thickener system provides. At low temperatures, the multiphase system facilitates movement by preventing the characteristic crystallisation of paraffin waxes in the base oil and by allowing deformation to take place not only in the different phases but also in the boundary between the separate phases.

The consistency of grease prevents both fluid and solid contaminants from entering the system and also eliminates problems with dripping and leakage. In addition grease facilitates the choice of mechanical seals. All this allows more freedom in design possibilities. The consistency of grease and its ability to adhere to metal surfaces (tackiness and polarity), prevent reactive liquids from coming into contact with the metal surfaces and causing corrosion. Lubricating greases can also absorb considerable quantities of water (or cutting fluids, emulsions, industrial coolants, etc.) and still perform satisfactorily as a lubricant.. Corrosion inhibition can be further enhanced by the inclusion of special additives.

Grease can also perform as a carrier for certain insoluble substances such as molybdenum disulphide and graphite. This is possible because the base material is solid and there is no problem with sedimentation, solvency or compatibility. These additives can be evenly dispersed into the multiphase system.

All these factors contribute to grease being an excellent lubricant and it is often quite simple to choose the right product for any given application. However, in our modern highly technological society, there are many factors to be considered and the art of choosing the right grease may well become a complex problem.

A modern high quality multipurpose grease can be said to cover up to 80% of all applications where grease lubrication is required. For the remaining 20%, special greases are necessary depending on three very important parameters: the component itself, the temperatures involved and the surrounding environment. A conventional lithium-based multipurpose NLGI 2 grade grease is often adequate for the lubrication of plain and rolling bearings working under normal conditions within a temperature range of, say, −20 °C to +120 °C. Special greases are required for special applications such as different types of machine elements (gears, instruments, couplings, slides, etc.), high temperatures, low temperatures, a wide temperature range, high loads, high speeds, low speeds, marine applications, vibrations, centralised lubrication systems (especially on automotive units) and in special environments such as in the foodstuffs industry, or where there is contact with radiation, oxygen or other highly reactive chemicals, or if there are high filtration requirements. The ways of solving specific lubrication problems vary in different parts of the world and from company to company depending on the technology available.

11.5 Future developments

Looking forward into the next century, the need for more advanced science in grease technology is essential. The design of special components is becoming more and more complicated and machines are becoming much smaller and lighter in weight, and are required to run faster and withstand heavier loads. To be able to develop the optimal lubricants for these new conditions, the mechanism behind grease lubrication must be studied and understood. There will be an increasing specialisation in both products and markets and the survival of individual lubricant companies will depend on their ability to adapt to a changing situation. Not only new machines but also new materials will affect the development of greases. Already plastics and ceramics are becoming more common in designs and newer, unconventional lubricants are being developed to meet the technical requirements and, at the same time, the increasing demands of health and the environment.

The increasing internationalisation of Europe will also be an important factor since the concept of a home market will no longer exist. In addition, many new international organisations are being formed to try to agree on topics such as standards and specifications, environmental legislation and quality assurance. The foundation of the ELGI (European Lubricating

Grease Institute) in 1989 was the European grease industry's reaction to this effect and the role of this new organisation is presumed to be as an advisory body to the different authorities and committees when it comes to matters concerning lubricating greases. A number of working groups are already active in fields such as health and the environment, foodstuffs lubrication, testing methods and fundamental research. The advent of the quality assurance system ISO 9000 in grease manufacturing is also expected to have a considerable effect and companies not conforming to this standard will eventually have to close down. A decrease in the total number of grease plants is forecast and those remaining are expected to enter into a network of co-operative activities perhaps involving take-overs and mergers; the survival of the fittest so to speak.

11.6 Conclusions

The new image for lubricating grease in the 21st century must abandon the concept of a cheap and relatively simple commodity. Grease is an exceptionally complex product incorporating a high degree of technology in chemistry, physics, rheology, tribology and the environmental sciences. It is a basic machine component, just as important as any other part of the machine, and should be considered right from the start of any development project. To achieve this, a great deal of attention has to be paid to education and training not only by the individual grease manufacturers but on a much wider scale. The ELGI hopes to be able to provide a basic course in grease technology in the not too distant future. Quality is expected to be an important property of the new grease image, encompassing not only product quality but quality throughout the range of operations.

The art of grease lubrication has become a complicated science. The dilemma facing the grease trade is that art is often much more profitable than science.

Acknowledgements

SKF Engineering and Research Centre BV in Holland for providing the SEM photographs used in Figure 11.1.

References

ASTM (1961) *Annual Book of ASTM Standards*. American Society for Testing and Materials.

Åström, H. and Höglund, E. (1990) *Rheological Properties of Six Greases and their Two Base Oils*. Technical report, Luleå University of Technology, Sweden.

Boner, C.J. (1954) *Manufacture and Application of Lubricating Greases*. Reinhold Publishing Corp.

Boner, C.J. (1976) *Modern Lubricating Greases*. Scientific Publications (GB) Ltd.

Cantin, R. (1981) Future developments of a constant pressure viscometer. *NLGI Spokesman* **XLV**(1) 20–27.

Cheng, D.C-H. (1989) The art of coarse rheology. *The British Soc. of Rheol. Bull.* **32**(1) 1–21.

Dorinson, A. and Ludema, K.C. (1985) *Mechanics and Chemistry in Lubrication*. Elsevier Science Publications, London.

Erlich, M. (ed.) (1984) *NLGI Lubricating Grease Guide*. National Lubricating Grease Institute.

Frand, E. (1988) *Product and Marketing Development in Mature Industries*. Paper presented at the 55th Annual Meeting of the NLGI, Wesley Chapel, Florida, USA.

Gow, G.M. (1988) Judges 5:5. *NLGI Spokesman* **LII**(9) 415–423.

Gow, G.M. (1990) *The CEY to Grease Rheology*. AB Axel Christiernsson, Sweden.

Hamnelid, L. (1990) *Amazing Grease*. AB Axel Christiernsson, Sweden.

Höglund, E. (1989) The relationship between lubricant shear strength and chemical composition of the base oil. *Wear* **130** 213–224.

Jacobsson, B. (1984) A high pressure–short time shear strength analyser for lubricants. *Trans ASME. J. Tribology*.

Jenks, G.R. (1985) You've come a long way from axel grease *NLGI Spokesman* **XLVIII**(12) 430–433.

Klamann, D. (1984) *Lubricants and Related products: Synthesis, Properties, Applications, International Standards* Verlag Chemie, Weinheim.

Klemgard, E.N. (1937) *Lubricating Greases, Their Manufacture and Use*. Reinhold Publishing Corp.

Lansdown, A.R. (1982) *Lubrication, a Practical Guide to Lubricant Selection*. Pergamon Press.

Powell, T.W. (1982) Activators for organophillic clays in lubricating greases. *NLGI Spokesman* **XLVI**(8) 269–277.

Sinitsyn, V.V. (1974) *The Choice and Application of Plastic Greases*. Khimiya, Moscow.

Sorli, G.E. (1980) Beyond the looking glass, grease in the 90s. *NLGI Spokesman* **XLIV**(2) 64–66.

Stephens, R.W. *et al.* (1985) Precautionary labelling of petroleum products in packages—a labelling system. *Concawe Report no 1/85*, Brussels, Belgium.

Vinogradov, G.V. (1989) *Rheological and Thermophysical Properties of Grease*. Gordon and Breach Science Publications.

Winterburn, G. (1988) Cooperative test results of apparent viscosity determination at low shear rates using the constant pressure viscometer. *NLGI Spokesman* **LII**(8) 365–372.

12 Extreme-pressure and anti-wear additives

A.R. LANSDOWN

12.1 Introduction

To understand the nature and function of extreme-pressure and anti-wear additives, it is useful to consider again the fundamental nature of lubrication by fluids, and of wear. In an ideal situation, lubricated surfaces are separated by a thick film of lubricant, and all the forces between the surfaces are transmitted by the lubricant. If, for any reason, the thickness of the lubricant film decreases, a point will be reached such that the contact stresses are carried increasingly by direct solid/solid contact between the surfaces or, at best, by interposed films of lubricant or other materials which are so thin that they behave as if they were solid.

The situation can be explained most simply in terms of the diagram in Figure 12.1, usually known as a Stribeck curve, but in fact first presented in this form by McKee and McKee (1929). The diagram was first used in connection with a plain journal bearing, but is equally appropriate for any bearing in which a 'pressure wedge' contributes to the formation of the lubricant film. The figure shows the relationship between the friction in the bearing and the expression $\eta N/P$, which is sometimes called the Hersey number. η represents the lubricant viscosity, P the bearing pressure, and N the speed of rotation of the shaft or, more generally, the relative speed of movement of the bearing surfaces. The zone in which a thick film of lubricant

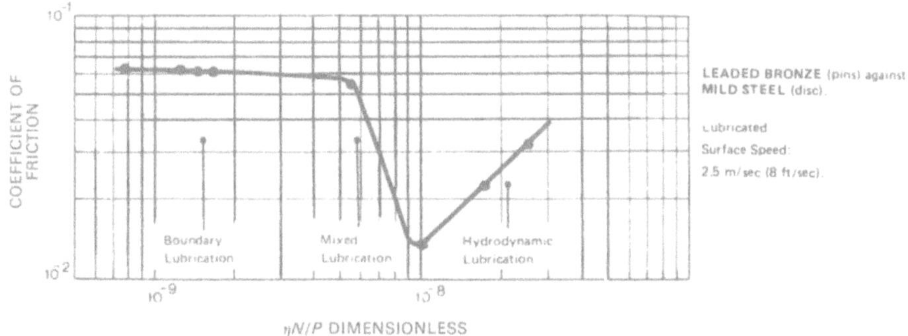

Figure 12.1 Relationship between lubrication regime and Hersey number. Reproduced with permission from A.R. Lansdown.

is present is termed 'hydrodynamic lubrication'. In this zone the lubricant film is thick enough to prevent any solid/solid contact between the bearing surfaces, and the friction is entirely due to the viscosity of the lubricant. The bearing load is carried more or less uniformly over the whole of the load-bearing area, and the stresses are therefore low. Since there are no high local stresses on the asperities of the bearing surfaces, there is no wear.

Three factors can lead to a reduction in the thickness of the lubricant film. These are: (i) a reduction in the lubricant viscosity, for example due to increasing temperature; (ii) a reduction in speed; or (iii) an increase in bearing load. Any of these changes causes a decrease in $\eta N/P$ and, therefore, a move towards the left-hand side of the diagram.

As the lubricant film becomes thinner, it reaches a stage at which its thickness is similar to the height of the microscopic roughness, or asperities, on the bearing surfaces. At this stage the thickness of the film and the load distribution are no longer uniform, but vary with the shape of the asperities. Where asperities on the opposing surfaces come close together, the film thickness will be low and the stress high. There will also be lateral stresses on the asperities because of the pressure variations, and these lateral stresses will begin to contribute to the friction. If the combined perpendicular and lateral stresses on an asperity are high enough, then repeated stress cycling can cause fatigue. This leads initially to surface or sub-surface microcracking and, eventually, to a particle breaking away from the surface: this is the first manifestation of wear, called *fatigue wear*.

Any further progressive reduction in $\eta N/P$ will further decrease the film thickness until there is actual solid/solid contact between the bearing surfaces. In the earliest stages this contact will only take place at the higher asperities, while a high proportion of the bearing load continues to be carried by the fluid. With further reduction in film thickness, an increasing proportion of the load will be carried by solid/solid contact between the asperities. This situation is represented by the 'mixed lubrication' zone shown in Figure 12.1. In this zone the friction is increasingly due to the solid/solid interactions and decreasingly due to the fluid viscosity.

Finally, the situation is reached in which solid/solid interactions carry all the load and generate all the friction. This is the zone called 'boundary lubrication' in Figure 12.1.

When solid/solid interactions occur between asperities, the asperities tend to adhere and the friction thus produced is called *adhesive friction*. The stresses on the asperities are sufficient to remove material, and this material loss is called *adhesive wear*. If the asperities of metal surfaces were purely metallic, the adhesion between them would be very strong, and would cause high friction and severe wear. In practice, bearing surfaces are normally covered with a film of metal oxide. Adhesion between asperities that are oxide-coated tends to be much milder and leads to lower friction and mild wear. Mild wear generally represents removal of metal oxide and, if the process is taking place in the presence of oxygen, the surfaces will re-oxidise. This regeneration of the surface oxide film enables mild wear to continue, and reduces any tendency to severe wear. Consequently, oxygen is one of the most effective of all anti-wear substances.

However, if the sliding motion is too fast, or the contact stresses too high, there may not be enough time for adequate regeneration of the oxide film before the next asperity collision takes place. The collision will then be between unoxidised or, at best, under-oxidised asperities. Adhesion will be strong, with high friction, and will eventually lead to severe wear. If severe wear continues, a stage where there is gross welding or seizure of the surfaces will usually be reached.

Practical engineering surfaces are usually contaminated with other substances, even when they are nominally dry. Such contaminants can have an effect on contact adhesion—usually reducing it—and, in this way, produce a decrease in friction and wear. When a lubricant is present, the base oil and any other substances in the lubricant will also affect contact adhesion.

Some lubricants contain naturally-occurring substances that reduce mild wear, but most modern lubricants rely on special additives to control mild or severe wear. Additives that reduce mild wear are traditionally called *anti-wear additives*, while additives that reduce or prevent severe wear are called *extreme-pressure (EP) additives*. In practice, the distinction is not always clear and some substances can act as both anti-wear and EP additives. Before going on to discuss the technology of anti-wear and EP additives, it will be useful to mention two other forms of wear.

Cutting wear. When a sharp projection on a surface slides over a softer surface, cutting or ploughing will take place. Extreme cases of this mechanism are the action of a file or grinder, or metal-cutting with a lathe, miller, shaper, planer, etc. A special case is where hard particles rub against a surface, causing abrasion or erosion.

Although there may be some controversy, it is probably true to say that lubricants or additives have little or no direct effect on cutting wear as such. However, where the edges or sides of the projection or cutting tool rub along the sides of the cut, quite severe adhesion can take place. The effects of this in metal-cutting are undesirable, causing wear of cutting-tools or clogging of a file or grinding wheel. This problem is a form of severe adhesive wear, therefore the materials used to prevent it are similar to ordinary EP additives, although much more powerful forms may be used in machining. On the other hand, where unwanted abrasion is taking place, the presence of EP additives may actually make it worse.

Corrosive wear. This is a type of wear in which chemical attack on the wearing surfaces contributes to a higher overall wear rate. This will later be seen to be relevant to the discussion of some aspects of additives.

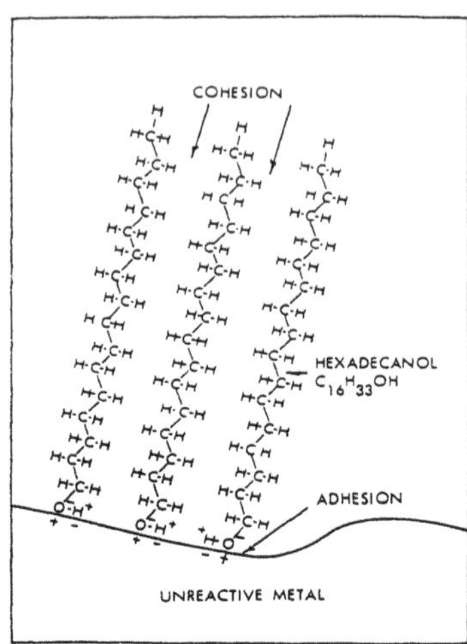

Figure 12.2 Adsorption of cetyl alcohol. Reproduced with permission from A.R. Lansdown.

12.2 Anti-wear additives

Many substances have an ability to reduce mild wear by forming a protective film on metal surfaces. Among these are the vegetable and animal oils or fats which, for many centuries, were mankind's first choice of lubricants.

It is now generally accepted that such mild, natural anti-wear additives act by adsorption on the surfaces. Most solid surfaces at low temperatures are completely covered with adsorbed films of contaminants, which may be one or more molecules thick. These adsorbed films lower the surface energy; the greater the reduction in surface energy, the more stable will be the adsorbed film. It follows that some substances will be adsorbed more strongly than others. The actual protection against mild wear is believed to be due to the ability of the adsorbed film to withstand the local contact pressures, while being able to shear readily with low frictional resistance when rubbed by the opposing surface.

The ideal molecule for anti-wear behaviour on a metal surface consists of a polar molecule for strong adsorption, with a long non-polar chain that will orientate itself perpendicular to the surface and thus create a thick film. This is shown in Figure 12.2, which represents a film of cetyl alcohol (hexadecanol) adsorbed on a steel surface. Other substances showing similar behaviour are the long-chain fatty acids such as palmitic and stearic acids, and their esters such as ethyl stearate. Vegetable oils and fats fall into the general category of long-chain esters, consisting mainly of esters of glycerol with long-chain fatty acids, and this is the probable reason for their good anti-wear properties.

Mineral oils produced from petroleum by normal refining processes also have useful anti-wear properties. This has generally been believed to be due to the presence of aromatic molecules or polar long-chain molecules containing oxygen, sulphur or unsaturated double bonds. This explanation seems to be only partly true, because Groszek (1967–68) showed that when a mineral oil was separated into several fractions, none of the fractions, not even that containing most of the trace elements like oxygen and sulphur, had as effective an anti-wear behaviour as the original oil.

Some mineral oils have always been recognised as having inferior anti-wear properties, and the technique of adding small amounts of natural vegetable oils or fats, or long-chain alcohols, acids or esters has been practised for many years. This was also done where the lubricated equipment was especially sensitive. These were the first substances that could be correctly described as anti-wear additives. Modern refining practices tend to remove the aromatics, unsaturates and sulphur-, oxygen- and nitrogen-containing compounds, resulting in base oils that have lower natural anti-wear performance. In most mineral oil lubricants, therefore, the addition of anti-wear additives has become normal practice.

The mildly adsorbed long-chain compounds described above are not very powerful anti-wear additives. They tend to be removed from the surfaces during strong rubbing and may be desorbed if the surface temperature rises much over 100 °C. There has therefore been considerable effort to find more

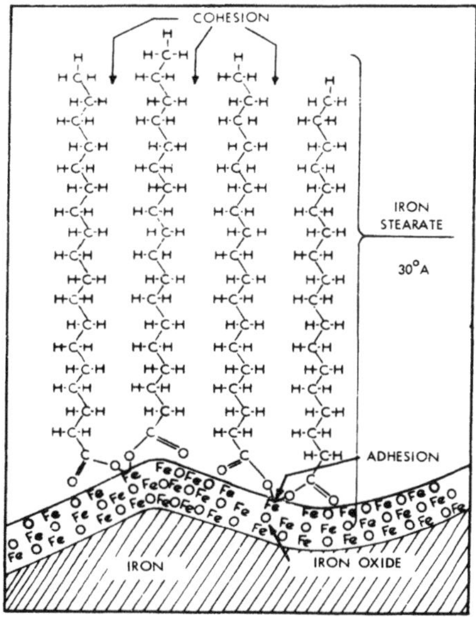

Figure 12.3 Chemisorption of stearic acid. Reproduced with permission from A.R. Lansdown.

powerful anti-wear additives that either adsorb more strongly or chemisorb onto the metal surfaces. Chemisorption (chemical adsorption) is a mechanism that takes place when electron transfer occurs between an adsorbed molecule and the surface. This produces either a covalent or an electrovalent bond, which is far stronger than the physical attraction in adsorption. Whereas adsorption is always reversible, chemisorption is believed to be generally irreversible, so that chemisorbed molecules are only removed when the element of surface to which they are attached is removed by wear.

The mechanism of chemisorption is shown diagrammatically in Figure 12.3, which represents chemisorption of stearic acid to a steel surface. It will be noted that although stearic acid was mentioned earlier as an adsorbed additive, it can also be chemisorbed. The probable mechanism is that a stearic acid molecule initially adsorbs on the surface but, if the temperature rises sufficiently, electron transfer can take place, producing an iron stearate salt.

Most of the common synthetic anti-wear additives are compounds containing phosphorus, such as zinc dialkyl(or diaryl) dithiophosphate (ZDDP), tricresyl phosphate (TCP), trixylyl phosphate (TXP) and dilauryl phosphate. One of the most widely used of all additives is ZDDP, which has anti-wear, antioxidant and corrosion-preventing properties. Like stearic acid, ZDDP initially adsorbs on the metal surface and, in this state, has mild anti-wear properties. Under the influence of heat, electron transfer can take place to produce a chemisorbed film; the heat itself is generated by sliding. Similar mechanisms occur with the other phosphorus compounds such as tricresyl and trixylyl phosphate.

Apart from ZDDP itself, there is also a variety of other dialkyl dithiophosphates. All are manufactured by a fairly simple process involving reaction of phosphorus pentasulphide with the relevant alcohol to produce an intermediate dialkyl dithiophosphoric acid. The intermediate acid is neutralised with zinc oxide. If the intermediate acid is neutralised with a different metal oxide, a different range of dialkyl dithiophosphates are formed. The exact range of properties of the product depends on the metal atom and the alkyl group.

12.3 Extreme-pressure additives

If the sliding conditions are more severe, then simple adsorbed or chemisorbed anti-wear additives will eventually be inadequate to prevent severe adhesion. The mechanism by which they fail may be thermal or mechanical. If the surface friction causes too great a temperature increase, the simpler adsorbed molecules will desorb so the protective film is no longer present to prevent adhesion of the surfaces. On the other hand, excessive mechanical stress will either tear away the surface films, or wear away the metal together with its adsorbed films. In either case, the result is a surface containing areas

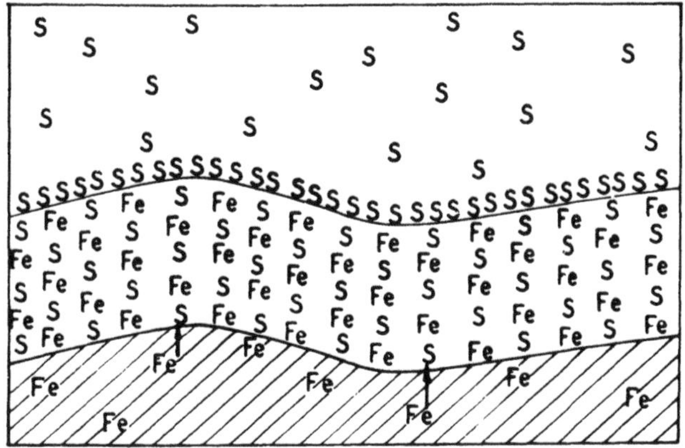

Figure 12.4 Reaction of sulphur with a steel surface. Reproduced with permission from A.R. Lansdown.

of unoxidised or under-oxidised metal, and more powerful films are required to protect it. These films are provided by reaction of various chemicals with the metal surfaces, and most of these chemicals contain sulphur, phosphorus or chlorine. These are extreme-pressure additives.

The simplest EP additive is elemental sulphur, which can be dissolved in a mineral oil to give a so-called sulphurised oil. The free sulphur will react with a metal surface, especially when the metal has been freshly exposed by wear, to give a surface layer of metal sulphide. Figure 12.4 shows an idealised picture of the surface film produced by free sulphur reacting with a steel surface. The reacted film will be worn away by rubbing against another bearing surface, and further sulphur will then react to regenerate the surface film. The repeated cycle of reaction and wear represents a form of corrosive wear. All EP additives are in fact corrosive to the surfaces which they protect. The rate of wear may not even be significantly less than that of the metal surface itself but, because the reacted surface has a lower shear strength, the sliding friction is markedly reduced, and the tendency to severe adhesion will also be reduced or even eliminated.

The primary target in the development and selection of EP additives is to maximise the protection against severe adhesion, while at the same time minimising the rate of corrosion. The two competing phenomena are affected by several factors.

There is an old established rule-of-thumb that the rate of a chemical reaction doubles for every 10 °C rise in temperature. This is a gross over-simplification, but is adequate for the present discussion. Friction theory suggests that when two asperities collide on lubricated rubbing surfaces, the local contact temperature may exceed the bulk temperature by 200 °C or more. The rate of any reaction at that point will then be 2^{20} or, in other words, over one million times faster than that at the bulk temperature.

Freshly-worn metal surfaces have a higher surface energy than unworn equilibrium surfaces. As a result, reactions will take place more readily at worn points than over the remainder of the surface. Morecroft (1971) has shown that, if the surface is completely free of oxide, the reactions tend to be more severe: hydrocarbons and long-chain fatty acids break down to give small fragments such as hydrogen, methane, carbon monoxide and carbon dioxide, which are of no value in wear prevention. However, partly-worn surfaces in which the surface oxide is reduced but not completely removed will have increased surface energy while reacting more conventionally with organic compounds.

It has also been shown that rubbing surfaces may catalyse reactions in the nearby liquid. For example Tabor and Willis (1969) have shown that, when in contact with rubbing surfaces, dimethyl silicones oxidise at temperatures as low as 95–100 °C, whereas in the absence of catalysts they do not oxidise until the temperature reaches more than 200 °C. It is not clear whether this effect is a true catalysis, or whether it may simply be a result of the temperature and surface energy effects described previously. In any case, the combination of local high temperature, increased surface energy, and any catalysis or other activation process certainly makes the action of extreme-pressure additives much more specific, taking place preferentially at the wear points where it is most needed.

In spite of these beneficial factors, it is very important to obtain the best balance between corrosiveness and protection for the wide range of systems in which EP additives are used. Many thousands of different compounds of phosphorus, sulphur and chlorine have been tested for EP activity, and Table 12.1 lists some of the most successful. They are listed roughly in order of increasing activity, from the mild phosphorus compounds to the reactive chlorine compounds. Missing from the table are the lead naphthenates, which are very useful EP additives in gear oils, but are becoming obsolete because of the general concern about toxicity of lead compounds. Also missing are the recently-developed organic boron–phosphorus compounds, which do not yet appear to be in widespread use.

It will be noticed that the table includes some compounds, such as tricresyl phosphate and ZDDP, which were previously mentioned as anti-wear additives. These compounds are valuable because they can act as anti-wear or EP additives, depending on the severity of the rubbing contact that is taking place.

For completeness some mention should perhaps be made of molybdenum disulphide. This is a solid lubricant but is quite often used dispersed in engine lubricating oils, where it has been shown to give a small improvement in fuel consumption. In laboratory tests it has also been found to give some improvement in the load-carrying capacity of oils, but it is not often used in

Table 12.1 Some successful extreme-pressure additives

Dilauryl phosphate
Didodecyl phosphite
Tricresyl phosphate (TCP)
Zinc dialkyl (or diaryl) dithiophosphates (ZDDPs)
Phospho-sulphurised fatty oils
Zinc dialkyldithiocarbamate
Mercaptobenzothiazole
Sulphurised fatty oils
Sulphurised terpenes
Sulphurised oleic acid
Alkyl and aryl polysulphides
Sulphurised sperm oil
Sulphurised mineral oil
Sulphur chloride treated fatty oils
Chlornaphtha xanthate
Cetyl chloride
Chlorinated paraffinic oils
Chlorinated paraffin wax sulphides
Chlorinated paraffin wax

this way. Its greatest benefit in load-carrying is in very high concentration, up to 60%, in greases and very viscous oils for anti-seize applications. Used in this way, it can prevent seizure at pressures high enough to produce plastic deformation of hardened bearing steels. It is probably not suitable for conventional EP use, such as in hypoid gears or metal-cutting, because these are in fact high-speed or high-temperature and not extreme-pressure requirements.

There are several other boundary lubricant additives that are neither anti-wear nor EP additives. These include the additives to control friction and stick-slip in automatic transmissions, limited slip differentials and lubricated clutches. The compounds used include some of the sulphur-containing EP additives and also some specialised ones such as *N*-acyl sarcosines and their derivatives.

12.4 Mechanisms of action of anti-wear and EP additives

It is not possible to describe a general mechanism for all anti-wear and EP additives, because they vary too much in their behaviour. Some, such as the long-chain alcohols, probably act only by adsorption. Others, such as the long-chain carboxylic acids, probably act only by adsorption and chemisorption. Yet others, such as the sulphurised or chlorinated oils, probably have no chemisorption behaviour, but go directly from adsorption to chemical reaction. The general relationship between severity of rubbing and the mechanisms of anti-wear or EP behaviour can, however, be illustrated by either ZDDP or TCP, which are generally believed to progress through all three stages of adsorption, chemisorption and chemical reaction.

When a solution of TCP in a lubricant base oil is in contact with a metal surface, the extent to which it adsorbs on the surface will depend on several factors. Overall adsorption is greater at lower temperatures, but the TCP will compete with all the other substances present for adsorption sites on the metal. On a metal or other polar solid, polar substances and substances that are least soluble will be preferentially adsorbed. It follows that, in order to ensure that the maximum amount of TCP adsorbs on the bearing surface, other additives present should be designed to be less polar and more soluble in the oil.

The adsorbed film may be one or more molecules thick. Since TCP is a relatively compact molecule, a monomolecular film will not be very thick. Films several molecules thick are weaker, so that pure adsorption with TCP is unlikely to give a high degree of protection, although adsorption must take place, and any adsorbed film will give some protection against wear in mild sliding. In this respect TCP would be likely to be inferior to adsorbed films of less-polar molecules, such as long-chain alcohols and esters. With all anti-wear or EP additives, adsorption is likely to have an additional benefit in increasing the concentration of the additive at the bearing surface.

As mentioned previously, adsorption is greater at lower temperatures, and adsorbed additives are therefore less effective when the surface temperature rises as a result of more vigorous sliding. There is no absolute temperature at which the adsorption of a particular compound ceases, but there is a significant reduction at a so-called critical desorption temperature, which is sometimes similar to the melting point. However, as the temperature rises, a point is reached where electron transfer can take place between the surface and the adsorbed molecules, so that a covalent or electrovalent bond is formed. In this condition, a new chemical compound can be considered to have been formed between the adsorbed molecule and the metal of the surface. It should perhaps be emphasised that in true chemisorption only electrons are transferred, and the basic structure of the chemisorbed molecule remains unchanged.

No direct proof of the existence of chemisorption with anti-wear or EP additives has been obtained. Support for the occurrence of chemisorption comes from the fact that, where the adsorbed compound has a suitable structure, the films formed have been shown to be more resistant to heat and sliding. In the case of TCP, Forbes *et al.* (1972) showed that films formed by sliding in the presence of TCP in mild conditions contained both phosphorus and suitable organic structures and, from this, inferred the existence of chemisorption.

The increased polarisation caused by chemisorption may well encourage the formation of two or more adsorbed layers, which would again increase the extent of protection against wear. The polarisation will certainly change the bond strengths and angles, and therefore affect the way in which the structure eventually behaves under the action of increased temperature or shear stress.

As the temperature rises further due to more severe frictional heating, a point at which the organic structures break down will be reached. It may be at this point that the thick viscous or solid films investigated by Cann *et al.* (1983) are formed. They argue that the probable structure of these films is polymeric or oligomeric, based on a repeated iron–oxygen–phosphorus–oxygen unit, with organic side-chains. They dealt specifically with ZDDP, but it seems probable that similar arguments apply to TCP.

At even higher temperatures of 300 °C or more, such as are likely to occur in asperity collisions at high load and speed, the remaining organic structures will decompose, and only inorganic compounds will be stable. Under these conditions, the final product from TCP will be iron phosphate or iron phosphide, depending on the availability of oxygen. Under similar conditions, ZDDP may generate iron phosphate, iron phosphide, iron sulphide, iron sulphate or zinc compounds. Other phosphorus-containing additives will also produce phosphates or phosphides, sulphurised additives will give sulphates or sulphides, and chlorinated additives will give chlorides.

It was mentioned earlier that if as a result of wear the oxide is completely removed, leaving a free metal surface, simple hydrocarbons or fatty acids will be severely broken down, giving fragments that provide no protection against wear. Obviously, if EP additives are present, similar reactions will release phosphorus, sulphur or chlorine, which can rapidly react with the bare metal surface to form effective protective films. All of these compounds have high thermal stability—high enough to withstand any flash temperatures generated in the sliding contacts. Mechanically they will adhere strongly to the metal substrates but will themselves shear more readily, therefore giving lower sliding friction. At the same time, they have only a very weak tendency to adhere to the opposite sliding surface.

12.5 Application of different classes of additive

Anti-wear and EP additives are chemically more or less reactive, and have lower resistance to heat and oxidation than most modern base oils. They will therefore reduce the overall thermal and oxidative stability of fully-formulated oils. The major exception is ZDDP, which acts as an antioxidant. In addition, the more powerful EP additives have some corrosive effect in contact with metals. In selecting anti-wear or EP additives, it is therefore a good primary rule to use them only where necessary, and to use the mildest type possible. However, there are two factors that make the use of some anti-wear protection generally desirable. The first is the increased complexity, and higher loads and speeds in modern machinery. The second is the increasing use of more highly-refined mineral oils and synthetic hydrocarbons, which have very little, or no, inherent anti-wear performance.

The mildest anti-wear additives, such as long-chain alcohols, fatty acids, esters and TCP are used in hydraulic fluids and turbine oils where the main anti-wear requirement is in the pumps and valves rather than in bearings or gears, and temperatures are low. Typical concentrations are 0.1% of stearic acid, or from 0.15 to 0.5% of TCP. For industrial gears or other machinery, where mild EP activity is needed at normal temperatures, the additives commonly used are the phosphorus-containing ones (such as di-lauryl phosphate, TCP and, especially, ZDDP) or the more stable sulphur compounds (such as phospho-sulphurised fatty oils and mercaptobenzothiazole). For vane pumps, in particular, the suitability of a lubricant has been said to correlate directly with the ZDDP content.

Automotive engines need EP properties particularly for various components of the valve train, especially the cams and tappets. The most successful additive is again ZDDP, but others used include di-lauryl phosphate, di-dodecyl phosphite, zinc dialkyl dithiocarbamates and sulphurised terpenes. With the advent of transaxles and combined engine and gearbox lubrication, the same types of additive have again proved adequate.

Among the most demanding machine components in their EP requirements are spiral bevel and, especially, hypoid gears, particularly in heavy trucks. These require powerful EP activity and the additives used are often mixtures of such compounds as chlorinated paraffin wax, chlorinated paraffin wax sulphides, chlornaphtha xanthate, sulphur chloride treated fatty oils, phospho-sulphurised fatty oils, sulphurised fatty oils, polysulphides, zinc dialkyl dithiocarbamate and, of course, ZDDP. Even more demanding, however, are metal-cutting operations, where the highly-loaded contact of a cutting tool against the freshly-exposed metal surface of the workpiece creates very severe adhesive conditions. Surface temperatures are also very high, but in this case the components are not high-precision bearings or gears, so that mild corrosion in the working area is acceptable. Chlorinated additives are therefore widely used, and include chlorinated wax and chlorinated mineral oil, together with the more active sulphur compounds such as sulphurised fatty oils, sulphurised mineral oil, sulphurised olefins, sulphur chloride treated fatty oils and benzyl polysulphides.

12.6 Future developments

Future developments in any field are most reliably predicted by extrapolation of recent and present developments. In the field of anti-wear and EP additives, recent developments have been dominated by the influence of environmental concerns. In recent years, the pressure for reduction of lead in the environment has led to the virtual disappearance of lead naphthenate additives. Similarly, concern about the conservation of the great whales has almost eliminated the use of sulphurised sperm oil, one of the most valuable of all EP additives.

Similar concerns are now leading to a search for substitutes for phosphorus and chlorine in EP additives. The justification for this is questionable, since neither element is listed as being hazardous to human health or plant growth by the British Government Interdepartmental Committee on the Redevelopment of Contaminated Land Paper TCRCL 59/83. Phosphorus is, of course, a contributor to harmful algal growth in inland waters, but the hundred tonnes used per annum in lubricants in the United Kingdom is trivial compared with the quantities used in agriculture. Nevertheless, it is likely that there will be a tendency to use sulphur-containing additives in preference to phosphorus and chlorine compounds.

For over 30 years, there has been a largely unsuccessful search for additives to use with synthetic, non-hydrocarbon base oils for high-temperature lubrication, especially in aerospace applications. The current interest in the so-called 'adiabatic' engines for automotive use has increased the prospect of much bigger markets for high-temperature lubricants. This may result in the successful development of effective anti-wear and EP additives for high-temperature applications, although the technical obstacles to the achievement of successful 'adiabatic' engines, and of their high-temperature anti-wear and EP additives are considerable.

In the long term there are three phenomena that may lead to useful new EP additives. These are the inherent rigidity of colloids, the anisotropic rheology of liquid crystals, and the variable rheology of ferro-fluids. However, considerable economic pressure is likely to be needed before any of these replaces the best present additives.

References

Cann, P.M., Spikes, H.A. and Cameron, A. (1983) Thick film formation by zinc dialkyl dithiophosphate. *ASLE Trans.* **26** 48–52.

Forbes, E.S., Upsdel, N.T. and Battesby, J. (1972) Current thoughts on the mechanism of action of tricresyl phosphate as a load-carrying additive. *Proc. Trib. Conf. C.64*, **172** 7–13.

Groszek, A.J. (1967–68) Wear reducing properties of mineral oils and their fractions. *Proc. Instn. Mech. Engrs.* **182** (3N) 160.

McKee, S.A. and McKee, T.R. (1929) Friction of journal bearings as influenced by clearance and length. *Trans. Am. Soc. Mech. Engrs.* **51** 161–171.

Morecroft, D.W. (1971) Reaction of *N*-octadecane and decoic acid with clean iron surfaces. *Proceedings of the Conference on the Chemical Affects at Bearing Surfaces*, Swansea, January 1971, p. 225.

Tabor, D. and Willis, R.F. (1969) The formation of silicone polymer films on metal surfaces at high temperatures and their boundary lubricating properties. *Wear* **15** 137.

13 Lubricants and their environmental impact

C.I. BETTON

13.1 Introduction

In 1985 a report by the European Oil Companies organisation CONCAWE (CONCAWE, 1985) showed that, of the total lubricant sold in the EC, over half was 'consumed' whilst some was recycled or burnt as fuel. The figures shown in Table 13.1 are for total lubricant production. A breakdown of lubricating oil sales for the US during 1981 (IARC, 1984) shows that automotive engine oils represent the major component of total sales.

The total percentage of lubricant that is successfully disposed of is that which is either recycled or burnt—32% (Table 13.1). The major proportion of lubricant enters the environment in one way or another, mainly due to lubricant use, i.e. 'consumption'. As far as automotive engine oils are concerned, the results of this consumption can be seen down the centre of any motorway lane, or in any car park, as a black coating of oil on the road.

Some lubricants are intended for use in such a way that they are 'lost' to the environment. Examples are two-stroke oils, greases and oils used on railways, chainsaw bar lubricants, rubber oils used in tyres, many white oils, and so on. Conversely, providing efficient management systems are in place, many industrial oils should be largely contained and should not escape into the environment. Accidents, however, inevitably occur and cause the environmental release of almost any and all types of lubricant. A gearbox seal that fails, high pressure hydraulic lines that fracture, and metal parts coated in residual oils all contribute to 'consumption' of lubricants and add to the environmental burden. In the EC alone, the total amount of lubricant that

Table 13.1 Total EC lubricant production

	Tonnes per year ($\times 10^3$)	%
Total EC lubricant sales	4500	100
Consumed	2350	50–55
Recycled	700	15
Burnt as fuel	750	17
Unaccounted for	600	13
Poured down drain deliberately	100	2

enters the environment each year can be said to be roughly the equivalent of twelve oil tankers or one Exxon Valdez per month.

Three major aspects of the environmental impact of lubricants can be addressed: (i) control via engineering to minimise losses; (ii) minimising the impact of these losses; and (iii) efficient collection and treatment of waste material. In terms of the lubricant, aspect (i) is largely a matter for the equipment designer and builder. The specialist lubricant company is affected by aspect (iii) and can affect aspect (ii). The collection and treatment of waste lubricants will now be described, and will be followed by consideration of the environmental impact of lubricants.

13.2 Collection of waste lubricant

Waste lubricant that is collected is usually disposed of either by burning as fuel or incineration as waste, or is reclaimed. Some lubricants, such as industrial oils, can be treated relatively easily and recycled, whereas automotive engine oils require more sophisticated treatment to produce re-usable base fluids of acceptable quality.

13.2.1 *Used industrial lubricants*

There are many potential sources of used industrial products but reprocessing is not an option for a large number of these synthetic and fatty oil based products. Some products, such as transformer oils and hydraulic oils, can be readily collected from large industrial concerns and segregated. Consequently, contamination can be avoided. These oils may be regenerated to a recognised standard and returned to the original source.

13.2.2 *Used automotive lubricants*

These will include mono- and multigrade crankcase oils from petrol and diesel engines, together with gear oils and transmission fluids. Used industrial lubricants that have been inadequately segregated will also be included. Apart from the degradation products from the in-service use of the oil, a wide range of contamination is possible. This includes:

- Water—combustion by-product, rain water/salt water ingress
- Fuels—residual components of gasoline and diesel fuel
- Solids—soot, additive and wear metals, rust, dirt, etc.
- Chemicals—used oil can be used as an unauthorised means of hazardous waste disposal
- Industrial oils—inadequate segregation of oil types can allow contamination by fatty or naphthenic products

There are many potential sources of used automotive oils, for example DIY motorists, service stations, company car or truck fleets and waste oil collection sites. There is also an equally wide range of commercial collectors. Good management of a used oil collection system is paramount to minimise both the variability of the material supplied to a re-refining plant, and the problems of contamination by toxic or hazardous materials.

13.3 Treatment of collected lubricant

13.3.1 *Production of fuel oil blending component*

Used lubricant is an excellent source of energy, but contains contaminants which, when burnt, can create environmental pollution and operational problems. Simple processing (e.g. settling and filtration/centrifugation) is sufficient to remove coarse solids and water, and a mild chemical treatment (e.g. caustic soda solution) may also be used to reduce concentrations of other contaminants. Disposal of waste products may be a problem.

13.3.2 *Reclamation of lubricating oils*

Some specific types of industrial oils can be readily segregated and are suitable for relatively simple reprocessing before being returned to their original service. Typical processing methods involve filtration and removal of water or volatile decomposition products under vacuum.

Large industrial customers can arrange for on-site reprocessing (e.g. reconditioning of transformer oils at power generation plants), or can collect specific drain oils for off-site reprocessing and return (e.g. reclamation of railroad diesel engine oils). These types of customer-specific reprocessing can be cost-effective compared to the expense of new oils.

13.3.3 *Production of re-refined lubricant base oils*

When dealing with oils from multiple sources, complex processes are needed to remove the wide range of contaminants and additives that will be present. The objective is to produce base oils that can then be used as substitutes or alternatives to virgin mineral base oils. The main re-refining technologies are summarised in sections 13.3.3.1 to 13.3.3.4. Most of these are continuous rather than batch processes and rely on receiving a reasonably homogeneous used oil feedstock if the products are to be consistent. One means of achieving consistency is to mix a complete month's supply of used oil and to use that to feed a continuous process appropriate to the characteristics of the resultant mixture. There is a risk that contamination will spoil a large batch, however, and quality checks on the feedstock are very important.

13.3.3.1 *Acid/clay treatment* This has been the principal re-refining process in commercial use (e.g. Meinken process), but has the disadvantages that it generates large amounts of hazardous waste and cannot effectively remove the high concentrations of additives used in modern engine oils. It is now being superseded by other, newer technologies.

In the acid/clay process, the used oil is first treated with 98% sulphuric acid; the resulting acid sludge is then separated and the remaining oil is clay treated, neutralised and filtered. Process yields from modern used engine oils are low (50–60% of lube boiling range material), and large quantities of acid sludge and oil-soaked clay are generated. These wastes are difficult and costly to dispose of in an environmentally acceptable manner. Sometimes a thermal pre-treatment is used to degrade some of the additives and reduce the workload on the acid treatment stage. Base oils produced by the acid/clay process are usually dark in colour, have a noticeable odour and are somewhat inferior in quality to virgin mineral oils.

13.3.3.2 *Solvent extraction* Propane or other solvents can be used to selectively extract the base lube material from the used oil (e.g. IFP process). A high-boiling asphaltic residue, which contains most of the additives and other impurities, is recovered as a separate stream. (Note that this is not the same process as refinery furfural or phenol extraction, which uses solvent to selectively remove aromatics from lubricant fractions.) A further finishing treatment of the lube material is normally required and may involve low severity acid/clay treatment, clay treatment alone, or hydrofinishing. Solvent extraction plants are expensive to build and operate, and generate significant amounts of waste materials and hazardous by-products. Overall base oil yields are 70–80%.

13.3.3.3 *Distillation/clay treatment* Thin-film distillation under high vacuum (e.g. Luwa evaporators) allows the separation of gas oil, lube oil and an asphaltic residue containing most of the additives and contaminants from the used oil. The lube oil stream is finished by clay treatment. Overall base oil yields are again 70–80% and significant amounts of spent clay must be disposed of.

13.3.3.4 *Distillation/hydrotreatment* This technology is the most recent (e.g. KTI and Mohawk processes). After pre-treatment and thin-film distillation, the base oil fraction is hydrotreated under moderate conditions. A final distillation step yields a range of base oil streams of different viscosity. Process yields of base oil are high at 90–95%. By-products are a low boiling distillate (used as fuel in the re-refining plant), gas oil and a non-hazardous asphaltic residue. Emissions and waste streams are limited, therefore this process is capable of meeting very strict environmental controls. Base oils produced by this route tend to be of superior quality compared to the older technologies. The hydrotreatment process is comparatively sophisticated, capital intensive and requires skilled operation, although operating costs are not exceptionally high.

13.4 Re-refined base oil quality

Lubricant base oils must meet certain, specified, minimum performance characteristics; viscosity, viscosity index, pour point and volatility must all meet the required standards. When dealing with re-refined oils, additional characteristics such as colour and odour must also be taken into account. Dark colour or unusual odour are readily perceived by the customer as representing deficiencies in quality. Many re-refined oils, for example, have a definite, characteristic 'cracked or oxidised' odour, which may be totally unacceptable in some countries and markets. Table 13.2 provides quality guidelines for re-refined oil acceptance.

Due to the variety of potential sources of used lubricants, their differing histories of use, and their subsequent levels of possible contaminants, assessment of the quality of the resulting re-refined base oil is a major concern. Quality assurance must be applied, both to the used oil feedstock and, more particularly, to the product(s) of the re-refining process. Limits for the

Table 13.2 Guidelines for quality acceptance of re-refined base oil[a]

	Grade	
	150	500
Viscosity at 100 °C (cSt)	5.0 ± 0.2	11.0 ± 0.4
Viscosity index	90–110	90–110
Flash point, PMC(°C)[b]	210	230
Pour point (°C)[c]	−9	−9
Dialysis residue (%)[c]	0.1	0.1
Ring analysis, C_A(%)[c]	12	12
ASTM colour[c]	3	4.5
Total acid no. (mg KOH/g)[c]	0.1	0.1
Noack volatility(% loss)[c]	18	6
IP 154 copper corrosion[c]	1	1
Odour	No foreign odour in finished product	
Chlorine (ppm)[c]	10	10
Water (ppm)[c]	50	50
PAHs (ppm)[c,d]	3	3
PAHs (ppm)[c,e]	250	250
Individual metals (ppm)[c]	10	10
Sulphated ash (%)[c]	0.01	0.01

[a]Data compiled by W.H. Preston and J.Z. Trocki
[b]minimum
[c]maximum
[d]by IP 346
[e]PAHs from fluoranthene to coronene by Grimmer method

concentration of various contaminants in the finished base oil must be set and strictly adhered to.

13.4.1 *Possible contaminants in re-refined base oils*

13.4.1.1 *Fuel and solvent residues* These can be present in varying amounts depending on the efficiency of the re-refining processes employed. Flash point and Noack volatility values in excess of those given in Table 13.2 indicate unacceptable contamination.

13.4.1.2 *Polycyclic aromatic hydrocarbons (PAHs)* These are primarily present as a result of the combustion process during previous use of the lubricant, prior to re-refining. PAHs are found at varying concentrations in used oil, determined by factors such as drain interval, fuel type and sump size. Some PAHs are known to be carcinogenic, and the concentration of total PAHs in base fluids must be controlled. For more detailed discussion of PAHs, see section 13.5.

13.4.1.3 *Metals* Metals should not be present in re-refined base oils, either as wear metals or from previous use of metallic oil additives, or as clay or catalyst fines. Metals can be measured by determining total sulphated ash content or, individually, by atomic absorption spectroscopy.

13.4.1.4 *Non-metallic impurities* Viscosity index (VI) improvers or polychlorinated biphenyls (PCBs) can be determined by dialysis or chlorine analysis, respectively. Indication of the presence of VI improver is given by unusually high viscosity index.

13.4.1.5 *Water and untreated acid* These may be carried over from some processes. Total acid number and copper corrosion tests will indicate the presence of such contaminants in the re-refined oil.

13.4.1.6 *Sulphur content* In virgin base oils this is seen as an index of inherent antioxidant capacity. With re-refined oils, those hydrocarbons that were inherently oxidatively unstable will have been oxidised during previous use. The function of the sulphur content of a re-refined base oil is not clear; more work is required to determine the need or concentration of sulphur required to meet minimum performance standards.

13.5 Health and safety aspects of re-refined oil

Used engine oil is recognised as posing a carcinogenic risk to man. However, providing sensible precautions, such as good personal hygiene, are taken, the

Table 13.3 PAH analysis of re-refined oil[a]

Individual PAH	Virgin reference oil	PAH concentration of oil sample (mg/kg) 1	2	3	4	5	6	7	8	9
Fluoranthene	0.02	6.74	1.64	2.77	3.26	0.14	1.42	1.78	0.03	0.05
Pyrene	0.42	5.33	2.75	7.16	9.19	0.42	0.60	20.66	0.26	0.36
Benzo(*b*)naphtho(2,1-*d*)thiophene	0.34	0.89	1.02	2.29	0.60	0.07	1.24	0.38	0.04	0.03
Benzo(*ghi*)fluoranthene + benzo(*c*)phenanthrene	0.01	1.02	0.11	0.37	0.21	0.12	0.19	0.29	0.03	0.04
Cyclopenta(*cd*)pyrene	< 0.01	0.01	0.13	0.05	0.06	< 0.01	0.11	0.08	< 0.01	< 0.01
Benz(*a*)anthracene	0.08	5.39	1.05	1.62	2.84	1.13	0.30	0.34	0.07	0.08
Chrysene + triphenylene	2.10	9.47	2.89	4.16	12.65	1.73	6.65	6.26	1.23	1.36
Benzofluoranthenes ($b+j+k$)	0.28	6.84	0.98	1.20	12.32	4.37	2.43	4.93	4.53	5.36
Benzo(*e*)pyrene	0.41	3.56	1.04	1.41	14.46	6.28	0.65	6.51	11.25	11.20
Benzo(*a*)pyrene	0.01	0.83	0.53	0.60	1.65	2.82	0.13	0.31	0.47	0.47
Indeno(1,2,3-*cd*)pyrene	0.06	0.31	0.22	0.30	2.37	2.30	0.08	1.40	6.10	9.33
Dibenz(*a,h*)anthracene	< 0.01	0.26	0.05	0.09	0.12	0.76	0.04	0.04	0.22	0.22
Benzo(*ghi*)perylene	0.53	1.63	0.86	1.14	5.05	11.49	0.40	5.97	38.5	36.3
Anthanthrene	0.03	0.12	0.09	0.21	0.11	0.16	0.11	0.08	0.6	0.6
Coronene	0.13	0.19	0.10	0.35	1.19	2.30	0.09	0.24	11.4	12.6
Known or suspect carcinogen measured	2.7	29.9	6.15	8.95	33.52	15.69	10.13	13.97	24.65	30.06
Total PAH content (3–7 rings)	7.5	87.3	22.6	52.5	93.5	49.5	30.2	112.0	67.7	126

[a]Data compiled from results provided by the Biochemiches Institut für Umweltcarcinogene, Hamburg

actual risks to health can be essentially eliminated. This is borne out by epidemiological studies on groups of mechanics where there has been no evidence of an increase in skin cancer.

The main carcinogenic agents in used oil are polycyclic aromatic hydrocarbons (PAHs) with 3–7 rings such as benzo(*a*)pyrene, benz(*a*)anthracene and chrysene. These chemicals, and many others like them, are found in used lubricant, having been formed during the combustion cycle in the engine. PAHs are also present in crude oil, and can be present in unused base fluids leaving the refinery. Refinery technology is, however, able to remove these harmful PAHs by hydrogenation and solvent extraction. Strict limits of 3% on the DMSO extract of virgin base oil have been introduced by the oil industry; these limits are based on animal studies, which have shown that there is no evidence of carcinogenic activity at DMSO extract levels below 3% (see CONCAWE, 1985).

PAHs are found in re-refined oil samples at a higher level than those found in unused base oil. A series of re-refined oils from different countries and different re-refining processes was analysed for PAH content by the Biochemiches Institut für Umweltcarcinogene in Hamburg. The results are shown in Table 13.3. Interpretation of these results in relation to carcinogenic risk is not easy. The limits on total DMSO extractable materials (taken as a measure of PAH content) in virgin base stocks were set, based on a knowledge of PAH distribution that is not applicable to the used oil situation. The probability is that re-refined base oils will present no significant risk to health, provided that normal precautionary measures of personal hygiene and handling are taken. However, until further biological studies are undertaken, the definitive position with regard to the carcinogenic potential of re-refined oils cannot be determined.

13.6 Environmental considerations of waste lubricant

It is important to consider the total environmental costs and benefits of dealing with used lubricant. The previous sections have indicated that the quality of the re-refined product is determined by the nature of the feedstock, and by the treatment process. The total amount of energy consumed in the collection and re-refining of used oils to an acceptable standard must be balanced against the energy to be gained from utilising the used oil as a fuel. Clearly, the choice between no lubricant or an inferior basestock is also a major consideration.

Of the three basic methods of disposal of used oil, the most efficient in terms of energy conservation, in that it displaces an equivalent amount of oil, is the use of the material as a fuel or fuel supplement. In heating processes where the fuel is supplied through a burner, some limited pre-treatment of the waste oils and blending with conventional fuels are necessary. Pre-treatment can range from simple settling and filtration, to heating, emulsion breaking, chemical treatment and centrifugation—the amount of treatment will depend on the nature of the used oil and the required fuel specifications. The flue gas components of principal environmental concern are PCBs, PAHs, dioxins and heavy metals. Some treatment of the oil will be required to ensure that emission standards for these materials are not exceeded when waste oil is used as fuel. Should this not be possible, the oil must be considered a hazardous waste and treated accordingly. Waste oils can successfully be burnt in boilers, ranging in size from electricity generating plants and industrial boilers to specially produced small furnace and boiler installations. An EC limit of 1 MW has recently been introduced, below which it is not advisable to burn used oil. Subject to control over metal content, one of the most satisfactory uses for waste oil is in cement production. The advantage of this disposal method is that wastes that would otherwise be vented to the atmosphere via the flue gases, are absorbed and contained within the product with no adverse environmental consequences. There are, of course, limits on the amount of cement production available to use such a fuel—particularly when the building industry is in recession.

Disposal of waste oil as hazardous waste requires that it be incinerated at high temperature to ensure complete oxidation of PAHs, PCBs and PCTs (polychlorinated terphenyls). Such disposal is mandatory in many countries if the content of some contaminant in the used oil exceeds predetermined concentrations, e.g. 20 ppm of PCB in Germany.

With regard to re-refining, providing the used oil is of a suitable quality and the process is able to generate a product of sufficiently high quality, the principal environmental consideration concerns the disposal of the by-products of the refinery process. The effluents/waste products from some re-refining processes can be highly toxic and/or potentially carcinogenic. If these wastes are not disposed of correctly, they represent a potential hazard to the environment, which is at least as severe as that of untreated used oils. The cost of safe disposal of these wastes is usually high.

Energy balance in terms of percent weight hydrocarbon feedstock for the three alternative disposal methods is as follows:

(i) Disposal as toxic/hazardous waste 0% energy gain; some additional energy input required to achieve high temperature combustion. Net loss = −X%
(ii) Re-refining to produce base oils Net energy loss of 17%, assuming 70% recovery rate of lube oil
(iii) Use as fuel Net energy gain of 90–95%

The use of re-refined base oils in the production of lubricants is being increasingly considered on a worldwide basis. The driving force behind this is the concept of recycling of resources, thus minimising environmental impact.

These considerations, while commendable in principle, must not take precedence over the quality of the resulting lubricant. Full life-cycle analysis is required in each market to determine the true picture with regard to environmental impact. If lubricant quality is compromised, the results in terms of engine failure and loss of performance could lead to an overall environmental deficit when the cost of replacement parts and decreased efficiency are taken into account. The health consequences of increased PAH concentrations are also unknown, and must be considered if re-refined oils are used.

Quality control and process control should be able to produce a product of known and consistent quality, suitable for use as a lubricant basestock. In practice, however, oils that can meet the required performance characteristics tend to be expensive. For these reasons, the use of re-refined oil as a lubricant base fluid is not considered to represent a viable alternative to virgin base oils of consistent quality when these are readily available. In these circumstances, the most appropriate means of recycling used engine oil is as a fuel or fuel supplement. Recycling by re-refining should only be considered where alternative base oils are unavailable. In such cases, the use of re-refined oil in the production of lubricants represents the only viable alternative. Care must be taken to ensure that the re-refining process can produce an oil of acceptable quality.

13.7 Environmental impact of 'consumed' lubricant

Following the Exxon Valdez incident in Alaska, the time and effort spent on clean-up was considerable. One of the most spectacularly effective means of treating contaminated beaches was the addition, to the oil, of nutrients that allowed the indigenous bacteria to biodegrade the oil *in situ*. Land farming and oxygen addition to contaminated soil have also proved successful in dealing with hydrocarbon contamination. Oil does biodegrade in aerobic conditions, albeit rather slowly. If a lubricant is subjected to one of the standard OECD tests for ready biodegradability, however, it will most probably fail to show more than 20–30% degradation. There are many reasons for this; perhaps the most obvious, yet most often ignored, is that the test was probably designed to assess a single water-soluble chemical, probably a surfactant, and not a complex mixture of poorly soluble hydrocarbons.

The environmental degradation of lubricating oils is less easily demonstrated. One of the problems is the complex and varied mixture of the used material. Materials other than hydrocarbons can inhibit or influence the rate of degradation, which is of the greatest interest. Once released into the environment, there is a finite time before the waste lubricant is bound up in sediments or soils. Once there, due to the hydrophobic nature of the material, water is excluded and conditions are essentially anoxic. Anaerobic degradation of oils does not readily occur in nature—much to the relief of the oil exploration and refining industry. However, waste lubricant on the road, or in soil and sediments is not locked in place in the same way as crude oil in rock formations. Heavy rain can wash surfaces, churn-up river beds and release sediments, while the action of animals and plants can oxygenate soils. One means of lessening the burden on the environment, and minimising the impact of 'consumed' lubricant, would be to use readily biodegradable lubricants.

Lubricants that are susceptible to microbial degradation were developed in the 1970s for use in outboard engines. These lubricants were based on synthetic esters (see chapter 2), structurally similar to naturally occurring triglycerides, which increased their acceptability to the degrading organisms. In recent years, lubricants based on rape-seed oil have been developed for use in chain saws and the hydraulic systems of vehicles such as snowmobiles, which are used in environmentally sensitive areas. In order to develop a biodegradable lubricant, however, it is important to have a means of assessing biodegradability. This is by no means straightforward.

13.8 Biodegradation tests for oils

The concept of biodegradation is, at least superficially, a simple one. A compound is degraded by biological mechanisms, and so is removed from the environment. The practical assessment of biodegradability is less certain, however, as evidenced by the number and variation of the tests available. The relationship between test results and actual environmental persistence is well established for some classes of compounds (for example surfactants, which have been extensively investigated). For the majority of other materials, however, no such link has been established.

All the test methods currently employed were designed mainly for use with single chemical species that have a demonstrable water solubility. Oils and oil products are mixtures of hundreds (sometimes thousands) of different chemicals, which are generally of poor water solubility. Whilst some existing methods can be used with materials of poor solubility, such applications are at best a compromise. In the following sections, some of the main terms used in biodegradation tests are described. The differences between the commonly used test methods, and the difficulties of applying these methods to oil products are then outlined. Finally, an indication of the current applicability of test results to oil products is given.

13.8.1 *Terminology*

13.8.1.1 *Biodegradation* Biodegradation is the breakdown of a chemical by organisms. This is considered at two levels: (i) *primary biodegradation* whereby the loss of one or more active groups renders the molecule inactive with regard to a particular function; and (ii) *ultimate biodegradation*, which refers to the complete breakdown to CO_2, H_2O and mineral salts (PO_4, SO_4, NO_3, etc.,), a process known as *mineralisation*.

13.8.1.2 *Ready biodegradability* Ready biodegradability (or readily biodegradable) is an arbitrary definition whereby a compound achieves a 'pass' level in one of *five* named tests (OECD, Sturm, AFNOR, MITI, closed bottle).

13.8.1.3 *Inherent biodegradability* Inherent biodegradability (or inherently biodegradable) is another arbitrary definition. In this case, a compound shows evidence of degradation in *any* test for biodegradability.

13.8.1.4 *Biochemical oxygen demand (BOD)* This refers to the amount of oxygen consumed by microbes when metabolising or degrading a compound.

13.8.1.5 *Theoretical oxygen demand (ThOD)* The theoretical oxygen demand is the amount of oxygen required to completely oxidise a given chemical compound. This is calculated from the chemical formula.

13.8.1.6 *Chemical oxygen demand (COD)* Chemical oxygen demand refers to the amount of oxygen consumed during chemical oxidation with hot acid dichromate. It is used to give an indication of the amount of oxidisable material present, but only as a second best to ThOD. Not all chemicals are oxidised by dichromate, particularly oil products and lubricants.

13.8.1.7 *Dissolved organic carbon (DOC)* This is the amount of carbon present in a test compound that is in aqueous solution.

13.8.1.8 *Acclimatisation* Acclimatisation is the process by which microbes adapt and multiply to enable them to metabolise a compound with which they have not previously had any contact. This may take hours or weeks and, in some cases, may not happen at all, e.g. with DDT.

13.8.1.9 *Inoculum* The inoculum is the mixture of bacteria and other microorganisms added to the test solution. The source of the inoculum can significantly affect the results of the test; increased bacterial numbers and previous exposure to similar materials can increase the likelihood of degradation occurring. Common sources of inoculum are river water, sea water, soil extract, sewage effluent, sewage sludge and activated sludge. Sewage-based inoculum can be obtained from works treating primarily either domestic or

Table 13.4 Tests for determining 'ready biodegradability'

Test method	Concentration of test substance (mg/l)	Concentration of bacteria (cells/ml)	Measures	Pass level (%)	Test period (days)
Closed bottle	2–10	250	O_2	60	28
OECD	5–40	50–250	DOC	70	28
Sturm[a]	10 and 20	10 000–200 000	CO_2	60	28
AFNOR[a]	40	500 000	DOC	70	28
MITI[a]	100	200 000	O_2	60	28
CEC[b]	2 500	1 000 000	C–H(ir)	?	21

[a]These tests are currently being updated so that they all utilise 1 000 000 cells/ml.
[b]There is no 'pass level' for the CEC test. A value of 70% has been adopted by some authorities, e.g. Nordic Council.

industrial effluent. Regulatory authorities stipulate effluent from a works treating primarily domestic sewage as the source of inoculum.

13.8.2 *Current test method variations*

The purpose of any biodegradation test is to demonstrate a potential for a material to be degraded; it does not seek to show what will happen in the environment, indeed it cannot. The reason for this is quite simple: in a biodegradation test, the microbes utilise the test compound as an energy source and so degrade it. The test is set up so that all other nutrient requirements are met, except for the carbon/energy source. The microbes (bacteria, fungi, etc.) *must* utilise the test compound if they are to grow and it is this utilisation that is monitored, by measuring CO_2 evolution, O_2 depletion or DOC reduction. In the environment there are many thousands of alternative energy sources available and there is no compulsion for the microbes to degrade the test compound, especially when other 'easier' materials are available.

Five methods are currently available for determining 'ready biodegradability'. These methods are OECD standard tests and are recognised by the EC, the EPA and the ISO. Table 13.4 lists the salient points of these five tests (which all last for 28 days) in order of stringency, together with the CEC test for outboard two-stroke oils. The higher the concentration of test material (providing it is not toxic) and the greater the number of bacteria, then the greater the chance that some of the organisms in the test flask will be able to adapt to utilise the test material. Compounds that pass the MITI test, for example, could easily fail a closed bottle test, yet both tests would enable a substance to be classed as readily biodegradable.

The choice of a 'pass' level is arbitrary. Reference materials in closed test systems, such as the five ready biodegradability tests, do not show 100% removal, although most reference chemicals degrade to over 90% DOC removal. As the material is degraded and bacterial numbers increase, so the concentration of the test material decreases. The law of diminishing returns begins to apply, and the threshold concentration for a material to be absorbed and metabolised can be passed such that it is no longer economical for the bacteria to metabolise the small amount of test substance available. Consequently, they 'switch off' their metabolic processes and become dormant. The accumulation of 'staling' factors due to bacterial metabolism may also occur, preventing further growth. The concentration at which this occurs will differ on a case-by-case basis, but it is for this reason that it is practically impossible to demonstrate 100% degradability in short-term tests. In many cases, the concept of 100% removal or complete degradation can only be shown in environmental simulation experiments using radiolabelled chemicals. Such an approach is not possible, however, with complex mixtures such as oils.

13.8.3 *Problems with lubricants*

Aside from the insolubility of lubricants and the fact that the microbes responsible for degradation live in water, the complexity of the mixture referred to as 'oil' makes assessment in the existing standard tests difficult, particularly if the additives are also considered.

13.8.3.1 *Solubility* In natural systems, microbes produce emulsifiers that render insoluble materials accessible for degradation. All the systems currently available were developed for water-soluble materials and, although the use of artificial solvents or emulsifiers is permitted in the test guidelines, their use is restricted to 100 ppm in tests related to ecotoxicology and product classification, and neither are permitted in the MITI test. This approach is, in most cases, ineffective with oils since much higher concentrations are necessary to achieve solubility, which renders the conditions of the test doubtful.

13.8.3.2 *Complexity* In assessing a single chemical, the analytical determination of concentration is relatively straightforward and the process of breakdown can be followed by the identification of metabolites. In a complex mixture of oil in water the amount of each individual hydrocarbon in aqueous solution varies according to the ratio of oil to water. As degradation proceeds this ratio changes, therefore the pattern of hydrocarbons in aqueous solution changes. The ability of a single chemical (at a concentration representative of those used in the standard test methods) to induce acclimatisation in bacteria is well established. In a complex mixture, it is possible that not all of the individual components reach a concentration high enough to induce acclimatisation.

13.8.3.3 *Residuals* With a pure compound, if 70% degradation is demonstrated, only 30% is left, which may be pure compound or metabolites—this may be determined analytically. With a complex mixture, such as an oil, if 70% is removed by degradation, what is the remaining 30%? It may be that all of the hydrocarbon is removed leaving only the additives. It could be that 70% of everything has gone leaving 30% residue, similar to the situation with a pure compound. Using standard tests, it is not possible to demonstrate which parts of the product have been degraded.

13.8.4 *Current status and applicability*

The only oil product for which there is comprehensive data on environmental persistence is crude oil. These real environmental data have been generated as a result of the various oil spills that have occurred over the years. Such incidents have shown that oil does degrade. More recently, events in Alaska have shown that the addition of essential nutrients—PO_4, SO_4, NO_3, etc.—

can promote rapid degradation by indigenous microorganisms in a short time and at low temperature.

Given that the existing standard tests were developed to correlate with the actual environmental situation for particular types of material (surfactants), it can be seen that the application of existing test methods (where oil products invariably fail) to lubricants is not always appropriate. There is one test method, developed by the CEC for use with two-stroke outboard oils (see Table 13.4), that does correlate with the environmental situation. Although a 67% pass level has been applied to this test, which correlates with environmental data for two-stroke outboard oils, some authorities require 70% to be in line with other methods measuring DOC.

There is a need, therefore, for an internationally recognised test for the biodegradation of oil products that will reflect the true environmental situation, i.e. given an adequate supply of nutrients and oxygen, microbiological degradation of oils can and does occur. The European oil companies working through CONCAWE have set up a task force of biodegradation experts to develop just such a test. Until internationally recognised and accepted test methods that accurately reflect the environmental impact of oil and oil products are available, the oil industry will be obliged to work with current, limited test methods. The inapplicability of these methods to complex mixtures of poorly soluble compounds must be understood in order for the apparent failures of oil products in biodegradation tests to be placed in their correct environmental perspective.

13.9 Future trends

As mentioned earlier, the largest proportion of lubricant enters the environment by virtue of its being used. Legislation can reduce unauthorised disposal, but this is a minor problem in terms of the overall situation. Engineering can minimise leaks and also increase the dependence on authorised waste disposal specialists to eliminate 'dumping', for example by using dry sump lubrication in car engines, with the lubricant in a replaceable canister that can be changed only at a service station.

The largest factor in minimising environmental impact, however, will depend on the lubricant itself. As mentioned earlier, vegetable oil base fluids are being used in some applications, while readily biodegradable synthetic esters are used in outboard two-stroke engines. The choice of a biodegradable base fluid can aid the search for 'environmental friendliness'.

What, however, of the additive components? As described in earlier chapters, these materials are present in the formulation to perform a specific job. As engineering and fuel change, so does the need for additives. Does a low sulphur fuel (which is becoming the norm due to legislative pressure on acid emissions) need a lubricant with as good a dispersant additive as a high sulphur fuel? The metallic and halogenated components of additives may eventually be replaced by easily biodegradable compounds that perform as effectively, if not better, than those currently in use.

It is said that conflict is one of the most effective stimuli to scientific research and new technology. Mankind has always been in conflict with the environment; it is only now that this realisation has set in, and that acid rain, CFCs, pesticides, etc. are recognised as a major threat to the environment. The consequent need to minimise mankind's effects on the environment will drive technology forward at a pace undreamt of, even a few years ago.

References

CONCAWE (1985) The collection, disposal and regeneration of waste oils and related materials. *CONCAWE Report No. 85/53.*

IARC (1984) *IARC Monographs on the Evaluation of the Carcinogenic Risk of Chemicals to Humans. Volume 33 Polynuclear Aromatic Compounds, Part 2 Carbon Blacks, Mineral Oils and Nitroarenes.* IARC, Lyon, France.

Further reading

Central Statistics Office (1988) *Annual Abstract of Statistics, No. 124.* HMSO, London.

HMSO (1981) *Assessment of Biodegradability.* HMSO, London.

Index

GPSR Compliance
The European Union's (EU) General Product Safety Regulation (GPSR) is a set of rules that requires consumer products to be safe and our obligations to ensure this.

If you have any concerns about our products, you can contact us on

ProductSafety@springernature.com

In case Publisher is established outside the EU, the EU authorized representative is:

Springer Nature Customer Service Center GmbH
Europaplatz 3
69115 Heidelberg, Germany

www.ingramcontent.com/pod-product-compliance
Ingram Content Group UK Ltd.
Pitfield, Milton Keynes, MK11 3LW, UK
UKHW042019290726
14061UKWH00001BB/72

* 9 7 8 0 2 1 6 9 2 9 2 1 0 *